Foundations of Semantic Communication Networks

IEEE Press
445 Hoes Lane
Piscataway, NJ 08854

Foundations of Semantic Communication Networks

Edited by

Walid Saad
Virginia Tech, Arlington, VA, USA

Christina Chaccour
Ericsson, Inc., Plano, TX, USA

Christo Kurisummoottil Thomas
Virginia Tech, Arlington, VA, USA

Merouane Debbah
Khalifa University of Science and Technology
Abu Dhabi, United Arab Emirates

IEEE Press
WILEY

Published by John Wiley & Sons, Inc., Hoboken, New Jersey.
Published simultaneously in Canada.

Library of Congress Cataloging-in-Publication Data Applied for:

Hardback ISBN: 9781394247882

Cover Design: Wiley
Cover Image: © Tuomas A. Lehtinen/Getty Images

Set in 9.5/12.5pt STIXTwoText by Straive, Chennai, India

"To those who believe that semantics matter,"

Walid Saad

"To those who believe 'less is more', finding substance, depth, and beyond,"
Christina Chaccour

"To my parents,"

Christo Kurisummoottil Thomas

"To my wife Radja and my three kids Lyne, Ines, and Noor,"
Merouane Debbah

Contents

About the Editors

Walid Saad received his PhD from the University of Oslo, Norway, in 2010. He is currently a Professor at the Department of Electrical and Computer Engineering at Virginia Tech, where he leads the Network intelligEnce, Wireless, and Security (NEWS) laboratory. His research interests include wireless networks (5G/6G/beyond), machine learning, game theory, quantum communications/learning, security, UAVs, semantic communications, cyber-physical systems, and network science. Dr. Saad is a Fellow of the IEEE. He is also the recipient of the NSF CAREER award in 2013, the AFOSR summer faculty fellowship in 2014, and the Young Investigator Award from the Office of Naval Research (ONR) in 2015. He was the (co-)author of twelve conference best paper awards at IEEE WiOpt in 2009, ICIMP in 2010, IEEE WCNC in 2012, IEEE PIMRC in 2015, IEEE SmartGridComm in 2015, EuCNC in 2017, IEEE GLOBECOM (2018 and 2020), IFIP NTMS in 2019, IEEE ICC (2020 and 2022), and IEEE QCE in 2023. He is the recipient of the 2015 and 2022 Fred W. Ellersick Prize from the IEEE Communications Society, of the IEEE Communications Society Marconi Prize Award in 2023, and of the IEEE Communications Society Award for Advances in Communication in 2023. He was also a co-author of the papers that received the IEEE Communications Society Young Author Best Paper award in 2019, 2021, and 2023. Other recognitions include the 2017 IEEE ComSoc Best Young Professional in Academia award, the 2018 IEEE ComSoc Radio Communications Committee Early Achievement Award, and the 2019 IEEE ComSoc Communication Theory Technical Committee Early Achievement Award. From 2015 to 2017, Dr. Saad was named the Stephen O. Lane Junior Faculty Fellow at Virginia Tech and in 2017, he was named College of Engineering Faculty Fellow. He received the Dean's award for Research Excellence from Virginia Tech in 2019. He was also an IEEE Distinguished Lecturer in 2019-2020. He has been annually listed in the Clarivate Web of Science Highly Cited Researcher List since 2019. He currently serves as an Area Editor for the IEEE Transactions on Communications. He is the Editor-in-Chief for the IEEE Transactions on Machine Learning in Communications and Networking.

Christina Chaccour, a distinguished research scientist, engineer, and entrepreneur, currently serves as a Network Solutions Manager at Ericsson Inc. In her role, Dr. Chaccour seamlessly bridges strategy, product solutions, and research, spearheading developments in 5G Advanced, 6G networks, and AI-integrated solutions while ensuring the responsible and innovative use of AI, addressing both policy and technical dimensions. Dr. Chaccour actively represents Ericsson as a delegate in prominent industry bodies and councils including FCC CSRIC IX, 5G Americas, and Next G Alliance. With a PhD in Electrical Engineering from Virginia Tech, her academic journey yielded significant contributions, particularly in 6G systems at THz frequencies for next-gen XR and holographic systems, her work also laid the groundwork for pioneering research in AI-native networks and semantic communications. Her dissertation contributions and their subsequent impact were recognized with the Bill and LaRue Blackwell Award.

Dr. Chaccour has been recognized with several honors, including the Best Paper Award at the 10th IFIP Conference on New Technologies, Mobility, and Security (NTMS) in 2019 and the Exemplary Reviewer Award from IEEE Transactions on Communications in 2021, a distinction given to fewer than 2% of reviewers. Her paper in IEEE Communication Surveys and Tutorials was featured in the Top Access article listing from June to November 2022. In 2024, she was named among the "Top 100 Brilliant and Inspiring Women in 6G."

Dr. Chaccour serves on the editorial board of IEEE Transactions on Machine Learning in Communications and Networking, IEEE Transactions on Cognitive Communications and Networking, Wireless Personal Communications (Springer), and the guest editorial board of IEEE Communications Standards Magazine within the series "AI for Wireless." She is also on the advisory board of "The Data Science Conference."

In her entrepreneurial endeavors, Dr. Chaccour co-founded the startup "Internet of Trees," which has garnered numerous local and international awards.

Christo Kurisummoottil Thomas received his BS in Electronics and Communication Engineering from National Institute of Technology, Calicut, India, in 010, MS in Telecommunication Engineering from Indian Institute of Science, Bangalore, India, in 2012, and PhD from EURECOM, France, in 2020. He is currently a postdoctoral fellow at the Electrical and Computer Engineering Department at Virginia Tech. His research interests include semantic communications, statistical signal processing, and advanced machine learning using causal inference and neurosymbolic AI for wireless communications. From 2012 to 2014, he was a staff design engineer on 4G LTE with Broadcom communications, Bangalore, and from 2014 to 2017, he was a design engineer with Intel corporation, Bangalore. During November 2020 till June 2022, he was a staff engineer on 5G modems with wireless research and development division of Qualcomm Inc., Espoo, Finland. He was a recipient of the best student paper award at IEEE SPAWC 2018, Kalamata, Greece, and also received third prize for his team titled "Learned Chester" ML5G-PHY channel estimation challenge, as part of the ITU AI/ML in 5G challenge, conducted at NCSU, US, 2020. He has delivered multiple tutorials on variational Bayesian inference at IEEE conferences such as ICASSP and EUSIPCO.

Merouane Debbah is a professor at Khalifa University of Science and Technology in Abu Dhabi and founding Director of the KU 6G Research Center. He is a frequent keynote speaker at international events in the field of telecommunication and AI. His research has been lying at the interface of fundamental mathematics, algorithms, statistics, information, and communication sciences with a special focus on random matrix theory and learning algorithms. In the communication field, he has been at the heart of the development of small cells (4G, massive MIMO (5G), and large intelligent surfaces (6G) technologies. In the AI field, he is known for his work on large language models, distributed AI systems for networks and semantic communications. He received multiple prestigious distinctions, prizes, and best paper awards (more than 40 IEEE best paper awards) for his contributions to both fields. He is an IEEE Fellow, a WWRF Fellow, a Eurasip Fellow, an AAIA Fellow, an Institut Louis Bachelier Fellow, an AIIA Fellow, and a Membre émérite SEE.

List of Contributors

Jean-Claude Belfiore
Advanced Wireless Technology Lab.
Huawei Technologies
Boulogne Billancourt
France

Daniel Bennequin
Advanced Wireless Technology Lab.
Huawei Technologies
Boulogne Billancourt
France

and

UFR de Mathématiques
Universite Paris Cite
Paris
France

Mehdi Bennis
Centre for Wireless Communications
University of Oulu
Oulu
Finland

Chenghong Bian
Department of Electrical and
Electronic Engineering
Imperial College London
London
UK

Van Phuc Bui
Department of Electronic Systems
Aalborg University
Aalborg Øst
Denmark

Christina Chaccour
Ericsson, Inc.
Plano, TX
USA

Mingzhe Chen
Department of Electrical and
Computer Engineering
University of Miami
Coral Gables, FL
USA

Zhiyong Chen
Cooperative Medianet Innovation Center
Shanghai Jiao Tong University
Shanghai
China

Jinho Choi
School of Electrical and Mechanical Engineering
the University of Adelaide
North Terrace, Adelaide, SA
Australia

Jinhyuk Choi
School of Electrical & Electronic Engineering
Yonsei University
Seodaemun-Gu, Seoul
Korea

Wan Choi
Department of Electrical and Computer Engineering
Seoul National University
Seoul
Republic of Korea

Merouane Debbah
6G Center
Khalifa University of Science and Technology
Abu Dhabi
United Arab Emirates

Homa Esfahanizadeh
Radio Systems Research Group
Nokia Bell Labs
Murray Hill, NJ
USA

Rafael Gregorio Lucas D'Oliveira
School of Mathematical and Statistical Sciences
Clemson University
Clemson, SC
USA

Deniz Gunduz
Department of Electrical and Electronic Engineering
Imperial College London
London
UK

Ekram Hossain
Department of Electrical and Computer Engineering
University of Manitoba
Winnipeg, Manitoba
Canada

Ye Hu
Department of Industrial and Systems Engineering
University of Miami
Coral Gables, FL
USA

Yoon Huh
Department of Electrical and
Computer Engineering
Seoul National University
Seoul
Republic of Korea

Tomás Huttebraucker
CEA-Leti
Université Grenoble Alpes
Grenoble
France

Seong-Lyun Kim
School of Electrical & Electronic
Engineering
Yonsei University
Seodaemun-Gu, Seoul
Korea

Seung-Woo Ko
Department of Smart Mobility
Engineering
Inha University
Yeonsu-Gu, Incheon
Korea

Emrecan Kutay
Department of Electrical and
Computer Engineering
INSPIRE@OhioState Research Center
The Ohio State University
Columbus, OH
USA

Rongpeng Li
College of Information Science and
Electronic Engineering
Zhejiang University
Hangzhou
China

Yingyu Li
School of Mechanical Engineering and
Electronic Information
China University of Geosciences
Wuhan, Hubei
China

Yiwei Liao
School of Electronic Information and
Communications
Huazhong University of Science and
Technology
Wuhan, Hubei
China

Yuanwei Liu
Department of Electrical and
Electronic Engineering
The University of Hong Kong
Hong Kong, Island

Zhilin Lu
College of Information Science and
Electronic Engineering
Zhejiang University
Hangzhou
China

Xinlai Luo
Cooperative Medianet Innovation Center
Shanghai Jiao Tong University
Shanghai
China

Alexander Mariona
Research Laboratory of Electronics
Massachusetts Institute of Technology
Cambridge, MA
USA

Muriel Médard
Research Laboratory of Electronics
Massachusetts Institute of Technology
Cambridge, MA
USA

Mattia Merluzzi
CEA-Leti
Université Grenoble Alpes
Grenoble
France

Xidong Mu
School of Electronic Engineering and Computer Science
Queen Mary University of London
London
UK

Hyelin Nam
School of Electrical & Electronic Engineering
Yonsei University
Seodaemun-Gu, Seoul
Korea

Shashi Raj Pandey
Department of Electronic Systems
Aalborg University
Aalborg Øst
Denmark

Jihong Park
Information Systems Technology and Design Pillar
Singapore University of Technology and Design
Singapore

H. Vincent Poor
Department of Electrical and Computer Engineering
Princeton University
Princeton, NJ
USA

Petar Popovski
Department of Electronic Systems
Aalborg University
Aalborg Øst
Denmark

Walid Saad
Department of Electrical and Computer Engineering
Virginia Tech
Arlington, VA
USA

Yalin Evren Sagduyu
Nexcepta
Gaithersburg, MD
USA

Mohamed Sana
CEA-Leti
Université Grenoble Alpes
Grenoble
France

Hyowoon Seo
Department of Electronics and Communications Engineering
Kwangwoon University
Seoul
Republic of Korea

Yulin Shao
State Key Laboratory of Internet of Things for Smart City and Department of Electrical and Computer Engineering
University of Macau
Macau
China

Guangming Shi
School of Electronic Engineering
Xi'an University of Electronic Science and Technology
Xi'an, Shaanxi
China

Heekang Song
School of Electrical Engineering
Korea Advanced Institute of Science and Technology
Daejeon
Republic of Korea

Emilio Calvanese Strinati
CEA-Leti
Université Grenoble Alpes
Grenoble
France

Zijian Sun
School of Electronic Information and Communications
Huazhong University of Science and Technology
Wuhan, Hubei
China

Meixia Tao
Department of Electronic Engineering and the Cooperative Medianet Innovation Center
Shanghai Jiao Tong University
Shanghai
China

Christo Kurisummoottil Thomas
Virginia Tech Research Center
Virginia Tech
Arlington, VA
USA

Sennur Ulukus
Department of Electrical and Computer Engineering
University of Maryland
College Park, MD
USA

Zhongwei Wang
Cooperative Medianet Innovation Center
Shanghai Jiao Tong University
Shanghai
China

Haotian Wu
Department of Electrical and Electronic Engineering
Imperial College London
London
UK

Yong Xiao
School of Electronic Information and Communications
Huazhong University of Science and Technology
Wuhan, Hubei
China

Aylin Yener
Department of Electrical and Computer Engineering
INSPIRE@OhioState Research Center
The Ohio State University
Columbus, OH
USA

Honggang Zhang
College of Information Science and Electronic Engineering
Zhejiang University
Hangzhou
China

and

Zhejiang Lab
Hangzhou
China

Zhifeng Zhao
College of Information Science and Electronic Engineering
Zhejiang University
Hangzhou
China

and

Zhejiang Lab
Hangzhou
China

Preface

The wireless industry is on the brink of a revolution with the emergence of connected intelligence applications such as the metaverse, connected autonomy, holographic societies, haptics, telehealth, smart manufacturing, and smart cities. This revolution has brought forth many concerted efforts that focus on various design challenges of wireless networks, in general, and 6G systems and beyond, in particular. However, those efforts remain focused primarily on incremental, but important advances to conventional communication system technologies such as massive MIMO, reconfigurable intelligent surfaces, network slicing, and the likes.

In this rapidly evolving landscape, the integration of connected intelligence applications highlights the pressing need for networks to gain intelligence in a non-siloed and ad hoc manner. The traditional incremental approach to network design is no longer sufficient to support the diverse and dynamic requirements of these emerging applications. This necessitates a paradigm shift toward more intelligent and adaptive network architectures.

The genesis of this book stems from our recognition of semantic communication as an emerging cornerstone in the telecommunication industry. We believe that exploring this concept through diverse viewpoints – encompassing information theoretic, physical layer, networking, and security perspectives – is paramount in unraveling its complexities and harnessing its potential. While our exploration may traverse differing schools of thought and encounter opposing views on granular concepts, we view this diversity as essential in shaping a comprehensive and consensus-driven understanding of semantic communication systems.

This book represents our endeavor to establish a staple resource in the realm of semantic communication. We envision it as a guiding beacon for researchers, practitioners, and enthusiasts alike, navigating the intricate pathways of this burgeoning field.

To achieve our objectives, this book is structured into four parts, each delving into fundamental aspects of semantic communications and networking. The introductory chapter sets the stage, providing an overarching framework for subsequent discussions. Each part is meticulously crafted with specific goals in mind, offering readers a comprehensive exploration of the subject matter.

As editors and authors, we are immensely grateful to the contributors whose expertise and dedication have enriched this volume. We extend our heartfelt thanks to the reviewers whose insights and feedback have helped refine the content.

Ultimately, we believe that this book will serve as a cornerstone in the ongoing discourse surrounding semantic communication. It is our sincere hope that it will inspire further exploration, innovation, and collaboration in this dynamic field.

March, 2024
United States of America

Sincerely,
Walid Saad
Christina Chaccour
Christo Kurisummoottil Thomas
Merouane Debbah

Acknowledgment

This effort was supported by the Office of Naval Research (ONR) under MURI Grant N00014-19-1-2621.

Acronyms

AI	artificial intelligence
AoI	age-of-information
AP	affinity propagation
APC	average power constraint
AR	augmented reality
AR	augmented reality
AUC	area under the curve
AWGN	additive white Gaussian noise
BER	bit error rate
BERT	bidirectional encoder representations from transformers
BIM	basic iterative method
BIP	basic invariance principle
BLER	block error rate
BLEU	bilingual evaluation understudy
BPG	better portable graphics
BPSK	binary phase-shift keying
B-RNNs	bi-directional recurrent networks
BS	base station
BSC	binary symmetric channel
CA	channel attention
CDF	cumulative distribution function
CLT	central limit theorem
CNN	convolutional neural network
CVaR	conditional value at risk
DeepJSCC	deep joint source-channel coding
DL	deep learning
DMC	discrete memory channel
E2E	end-to-end
EnSC	encrypted semantic communication

FC	fully connected
FGSM	fast GradientSign method
FL	federated learning
GAN	generative adversarial network
GenAI	generative artificial intelligence
GO	goal oriented
IB	information bottleneck
iCR	inverse contextual reasoning
IID	independent and identically distributed
IoT	Internet of Things
IoU	intersection over union
ISS	image-to-graph semantic similarity
JSCC	joint source-channel coding
KL	Kullback–Liebler
LCR	linearized contextual reasoning
LDPC	low-density parity check
LLM	large language model
LPIPS	learned perceptual image patch similarity
LSC	language-oriented SC
MAB	multi-arm bandits
MAC	multiple access channels
MCMC	Markov chain Monte Carlo
MIM	momentum iterative method
MIMO	multiple-input and multiple-output
ML	machine learning
MSE	mean squared error
NE	nash equilibrium
NLP	natural language processing
NOMA	non-orthogonal multiple access
OMA	orthogonal multiple access
O-RAN	open-RAN
PGD	projected gradient descent
PPC	peak power constraint
PSNR	peak signal-to-noise ratio
PSR	perturbation-to-signal ratio
QPSK	quadrature phase shift keying
RHS	right-hand side
RL	reinforcement learning
ROC	receiver operating characteristic
RS	reed Solomon
RX	receiver

SBERT	sentence-BERT
SC	semantic communication
SCM	structural causal model
SER	symbol error rate
SGD	stochastic gradient descent
SKD	semantic knowledge distillation
SL	split learning
SNC	semantics-native communication
SNR	signal-to-noise ratio
SSCC	separate source and channel coding scheme
SV	Shapley value
SVD	singular-value decomposition
T2I	text-to-image
T2T	text to text interpreter
TV	total variation
TX	transmitter
UAV	unmanned aerial vehicle
UCB	upper confidence bound
UE	user equipment
UMAP	uniform manifold approximation and projection for dimension Reduction
V2X	vehicle-to-everything
VD-ERL	value decomposition entropy-maximized multi-agent RL
VG	visual genome
ViT	vision transformer
VMAF	video multimethod assessment fusion
VQ AE	vector quantized autoencoder model
VQ-VAE	vector-quantized variational autoencoder
VR	virtual reality
VR	virtual reality
WSN	wireless sensor networks
XR	extended reality

1

Introduction to Semantic Communications

Christina Chaccour[1], Christo Kurisummoottil Thomas[2], Walid Saad[3], and Merouane Debbah[4]

[1] *Ericsson, Inc., Plano, TX, USA*
[2] *Virginia Tech Research Center, Virginia Tech, Arlington, VA, USA*
[3] *Department of Electrical and Computer Engineering, Virginia Tech, Arlington, VA, USA*
[4] *6G Center, Khalifa University of Science and Technology, Abu Dhabi, United Arab Emirates*

1.1 From Information Streams to Streams of Understanding: The Rise of Semantic Communication Networks

Throughout history, networks have mirrored the evolution of our relationship with information. Initially, they facilitated the transmission of human voices, carrying the ephemeral nature of conversation across vast distances. Subsequently, they evolved to transport the richness of media, bringing the world into our living rooms. As technology progressed, networks connected a multitude of smaller devices, forming the nascent internet of things. However, with the dawn of 6G and beyond, and the promise of a hyper-connected future, the focus of our network infrastructure shifts from mere information transport to the interconnection of intelligence.

This fundamental shift necessitates a radical transformation in how we communicate. As we envision a world of holographic teleportation, the immersive metaverse, and the sophisticated automation of Industry 5.0, the limitations of traditional networks become increasingly apparent. These networks, acting as mere "bit pipes," rely on brute force, pushing ever-increasing volumes of data through ever-widening bandwidths, simply cannot sustain the demands of this intelligent future. Complexity would spiral out of control, and the very infrastructure that underpins this progress would become its own Achilles' heel.

Foundations of Semantic Communication Networks, First Edition.
Edited by Walid Saad, Christina Chaccour, Christo Kurisummoottil Thomas, and Merouane Debbah.

To address this challenge effectively, we need to move away from makeshift fixes and fundamentally rethink the structure of our network. This means giving greater importance to the connections between network nodes. These connections shouldn't just pass on data blindly; they should possess the capacity to comprehend the significance of the information they carry. This transformation, often termed as the "humanization" of the network, forms the core of semantic communication – a paradigm shift that transcends mere data transmission to facilitate the exchange of meaningful information.

It is worth emphasizing that semantic communication does not aim to replicate the full complexity of human interaction in transmitting data. Rather, its focus lies in conveying the essence of the data along with pertinent metadata, irrespective of specific data formats, and functions at the network's foundational layers. This approach starkly contrasts with traditional wireless networks, where communication relies solely on raw, unprocessed bits. As we navigate the increasingly intricate landscape of future applications, our reliance on spectrum and coverage alone becomes unsustainable. Semantic communication represents a shift in networking philosophy toward a deeper understanding and intelligent handling of data traffic.

With the advancement of applications and the proliferation of consumer AI, the intelligence embedded within AI systems is evolving rapidly. Tasks as simple as a Google search now translate into dynamic generative pre-trained transformer (GPT) conversations, while a single recorded video can spawn numerous generated videos through AI algorithms. Moreover, the landscape is expanding with complex applications like emerging services such as next-generation extended reality (XR), full and others. In this context, it becomes imperative for networks to adapt their approach to data and traffic. Rather than adhering to the traditional "bit pipe" model, networks must mirror the intelligence embedded within applications. We envision a future where traffic continues to escalate, but instead of exponentially increasing to the point where new spectrum allocation becomes unfeasible, the concept of semantic communications intervenes to transform networks into truly AI-native entities (as shown in Figure 1.1). In this future, traffic is managed and controlled through semantic understanding, alleviating the relentless demand for spectrum that is becoming increasingly scarce.

However, by embracing semantic communication, we unlock a powerful new paradigm. We can leverage our computing resources as "in-memory" networks, reducing the strain on spectrum and enabling efficient handling of complex data. Additionally, the growing ubiquity of AI, both in consumer devices and across diverse applications, necessitates a network that can keep pace with the rising tide of data volume and intelligence. Semantic communication networks represent the crucial leap forward, enabling us to navigate this exciting future where machines not only share information but also share understanding.

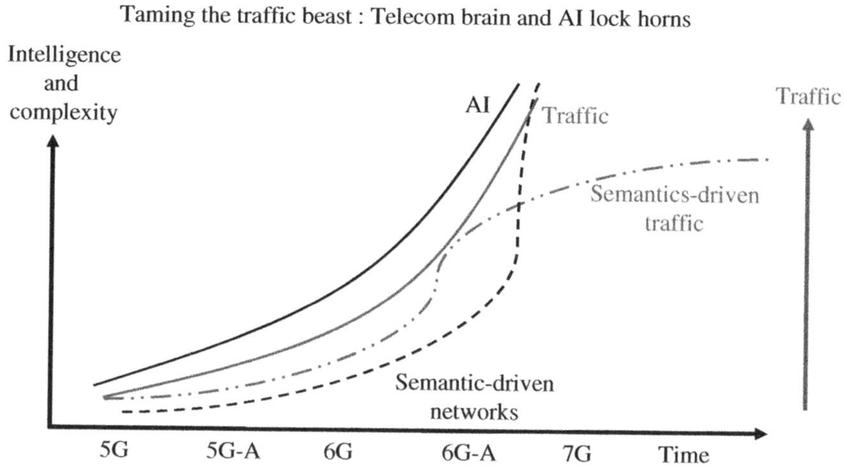

Figure 1.1 Hypothesis for Taming traffic: Telecom brain and AI lock horns.

1.1.1 How Does It Work?

Semantic communication necessitates a fundamental reconsideration of the communication quandary, as delineated by Weaver's three-tier framework [Weaver, 1953]. Following Shannon's inception of information theory, Weaver delineated communication hurdles across three levels [Weaver, 1953]: (i) *Level A*, pertaining to the technical precision of communication symbols to be transmitted; (ii) *Level B*, addressing the semantic accuracy of transmitted symbols vis-à-vis intended meaning; and (iii) *Level C*, evaluating the impact of received meaning on overall system conduct. Traditional communication systems primarily tackled Level A challenges. Nonetheless, a judiciously crafted semantic communication system can leverage AI advancements and computational prowess to potentially transcend Weaver's framework, incorporating a *reasoning plane* [Chaccour et al., 2024], achieving more with less. This entails a paradigm shift from conventional transmitter–receiver pairs to what we propose as *teacher and apprentice* nodes, endowed with the following capabilities:

- **Transitioning from a bit-driven transmitter to a knowledge-driven teacher:** The conventional transmitter model needs to evolve beyond merely serving as a conduit for bits of data. Instead, it should metamorphose into a *teacher* endowed with the ability to decipher multiple *semantic content elements* embedded within the source data. This entails the extraction of distinct meanings, or semantics, from the message. Subsequently, for each identified semantic content element, the teacher must generate a *semantic representation* possessing desirable attributes. Fundamentally, the semantic content represents the essence of the data, while the semantic representation encapsulates this essence in a minimal form – akin to how individuals meticulously select words to articulate their thoughts. Moreover, various semantic content elements may correspond to different modalities within the data. For instance, in an audio recording, the tone of voice and the spoken words constitute distinct semantic content elements. Human cognition effortlessly disentangles and comprehends these elements, a capability lacking in current communication system transmitters. Hence, there is a pressing need to reconfigure transmitters to emulate human reasoning capabilities to the best of their ability. This necessitates *reasoning* at the transmission end – the capacity enabling the transmitting agent to identify, differentiate, and efficiently represent each semantic content element within the data. This stands in stark contrast to the conventional approach in today's networks, where transmitters treat input as a random, uncertain string of information transmitted through a bit-pipeline, devoid of semantic understanding. Furthermore, the shift from a data-driven to a reason-driven approach is showcased in Figure 1.2.

- **Shifting from a bit-driven receiver to a knowledge-driven apprentice:** Similarly, the receiver's role should evolve from merely processing bits to embodying an apprentice endowed with *reasoning* abilities. This transformation enables the receiver to *comprehend the minimal semantic representation*

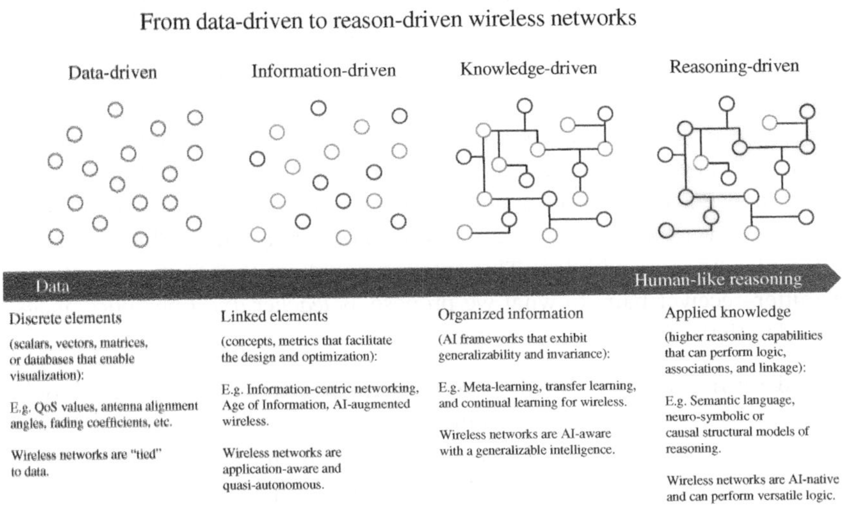

Figure 1.2 The evolution of wireless networks from data-driven ones to reasoning-driven ones [Chaccour et al., 2024].

utilized by the teacher, thereby mapping it back to its corresponding semantic content element. Additionally, the apprentice must harness its computational resources to accurately recreate the semantic content element derived from the transmitted semantic representation with utmost fidelity. For instance, if a holographic element is transmitted, the apprentice must possess the capability to reproduce it with identical resolution as transmitted by the teacher. Furthermore, the developed reasoning capabilities empower the apprentice to utilize both causal and associational (statistical) logic across the networking stack. These logic frameworks, derived from an evolving knowledge base, enable the apprentice to undertake diverse projections and decisions concerning the received semantic representation.

- **Transitioning from a bit-pipeline to a semantic language:** In the realm of semantic communications, the fundamental unit of meaning is encapsulated within a semantic representation. These representations, when organized sequentially, form what we term a *semantic language*. Unlike natural languages, semantic languages prioritize automation by minimizing emphasis on syntax and pragmatics. Moreover, the semantic representations within a semantic communication language must adhere to three crucial properties:
  1. **Minimalism**: This property entails characterizing information structure using the fewest possible language elements (and their corresponding bits). The aim is to reduce the number of exchanged messages over time while maintaining a comprehensive understanding of the underlying data structure.
  2. **Generalizability**: A critical aspect of semantic representations is their ability to represent or comprehend underlying structures invariant to changes in distribution, domain, and context. Context, in this context, refers to the thematic umbrella under which various semantics coexist (e.g., steering actions on a tennis court for a robot). Generalizability enables a node to apply learned representations across diverse domains, distributions, and contexts, akin to how words in natural language generalize their meaning to describe various events.
  3. **Efficiency**: The apprentice's proficiency in regenerating information with high fidelity in minimal time is crucial. This means that the generated data must match or exceed the resolution achievable by a traditional receiver. For instance, if reconstructing an audio clip, the apprentice must replicate the intended resolution precisely. It is worth noting that while efficiency can be quantified through concrete metrics (as defined in Chapter 11), evaluating the efficacy of the system is more intricate. Thus, for brevity, we consider "efficiency" in this tutorial to encompass both efficacy and efficiency.

1.1.2 Why Now? What Factors Contribute to Our Ongoing Reliance on Traditional Communications?

Since the inception of digital communication, Shannon's formulation of the "reconstruction problem" has been at the core of its challenges. This problem is primarily rooted in the limitations of computing power, which consequently restricts the intelligence of AI-guided tasks within communication systems. Traditionally, the focus of communication has revolved around replicating transmitted messages at another point, relying heavily on compression, transmission, and decompression techniques. However, these techniques predominantly address stochasticity arising from the source, channel, and destination, often overlooking meaningful information about the message's significance or context.

Within this traditional framework, various shortcomings emerge in efficiently conveying the intended meaning of transmitted messages. These shortcomings manifest in multiple issues within end-to-end (E2E) communication system design. Firstly, there's a prevalent need for repetitive transmissions in many cases to adequately transfer application information. This is primarily because transmitters lack automation in message generation at the receiver and fail to leverage memory or historical patterns in the data. Secondly, the asymmetrical nature of communication, characterized by passive receivers and active transmitters, renders receivers vulnerable to channel conditions and hardware errors.

Introducing symmetry into communication, where receivers can learn message structures and leverage historical context, holds the potential to enhance robustness against irregularities. Semantic communication, emerging as a transformative approach, facilitates radio nodes in extracting meaning from datastreams, thereby fostering shared understanding and efficient communication, particularly in scenarios with unreliable links [Lan et al., 2021].

Addressing intermittent behavior through receiver generative capabilities, semantic communication minimizes back-and-forth communication for a reliable link. By delving into low–level data analysis, including causal roots of events, semantic systems empower communication networks to efficiently achieve specific goals.

The transformation of current systems into reasoning-driven semantic communication platforms presents a promising avenue for enhancing efficiency and intelligence in wireless networks. This paradigm shift, leveraging computing resources and emphasizing languages, reasoning, and causality, offers significant potential to achieve more with less. Subsequent sections will delve deeper into exploring and understanding this transformative journey.

1.1.3 What Is NOT Semantic Communications?

Semantic communications might appear as a mere evolution of existing approaches and methods. However, in this section, we aim to clarify this misconception by emphasizing the core distinctions between conventional techniques and semantic communications. This is further illustrated in Figure 1.3.

1.1.3.1 Semantic Communications Is Not Data Compression

In a traditional communication framework, as described in information theory by Shannon [1948], data compression, also called source coding, involves encoding information using fewer bits than the original data stream. This compression exploits statistical redundancies within the data bits, allowing for data representation without sacrificing any information, thus ensuring reversibility at the receiver end. While both data compression and semantic communications share the objective of minimizing the size of transmitted data, they are fundamentally distinct concepts:

- Data compression achieves minimalism, i.e., shrinking the size of a particular data stream, by identifying and eliminating statistical redundancy. For instance, the most prominent lossless compressors employ probabilistic models such as prediction by partial matching [Zhang and Adjeroh, 2008]. Notably, there is a close connection between data compression and machine learning (ML), in that they both specifically attempt to predict the posterior probabilities of a sequence given its history. However, data compression does not bear any learning ability that contributes to a particular training memory or a trained model. Put simply, from an ML perspective, data compression techniques tend to overfit as they solely aim to reduce the current data stream without considering futuristic ones. They lack learning abilities and reasoning. From a minimalist viewpoint, data compression might offer short-term benefits over semantic communications. However, it fails to provide contextual information and knowledge-driven memory to the receiver.
- In contrast to data compression, semantic communications focus on identifying, learning, and representing patterns that map to structure and semantic content instead of merely compressing data redundancies. While data compression aims to overfit statistical characteristics of data streams, semantic communications prioritize characterizing structure rather than pure randomness exhibited in data. Random data points are often better transmitted conventionally. Semantic communication achieves "minimalism" through semantic representations that not only comprise fewer total bits but also teach the receiver to

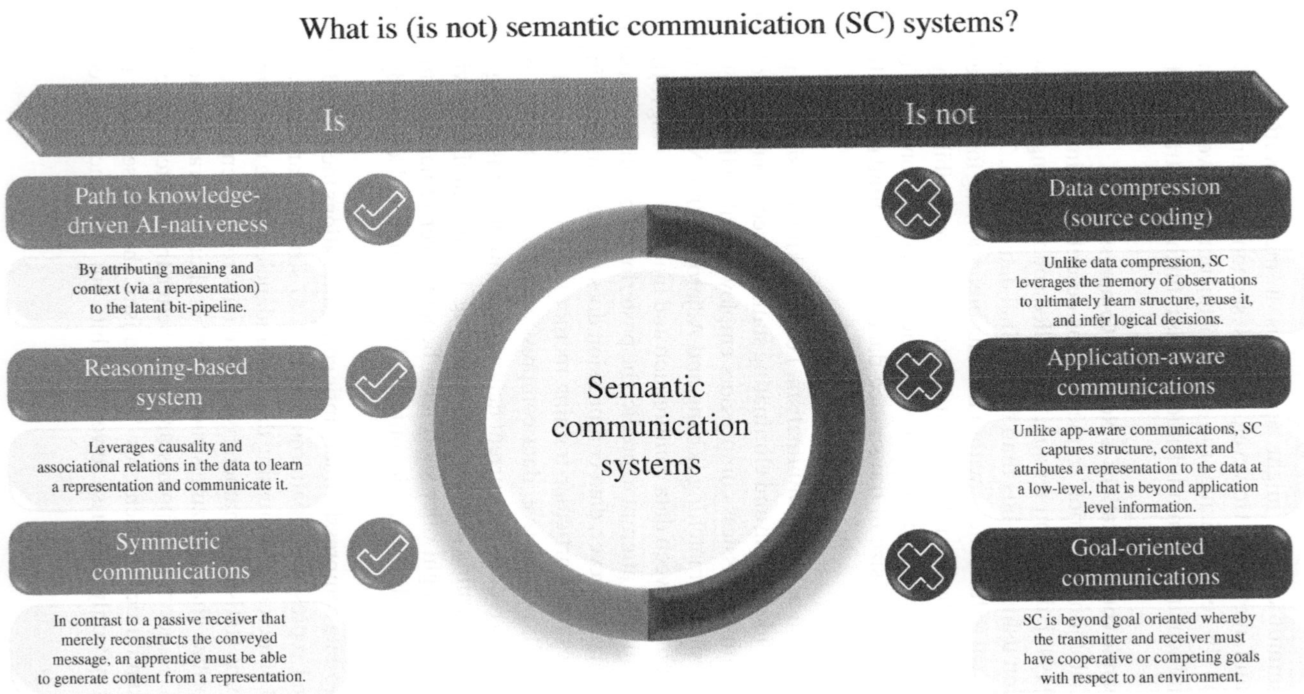

Figure 1.3 What is (is not) semantic communications [Chaccour et al., 2024]?

learn, generate, and automate tasks or messages. This is enabled by the characteristics of the adopted representation and the acquired reasoning capabilities, which minimize the number of bits per transmission and the total number of transmissions required to convey a message or accomplish a task. Semantic communication networks achieve minimalism through mechanisms that surpass simple bit compression. Moreover, when radio nodes operate based on organized knowledge, they can make more informed logical conclusions across the networking stack, a capability unique to reasoning-driven semantic communication networks.

1.1.3.2 Semantic Communications Is Not Only an "AI for Wireless" Concept

AI has played a significant role in addressing various wireless-related challenges, including channel estimation, beamforming, network management, and receiver design. While AI has indeed enhanced the accuracy, precision, or key performance indicators (KPIs) relevant to these tasks, the fundamental functionality and dynamics of the corresponding wireless tasks have largely remained unchanged. For instance, AI-enabled channel estimation represents an advanced method for conducting classical channel estimation, albeit with the same core task. In contrast, semantic communication systems do not simply add an AI layer atop existing tasks; they represent a fundamental shift in communication paradigms. Semantic communications do not merely improve the air interface with AI; instead, they introduce a novel approach to communication tasks. With semantic communications, user equipment (UEs) and base stations (BSs) no longer rely solely on continuous channel probing. In scenarios where radio nodes have amassed substantial knowledge, the necessity for frequent channel estimation diminishes. Furthermore, rather than relying on bit-driven signaling messages to gauge the channel, semantic communications enable a continual understanding of contextual information from previous messages. This contextual awareness can facilitate learning channel characteristics while utilizing computational resources efficiently. Essentially, semantic communications fundamentally alter the mechanism of communication tasks, as exemplified by the transformation in channel estimation.

1.1.3.3 Semantic Communications Is Not Only Goal-Oriented Communications

A goal-oriented communication system involves a number of agents that interact and exchange messages to achieve a joint goal or separate goals that include the same environment. For example, two robots can interact with each other to execute a common mission. Here, in contrast to sending the information gathered by sensors bit-by-bit, the robots can exchange multiple feedback messages of their current semantic action, their next expected outcome, all while achieving a unique joint goal. In a goal-oriented framework, the nodes, e.g., the teacher and apprentice, can also be achieving two separate goals. Much of the early-on work on semantic communication has equated it with such goal-oriented communication systems [Xie et al., 2022; Zhang et al., 2022; Farshbafan et al., 2023]. However, there are fundamental differences between the two concepts. In some sense, goal-oriented communications falls under the umbrella of semantic communications. For instance, in every goal-oriented communication system, the nodes will have to embed semantic representations to ultimately achieve a particular goal. In contrast, under the broader auspices of a semantic communication system, the generation and communication of semantic representations is not necessarily done for the purpose of serving a system-wide goal. In this regard, limiting the concept of semantic communications to the confines of goal-oriented systems will therefore unnecessarily limit its use to a subset of use cases that have a competitive or cooperative nature. Meanwhile, there are many instances in which the teacher and the apprentice do not necessarily share any joint goals nor interact with a common environment. For instance, the teacher can be a server that is transmitting highly data-intensive content (e.g., XR content) to a particular user. Here, every standalone content transmitted can have an entirely different goal, and there are no cooperative or competing goals between the teacher and the apprentice. Yet, in this case, semantic communications can still be used to: (i) rely less on the channel to transfer massive information content and (ii) empower radio nodes with reasoning to make versatile decisions, which can enhance network's capability in meeting the stringent requirements of future applications.

1.1.3.4 Semantic Communications Is Not Only Application-Aware Communications

Implementing the context of information within the transmission of messages may seem at first glance similar to the traditional concept of application-aware communication. In fact, there are many prior works (e.g., Shajaiah et al. [2019], Chen et al. [2019], Temdee and Prasad [2018], and Chen et al. [2020]) that have fine-tuned the network optimization process to address application-level requirements. For example, in XR applications, the XR content transmitted by users may exhibit a particular correlation. Here, some works such as Chen et al. [2019] exploit this correlation to ultimately improve the management of uplink and downlink wireless transmissions. Notably, it is important to distinguish between the "context-awareness" concept defined by such frameworks and the one granted with semantic communications systems, as the former is a mere application and use case-specific awareness. In contrast, in semantic communication systems, "context" is a concept defined with respect to the low-level structure of exchanged data streams between the transmitter and the receiver. Such low-level intelligence opens the door for an inter-application, intra-application, and out-of-domain generelizability. In other words, a radio node can leverage the meaning attributed to low-level data corresponding to service A by using it to improve the E2E performance for service B.

1.2 Reimagining Future G Applications with Semantic Communications

In this section, we will explore how the adoption and deployment of semantic communication will fundamentally transform emerging applications like XR, digital reality, and concepts such as sustainable networks.

1.2.1 Semantic Communication for Next-Generation XR

The realm of XR, encompassing augmented reality (AR), virtual reality (VR), and evolutionary merged reality concepts like holographic teleportation, promises to revolutionize the way we interact with the world around us. However, current XR experiences face significant challenges, particularly in the realm of data transmission. The vast quantities of data required to render realistic environments create bottlenecks, hindering immersion and user experience [Chaccour et al., 2022]. Essentially, XR experiences often require extremely high data rates as well as high reliability, availability, resilience, and low latency. This combination is often difficult to achieve due to physical layer limitations. Strategies include advanced compression, edge computing, network slicing, adaptive streaming, and foveated rendering. These strategies could gain added flexibility with semantic communication as the backbone instead of classical methods.

1. **From raw data to semantic meta-data:** Current XR systems transmit vast amounts of data to represent complex environments [Chaccour et al., 2022]. Semantic communication breaks this paradigm by prioritizing the transmission of critical metadata, such as object type, location, and attributes as well as their inherent structure. This metadata carries the semantic meaning, significance, or structure of the scene, allowing for efficient reconstruction at the receiving end. For instance, an AR system utilizing semantic communication might transmit the "control" semantic information "fire exit" associated with a specific door, rather than the entire door image. This significantly reduces data volume and bandwidth requirements, leading to a more lightweight and efficient XR experience.
2. **Reliability and error correction in semantic XR:** While advancements like massive multiple-input and multiple-output (MIMO) and higher carrier

frequencies promise high data rates for future XR, semantic communication offers a more nuanced approach. By focusing on transmitting critical metadata, via semantic representations, the overall data volume shrinks. This reduction in data volume translates to more robust connections, even in less-than-ideal network conditions. Additionally, with the rise of AI-powered error correction techniques at receiving devices, the need for extremely high transmission reliability lessens for established knowledge bases. However, for novel and complex transmissions, ensuring reliable metadata delivery remains crucial. Here, novel reliability, resilience, and robustness studies are encouraged to be developed by the research community, to address potential underlying challenges in these different scenarios (nodes with established knowledge bases versus nodes in a transient language learning phase).

3. **AI and context-aware communication for personalized XR:** The power of semantic communication extends beyond data reduction. When coupled with AI, it unlocks a new level of personalization, contextualization, and situational awareness in XR experiences. Semantic encoding of sensor data, rich with contextual information, allows AI to understand user intent, future goals and behaviors as well as underlying implicit preferences. This can be leveraged to deliver highly personalized AR experiences, such as an AR assistant that analyzes user behavior, location, and gaze direction to highlight relevant products with semantic overlays.
4. **Sensor fusion and joint communication for seamless XR:** XR heavily relies on sensor fusion, where data from various sensors (cameras, LiDAR, heterogeneous multi-band networks) is combined to create a complete picture of the environment. Semantic communication beautifully complements this approach. When sensor data is processed so as a semantic representation is captured, it undergoes a transformation that enhances its interoperability across various types of sensors and modalities. By translating raw data into a semantic domain, the information becomes more universally understandable and easier to integrate with different systems. This interoperability among heterogeneous data types facilitates a tighter cohesion in sensor fusion, where inputs from different sensors are combined to provide a comprehensive understanding of the environment. Moreover, semantic encoding helps minimize potential lag in data transmission and processing. By reducing the overhead associated with transmitting and interpreting raw sensor data, real-time synchronization among different sensors is improved, leading to a smoother and more immersive experience in XR applications. Essentially, this could potentially leverage the concept of semantic similarity, which allows different pieces of information pertaining to the same angle or subject to be consolidated into a cohesive worldview.

5. **Computation over communication: Edge processing for challenging environments:** Advancements in on-device processing power pave the way for the integration of edge computing with semantic communication. By transmitting only the essential semantic data, the receiving device can leverage edge computing resources to reconstruct the intended information. This is particularly beneficial in challenging environments with high signal reflection, where traditional communication methods struggle.

1.2.2 Digital Reality and Massive Twinning: Speaking the Same Language

Analogous to the paradigm shift in XR with semantic communication, digital twins stand to gain significant advantages from the establishment of a common language, or vocabulary. This chapter explores the concept of a digital twin language, focusing on its potential to bridge the communication gap between the physical system (the "physical twin") and its corresponding digital representation (the "digital twin"). Imagine a scenario wherein a connected autonomous robot (CAR) and its digital twin engage in constant communication. By establishing a standardized set of definitions and structures for their interactions, these entities can achieve substantial benefits:

1. **Minimized synchronization requirements:** A shared language eliminates the need for the physical and digital twins to transmit real-time details. Communication can focus on exchanging critical information like the physical twin's location, sensor readings, and intended actions. This streamlined approach minimizes the frequency of required synchronization, thereby reducing the overall communication burden, data rate, reliability, and synchronization requirements.
2. **Extrapolation despite latency:** Tight synchronization requirements in digital twins, especially when they scale to massive twinning systems inherently involve a lag. However, if the digital twin comprehends the "language" of the physical twin's data, it possesses the ability to potentially predict the physical twin's future actions based on past information and the current knowledge base and context. This capability allows the digital twin to maintain relevance even in the presence of some lag, ultimately enhancing system performance.
3. **Semantic showers and intent recognition:** Now consider a situation where the physical twin, our CAR, encounters an unforeseen obstacle. Rather than inundating the digital twin with a deluge of raw sensor data, the CAR could transmit a high-level semantic representation that entails "unidentified obstacle detected, rerouting." This concise semantic communication enables the digital twin to grasp the CAR's intent (avoiding the obstacle) and subsequently update its own simulation accordingly.

This approach eliminates the need for continuous, low-level data transmission. Consequently, communication becomes more efficient, allowing the digital twin to concentrate on the broader objective – understanding the robot's goals and adapting its simulation to support them.

4. **Beyond efficiency: A broader impact:** A shared language between physical and digital twins extends beyond simply reducing communication load. Essentially, semantic communication enables a better planning, reasoning, and conclusion drawing.
   - **Cognitive synergy**: The symbiotic alliance between physical and digital twins unleashes unprecedented cognitive synergy. By capturing the structural and semantic situational awareness of humans in their environment, this synergy enables the digital twin not only to comprehend the state and intentions of its physical counterpart but also to engage in collaborative decision-making. This collaborative process seamlessly blends data-driven insights with human expertise, facilitated by a confluence of semantics and imitation learning techniques [Hua et al., 2021].
   - **Proactive insight generation**: Drawing upon its extensive knowledge base and profound understanding of the behavior of the physical twin, the digital twin proactively anticipates potential scenarios and generates insightful predictions, fostering proactive problem-solving and ultimately enhancing system performance and resilience. Essentially, if the knowledge base is imbued with causal and associational logic, this insight generation will be equipped with granular planning and robust meta-decision-making capabilities.

1.2.3 Semantic Communication and Sustainable Networks: A Convergence for Efficiency

The convergence of communication and computation at the network edge takes on a new dimension with the emergence of semantic communication. This transformative approach transcends the mere placement of processing power at the edge. It signifies the birth of a network infused with AI, capable of understanding and reasoning about the data it transmits. Semantic communication aims to revolutionize how information travels, focusing not just on the raw data itself, but on the meaning and intent it carries.

However, unlocking this potential necessitates a paradigm shift. While current energy efficiency efforts focus on optimizing power amplifiers in the Radio Access Network (RAN) through on/off switching, a truly sustainable future demands a more comprehensive approach. To achieve this, we must transition from traditional, resource-intensive AI in the telecom space toward lighter-weight, scalable, and learning-optimized models specifically designed for the network edge. These new AI models will be the cornerstone of semantic communication, enabling efficient and intelligent communication at the edge. This shift necessitates advancements on multiple fronts:

1. **Novel hardware:** Energy-efficient hardware specifically designed for AI workloads at the network edge is crucial. Emerging technologies like neuromorphic computing, which mimics the structure and function of the human brain, hold immense promise. Neuromorphic chips, with their significantly lower power consumption compared to traditional processors while delivering excellent performance for specific AI tasks, are a prime example [Shastri et al., 2021]. Additionally, exploring technologies like *quantum computing* offers the potential for revolutionary advancements in AI algorithms for communication networks. Quantum computers, with their ability to perform certain calculations exponentially faster than classical computers, could unlock entirely new paradigms for network optimization and resource allocation.
2. **Advanced architectures:** Beyond hardware, novel software architectures are needed to optimize AI models for resource-constrained edge devices. Techniques like *knowledge distillation*, where a complex model's knowledge is transferred to a smaller, more efficient model, offer significant potential. Additionally, exploring other compression and pruning techniques like *structured pruning* (removing entire filters or neurons) and *magnitude pruning* (reducing the weights of specific connections) will be essential in creating lightweight AI models suitable for edge deployments. Furthermore, leveraging distributed learning architectures like *federated learning*, *split learning*, and *sequential learning* can be instrumental. Federated learning allows training models on distributed datasets residing on user devices without compromising user privacy, while split learning focuses on training different parts of the model on different devices, further reducing computational burden at the edge. Sequential learning techniques, adept at handling ordered data streams, can be particularly useful in network scenarios where data arrives in a specific sequence, such as traffic patterns or user behavior over time.
3. **Optimization frameworks**: Developing robust optimization frameworks specifically tailored for the unique constraints of the network environment is paramount. These frameworks should address factors like latency, bandwidth limitations, and energy consumption during the training and deployment of AI models. Frameworks should incorporate techniques like *dynamic model selection* based on network conditions and user requirements and *adaptive learning algorithms* that can continuously optimize model performance while minimizing resource consumption. In addition, integrating different types of logic, such as causal and associational types, along with common sense, into the mix could reduce the amount of processing needed in compute. By leveraging causal reasoning, the system can infer cause-and-effect relationships, allowing

for more efficient decision-making and resource allocation. Associational logic enables the recognition of patterns and correlations, enhancing the system's ability to interpret and respond to complex data inputs. Furthermore, incorporating common sense knowledge helps contextualize information and guide reasoning, streamlining the processing required for accurate decision-making. This integration not only enhances the intelligence of the system but also optimizes computational resources by enabling more targeted and efficient processing.

By embracing these advancements, the telecommunications industry can harness the full potential of AI at scale while ensuring sustainability. This native AI approach prioritizes lightweight models, efficient hardware, and distributed learning techniques, setting the stage for seamless integration of transformative technologies. With a focus on semantic communications, this approach paves the way for scalable advancements. As research advances, particularly in areas like quantum computing, the prospect of sustainable and powerful communication networks anchored in semantic communication becomes increasingly feasible.

1.3 Structure and Path of the Book

This book serves as a comprehensive reference volume for anyone looking to deepen their understanding, conduct research, or innovate in the field of semantic communications. Divided into four meticulously curated parts, each section explores distinct facets of semantic communication, offering invaluable insights, theoretical foundations, practical applications, and innovative solutions to address emerging challenges in this rapidly evolving domain.

- Part I of this book serves as the bedrock upon which our exploration of semantic communication is built. We begin by establishing the theoretical underpinnings, delving into information theory to elucidate how semantic information augments classical notions. Moving beyond theoretical discourse, we dissect the very essence of semantic information, employing mathematical constructs such as topology and category theory to provide a rigorous foundation. Our journey then advances into the realm of communication theory, where we unveil the principles driving semantic communications, optimizing transmission processes for robust wireless systems. By selectively encoding and transmitting vital information while gracefully degrading less crucial data, we pave the way for efficient semantic communication networks. Finally, we transcend individual systems to design methods that foster coexistence and interoperability among diverse generations of communication technologies, ensuring a seamless transition to 6G semantic networks.

- In Part II, we delve into the practical applications and networking features of semantic communication. Here, our focus shifts from theoretical constructs to real-world scenarios, exploring multiple access and networking challenges inherent in semantic communication systems. Through collaborative frameworks optimizing image transmission and a meticulous examination of multiple access schemes, we strive to enhance transmission efficiency and coexistence with legacy systems. Moreover, by harnessing contextual reasoning techniques and delving into the critical aspect of interoperability, we aim to unlock the full potential of semantic communication networks, laying the groundwork for seamless integration into existing infrastructures.
- As we transition into Part III, our exploration takes us deeper into the realms of AI and ML that underpin semantic communication systems. Here, we embark on a journey to understand the transformative role of deep learning in reshaping semantic communication networks. By advocating for a departure from data-driven paradigms to knowledge-driven AI-native networks, we seek to construct a semantic language grounded in causality, transcending mere associational logic. Furthermore, by confronting scalability challenges and addressing the "semantic gap" through innovative supervision mechanisms, we strive to bridge the divide efficiently during the learning process. Our exploration culminates in a discussion on concealed information and the strategic valuation of data, shedding light on unique challenges and potential solutions in semantic and goal-oriented communication frameworks.
- In the fourth and final section of our book, we confront the paramount challenge of security in semantic communication. Here, we undertake a comprehensive examination of the multi-domain security risks inherent in deploying deep neural networks. Through meticulous analysis of adversarial machine learning and the introduction of encrypted, privacy-preserving semantic communication systems, we endeavor to fortify the integrity and reliability of semantic communication networks. By deploying robust adversarial training methodologies and embracing a data valuation perspective, we pave the way for a new era of trust and reliability in semantic communication, safeguarding privacy and ensuring the secure exchange of information in an increasingly interconnected world.

Bibliography

Christina Chaccour, Mehdi Naderi Soorki, Walid Saad, Mehdi Bennis, and Petar Popovski. Can terahertz provide high-rate reliable low-latency communications for wireless VR? *IEEE Internet of Things Journal*, 9(12):9712–9729, 2022.

Christina Chaccour, Walid Saad, Merouane Debbah, Zhu Han, and H. Vincent Poor. Less data, more knowledge: Building next generation semantic communication networks. *IEEE Communication Surveys and Tutorials*, 2024.

M. Chen, W. Saad, C. Yin, and M. Debbah. Data correlation-aware resource management in wireless virtual reality (VR): An echo state transfer learning approach. *IEEE Transactions on Communications*, 67(6):4267–4280, Feb. 2019.

X. Chen, T. Tan, G. Cao, and T. F. L. Porta. Context-aware and energy-aware video streaming on smartphones. *IEEE Transactions on Mobile Computing*, 21(3):862–877, Aug. 2020.

M. K. Farshbafan, W. Saad, and M. Debbah. Curriculum learning for goal-oriented semantic communications with a common language. *IEEE Transactions on Communications*, 71(3):1430–1446, Mar. 2023.

Jiang Hua, Liangcai Zeng, Gongfa Li, and Zhaojie Ju. Learning for a robot: Deep reinforcement learning, imitation learning, transfer learning. *Sensors*, 21(4):1278, 2021.

Qiao Lan, Dingzhu Wen, Zezhong Zhang, Qunsong Zeng, Xu Chen, Petar Popovski, and Kaibin Huang. What is semantic communication? A view on conveying meaning in the era of machine intelligence. *Journal of Communications and Information Networks*, 6(4):336–371, Dec. 2021.

H. Shajaiah, M. Ghorbanzadeh, A. Abdelhadi, and C. Clancy. Application-aware resource allocation based on channel information for cellular networks. In *Proceedings of IEEE Wireless Communications and Networking Conference (WCNC)*, Marrakesh, Morocco, Apr. 2019.

C. E. Shannon. A mathematical theory of communication. *The Bell System Technical Journal*, 27(3):379–423, Jul. 1948.

Bhavin J. Shastri, Alexander N. Tait, Thomas Ferreira de Lima, Wolfram H. P. Pernice, Harish Bhaskaran, C. David Wright, and Paul R. Prucnal. Photonics for artificial intelligence and neuromorphic computing. *Nature Photonics*, 15(2):102–114, 2021.

P. Temdee and R. Prasad. *Context-Aware Communication and Computing: Applications for Smart Environment (Springer Series in Wireless Technology)*. Springer, 2018.

Warren Weaver. Recent contributions to the mathematical theory of communication. *ETC: A Review of General Semantics*, 10(4):261–281, 1953.

H. Xie, Z. Qin, X. Tao, and K. B. Letaief. Task-oriented multi-user semantic communications. *IEEE Journal on Selected Areas in Communications*, 40(09):2584–2597, Nov. 2022.

Y. Zhang and D. A. Adjeroh. Prediction by partial approximate matching for lossless image compression. *IEEE Transactions on Image Processing*, 17(6):924–935, Apr. 2008.

C. Zhang, H. Zou, S. Lasaulce, W. Saad, M. Kountouris, and M. Bennis. Goal-oriented communications for the IoT and application to data compression. *IEEE Internet of Things Magazine, Special Issue on Pervasive, Efficient and Smart Signal Processing for IoT*, 5(4):58–63, Dec. 2022.

Part I

Fundamentals of Semantic Communications

This part starts with a chapter whose goal is to lay down the information-theoretic foundations of semantic communications for text transmission that represents a simplified scenario, delving into fresh perspectives on how semantic information could push the boundaries of classical information theory. Moreover, in Chapter 3 of this part, an in-depth analysis of the theoretical aspects of defining semantic information will be conducted, from a topology and mathematical perspective. Here, new mathematical constructs that are indispensable for rigorously formulating semantic information, such as topology and category theory, will be explained and discussed. The subsequent chapter moves into the communication-theoretic principles of semantic communications by introducing the framework of semantic-driven joint-source channel coding, which enables link-level semantic communications by optimizing the transmission process and filtering out irrelevant information. By combining source coding techniques with channel coding strategies, this framework improves the robustness of wireless systems, enabling reliable communication even in challenging channel conditions. Selective encoding and transmission of crucial information, along with the removal of redundant and irrelevant data, enhance the overall efficiency and effectiveness of semantic communication networks. Furthermore, a novel perspective on semantic information is presented, which views information as a function of multiple data points. By adopting an information theoretic framework, this approach delves into semantic communication via the lens of invariance principle and function reconstruction. Finally, building on the previously developed information and communication theoretic foundations, the final chapter of Part I discusses novel methods for achieving coexistence and interoperability between 6G semantic, goal-oriented, and legacy systems.

Foundations of Semantic Communication Networks, First Edition.
Edited by Walid Saad, Christina Chaccour, Christo Kurisummoottil Thomas, and Merouane Debbah.

2

Semantic Compression and Communication: Fundamentals and Methodologies

Emrecan Kutay and Aylin Yener

Department of Electrical and Computer Engineering, INSPIRE@OhioState Research Center, The Ohio State University, Columbus, OH, USA

2.1 Introduction

In 1948, Shannon formulated the communication problem as the exact reconstruction of a message at another point. This seminal work abstracted messages irrespective of what they represented, i.e., their meanings [Shannon, 1948]. This abstraction thus represented messages as digital sequences, thereby formulating the problem of reliable communication as the reconstruction of these sequences. The resulting framework led to the development of syntactic information measures and error metrics. Specifically, source characteristics were utilized to establish bounds on compressibility, while noise characteristics were incorporated to determine the fundamental limit on the rate for reliable communication.

The next 75 years of research on information theory developed in a number of directions from lossy recovery to networked communication, and from codes that approach Shannon limit to application domains such as finance, machine learning, quantum, and security that benefit from the foundational approach of constructive and impossibility proofs that the original framework built.

Future applications of networked systems converge a number of traditionally distinct operations such as communications, computing, and sensing. Additionally, applications expected to materialize in the next generation such as digital twins and autonomous networked systems continue to be resource-intensive calling for solutions toward distilling information that is strictly necessary for a task and application at hand [Atzori et al., 2010; Jameel et al., 2019; Gunduz et al., 2020]. In these systems, considering the meanings of messages and designing systems that are semantic-aware are viable alternatives toward better utilization, in

Foundations of Semantic Communication Networks, First Edition.
Edited by Walid Saad, Christina Chaccour, Christo Kurisummoottil Thomas, and Merouane Debbah.

particular, in the wireless domain [Shi et al., 2021; Kutay and Yener, 2023; Yang et al., 2023].

It is worth noting that efforts to incorporate semantics into the communication problem dates back to shortly after Shannon's 1948 paper (see Shannon and Weaver [1949]). These efforts drew on concepts from information theory, graph theory, game theory, and, more recently, deep learning Wheeler and Natarajan (DL) [Wheeler and Natarajan, 2023]. In early work, Carnap and Bar-Hillel [1952] introduced the concept of semantic entropy, utilizing logical probability functions to quantify semantic information, resulting in contradictory sentences being assigned the highest information [Gunduz et al., 2023]. Using graph theoretic and zero-error interactive communications, Guler et al. [2013] formulated a semantic two-way communication problem, where each end aims to communicate its own facts to obtain a meaningful conclusion. Shifting to game theory, Guler et al. [2018] examined a problem involving an agent, which can be helpful or adversarial, and investigated optimal encoding–decoding strategies at equilibrium of the game between the agent and the system.

The universal approximation capabilities of neural networks and their excellent performance in a number of domains have motivated researchers to integrate them into (wireless) communication systems design [Hornik, 1991; He et al., 2015]. In pioneering reference O'Shea and Hoydis [2017], communication problem has been reframed as an end-to-end reconstruction task, structuring transmitter and receiver blocks as fully connected (FC) layers. In particular, the problem involves transmitting M possible messages, treating the communication task as a receiver-end classification. Results demonstrate that joint optimization of transmitter and receiver blocks, achieved through backpropagation of error, outperforms conventional techniques such as uncoded and Hamming-coded binary phase-shift keying (BPSK) in terms of block error rate (BLER). These results have sparked interest in autoencoder communication, where the encoder block performs channel coding and modulation, while the decoder handles demodulation and channel decoding [Sagduyu et al., 2023].

Recent semantic communication models build upon autoencoder communication and leverage DL to extract and convey underlying meaning. In Xie et al. [2021], the utilization of transformer neural networks with FC layers has been investigated to obtain the message semantic information in a point-to-point semantic text communication problem. For images, reference Lo et al. [2023] incorporates convolutional layers to extract image features for task-oriented communication in multiple access channels (MAC), where edge devices implicitly collaborate for an image retrieval task [Bourtsoulatze et al., 2019]. Motivated by autoencoder communication, recent DL-based models adopt a joint optimization approach involving end-to-end training for transmitter–receiver blocks.

In this chapter, we introduce the early work in semantic communications, reviewing the semantic index assignment problem, with a specific emphasis on constructing a codebook guided by a semantic distortion objective in Section 2.2. We briefly summarize the progress in DL since the introduction of this framework in Section 2.3. We reconsider the codebook construction problem with a task-oriented communications problem, leveraging tools from DL to address semantic distortion objectives in Section 2.4. As we showcase the advantages of integrating semantics, we also underscore the impact that DL has had on the landscape of semantic communication. The chapter concludes in Section 15.9.

2.1.1 Notation

In the remainder of this chapter, logarithms are taken to be base 2, $\lceil . \rceil$ is the ceiling operator, $\exists$ denotes there exists, $\forall$ denotes for all, $||X||_2$ denotes Euclidean norm of vector X. Given a set $\mathcal{W}$, $|\mathcal{W}|$ denotes its cardinality. For a matrix A, $A(i)$ refers to the ith row.

2.2 Semantic Index Assignment

Conventional communication models rely on syntactic error metrics, such as the bit error rate (BER) and symbol error rate (SER), for reconstruction of bit sequences representing the source message. These metrics have been optimized by employing optimal constellation diagrams and detection algorithms in modulation, along with error-correcting codes that introduce controlled redundancy. Naturally, there may be situations where rather than exact reconstruction of a message, conveying its meaning correctly would suffice, as is often the case in human communications. For instance, consider Alice transmitting the word "automobile" to Bob. If Bob reconstructs it as a "car," syntactic metrics treat it as an error despite the two words being synonyms, and the meaning is in fact correctly conveyed. The basic premise of semantic communications is to move from these syntactic metric to semantic metrics, in order to design systems that are more resource efficient, effectively reducing the uses of the channel (resource).

The definition of an end-to-end semantic metric dates back to Guler and Yener [2014]. Therein, the concept of semantic distortion is introduced by utilizing lexical taxonomies. In linguistics, lexical taxonomies are often organized in hierarchical order, allowing for the inclusion of structural information in the calculation of the semantic similarity of given words. Specifically, two main approaches exist for taxonomy-based semantic similarity calculation: node-based and edge-based [Jiang and Conrath, 1997].

Node-based similarity relies on the notion that words sharing more information in common are semantically more similar. In this context, information content (*IC*) is determined by the concept subsuming both words, leading to the calculation of similarity through concepts [Resnik, 1995]. By incorporating the definition of *IC*, the similarity between concepts c_i and c_j is expressed as follows [Resnik, 1995].

$$sim(c_i, c_j) = \max_{c \in S(c_i, c_j)} [IC(c)] = \max_{c \in S(c_i, c_j)} \left[\log \left(\frac{1}{p(c)} \right) \right]. \tag{2.1}$$

In Eq. (2.1), $S(c_i, c_j)$ represents the set of concepts subsuming both c_i and c_j, while $p(c)$ denotes the probability of the occurrence of concept c in a large document. Building upon Eq. (2.1), the semantic similarity between words w_i and w_j is defined as follows [Resnik, 1995].

$$sim(w_i, w_j) = \max_{c_i, c_j} [sim(c_i, c_j)]. \tag{2.2}$$

In Eq. (2.2), c_i and c_j represent the concepts of words w_i and w_j, respectively, reflecting different senses (meanings) of these words. In a tree-form taxonomy, this computation reveals that words demonstrate higher similarity when their common ancestor node (concept) is situated closer to the leaf edges.

The edge-based similarity approach utilizes distances within the taxonomy to calculate similarities by examining the relative path distance between words. In this method, words that are closer in the taxonomy are considered semantically more similar. However, relying solely on the distance calculation along the shortest path is insufficient, as links in a taxonomy lack uniformity [Rada et al., 1989]. To address this limitation, a refined approach involves assigning weights to edges in the taxonomy, considering factors such as node depth, link type, and hierarchical density [Jiang and Conrath, 1997].

The hybrid similarity metric, denoted as sim_{hybrid}, integrates concepts from both edge- and node-based approaches, utilizing a weighted edge structure. This approach incorporates IC, similar to the node-based approach, by integrating it into the calculation of edge weights [Jiang and Conrath, 1997]. Guler and Yener [2014] defines the semantic distortion between any two words, w_i and w_j, using a normalized version of the hybrid similarity, denoted as $sim_{hybrid}(w_i, w_j) \in [0, 1]$.

$$d(w_i, w_j) = 1 - sim_{hybrid}(w_i, w_j). \tag{2.3}$$

2.2.1 System Model

In conventional communication systems, codebook construction addresses syntactic optimization. For example, in a variable-length code, the objective is to minimize the expected codeword length while ensuring unique or instantaneous decodability [Cover and Thomas, 2006]. In the case of fixed-length coding, concepts such as Gray coding are utilized, wherein neighboring symbols have codewords with a Hamming distance of 1 [Proakis and Salehi, 2005]. These coding schemes focus directly on the probabilistic characteristics of the source and not its semantics.

The codebook construction problem has been reformulated in Guler and Yener [2014] to incorporate semantic distortion in Eq. (2.3) into the optimization process. This reformulation establishes a semantic-aware coding scheme and is also fitting to query-based semantic text communication (with context information). In this setup, the receiver initiates communication by sending a query to the transmitter. The transmitter's objective is to answer the given query, utilizing a subset of its symbol set $\mathcal{W}$, which consists of words. For example, the receiver may pose the query "What is the moving object approaching the building?" In response, the transmitter utilizes the associated subset of $\mathcal{W}$, such as {car, automobile, cat}, from the complete set $\mathcal{W}$ = {car, automobile, cat, table, school}. Within this framework, queries are assumed to originate from a predetermined set $\mathcal{Q}$ known by both ends in advance. The semantic index assignment, denoted as $\pi(\cdot)$, involves assigning codewords to words in $\mathcal{W}$ while considering semantic relations within $\mathcal{W}$ and between $\mathcal{W}$ and $\mathcal{Q}$.

This problem is addressed using two key concepts: semantic distortion minimization and graph coloring. The former aims to mitigate effects over noisy channels in a fixed-length coding scheme, while the latter is employed to reduce codeword length, as discussed in Sections 2.2.1.1, 2.2.1.2, and 2.2.1.3.

2.2.1.1 Minimization of Semantic Distortion

This section investigates the minimization of semantic distortion by optimizing the index assignment operator $\pi(w_i), \forall w_i \in \mathcal{W}$. Since a fixed-length scheme is employed, the codeword lengths are set to $n = \lceil \log_2(|\mathcal{W}|) \rceil$ bits during calculations. In particular, for the word w_i, $\pi(w_i)$ is transmitted through a binary symmetric channel (BSC) with a fixed crossover probability β. Semantic distortion is calculated using the codeword transition probabilities over the BSC [Guler and Yener, 2014].

$$p\left(\pi(w_i)\middle|\pi(w_j)\right) = (1-\beta)^{n-h(\pi(w_i),\pi(w_j))} \times \beta^{h(\pi(w_i),\pi(w_j))}. \tag{2.4}$$

In Eq. (2.4), $h(\cdot,\cdot)$ denotes the Hamming distance between given codewords. Using Eqs. (2.3) and (2.4), the average semantic distortion associated with the given coding scheme $\pi(\cdot)$ is defined as follows [Guler and Yener, 2014].

$$D(\pi) = \sum_{i=1}^{|W|} p(w_i) \left(\sum_{j=1}^{|W|} p(\pi(w_j)|\pi(w_i)) \times d(w_i, w_j) \right). \tag{2.5}$$

In Eq. (2.5), $p(w_i)$ represents the source probability of word w_i. The objective is to minimize the distortion defined in Eq. (2.5) to achieve the optimal index assignment, denoted as π_{opt}. This problem is a quadratic assignment problem, which is NP-complete [Guler and Yener, 2014].

$$\pi_{opt} = \arg\min_{\pi} D(\pi). \tag{2.6}$$

Minimizing Eq. (2.5) requires smaller semantic distortions for codewords with higher transition probabilities. Also, for $\beta < 0.5$, Eq. (2.4) indicates that codewords with smaller Hamming distances will exhibit higher transition probabilities. These implications suggest that words with similar meanings should be assigned to codewords with small Hamming distances.

2.2.1.2 Graph Coloring Problem

This section investigates the graph coloring problem based on the query-based nature of communication. Specifically, meaningful answers to each query $q_z \in \mathcal{Q}$ constitute a subset of $\mathcal{W}$, denoted as $\mathcal{W}_{q_z} \subseteq \mathcal{W}$. The graph is constructed by considering elements of $\mathcal{W}$ as vertices, and edges are defined based on the relations between $\mathcal{W}$ and $\mathcal{Q}$. More precisely, an edge is established between words $w_i \in \mathcal{W}$ and $w_j \in \mathcal{W}$ according to the following conditions [Guler and Yener, 2014].

- There exists a z such that $q_z \in \mathcal{Q}$ and $w_i \in \mathcal{W}_{q_z}$, $w_j \in \mathcal{W}_{q_z}$
- w_i and w_j are not synonyms

These conditions result in a graph $G = (\mathcal{W}, E)$, where E denotes the edges. In G, connected vertices represent meaningful answers to a given query, i.e., E represents sets $\mathcal{W}_{q_z}$, $\forall q_z \in \mathcal{Q}$. This construction of G enables the problem to be considered as graph coloring, i.e., vertices sharing an edge should be assigned to distinct colors. Thus, the minimum number of codewords corresponds to the chromatic number of G, i.e., the smallest number of colors satisfying the coloring problem. Colors used in this process are denoted as a set C, and codeword assignment involves assigning bit sequences to elements of C instead of $\mathcal{W}$. Since $\lceil \log_2(|C|) \rceil \leq \lceil \log_2(|\mathcal{W}|) \rceil$, the length of codewords decreases in parallel with our objective.

The established framework necessitates a graph G in which words, serving as vertices, must be assigned distinct colors if an edge connects them. Given the construction of G, this requirement becomes assigning different codewords to words that function with the same query except for the ones that are synonyms. The objective function for coloring can be expressed in terms of the codeword assignment as follows [Guler and Yener, 2014].

$$f(\pi) = \sum_{w_i, w_j \in \mathcal{W}} I_{E,\pi}(w_i, w_j). \tag{2.7}$$

In Eq. (2.7), $I_{E,\pi}(\cdot,\cdot)$ is the indicator function defined accordingly.

$$I_{E,\pi}(w_i, w_j) = \begin{cases} 1 \text{ if} (w_i, w_j) \in E \text{ and } \pi(w_i) = \pi(w_j) \\ 0 \text{ otherwise} \end{cases}. \tag{2.8}$$

In Eq. (2.8), $(w_i, w_j) \in E$ denotes that there exists an edge connecting w_i and w_j. The formulated coloring problem imposes a bijective codeword assignment on elements of sets $\mathcal{W}_{q_z}$, $\forall q_z \in \mathcal{Q}$, but not necessarily for the entire source symbol set $\mathcal{W}$. Given the receiver's possession of the query (side information), even if certain source symbols share the same codeword, the receiver can still differentiate between them. This principle forms the basis for the introduced efficiency.

2.2.1.3 Joint Graph Coloring and Index Assignment

In Sections 2.2.1.1 and 2.2.1.2, each problem is examined independently. In this section, the combined problem is formulated through a weighted multi-objective optimization problem [Guler and Yener, 2014].

$$\pi_{opt} = \arg\min_{\pi} \alpha_1 f(\pi) + \alpha_2 D(\pi) \tag{2.9}$$

In Eq. (2.9), α_1 and α_2 represent the weights of the associated functions. Simulated annealing is utilized for optimizing the formulated problems, i.e., Eqs. (2.6) and (2.9).

In addressing the semantic distortion problem formulated in Section 2.2.1.1, the message set is defined as illustrated in Table 2.1. Since this problem is considered independently, without incorporating graph coloring, codeword lengths are set to $n = \lceil \log_2(|\mathcal{W}|) \rceil = 4$.

Table 2.1 represents the calculated optimal assignment π_{opt}, revealing that semantically similar words are assigned to codewords with small Hamming distances, aligning with the intuition discussed in Section 2.2.1.1. For instance, word tuples like (car, automobile), (midday, noon), and (boy, lad) all have a Hamming distortion of 1, indicating their near synonymy. This facilitates robustness, as even if a bit error occurs, the receiver will be able to estimate the intended meaning.

In addressing the combined problem, i.e., Eq. (2.9), the weights are set to $\alpha_1 = 0.7$ and $\alpha_2 = 0.3$. The word and query sets $\mathcal{W}$ and $\mathcal{Q}$ are reconstructed as given below.

$$\mathcal{W} = \{\text{car, person, office, automobile, woman, building, vehicle, school}\},$$

$$\mathcal{Q} = \begin{Bmatrix} q_1 = \text{“What is the moving object approaching the building?”} \\ q_2 = \text{“Where is Bob working at?”} \end{Bmatrix}.$$

Table 2.1 Optimal codeword assignment for semantic distortion problem.

Words	Assigned Codewords
Car	0100
Monk	1010
Noon	1001
Automobile	0000
Wizard	1111
Lad	0111
Midday	1101
Jewel	0101
Magician	1011
Fruit	0010
Journey	1000
Voyage	1100
Food	0011
Brother	1110
Gem	0001
Boy	0110

Source: Guler and Yener [2014]/With permission of IEEE.

For these sets, $\mathcal{W}_{q_1}$ and $\mathcal{W}_{q_2}$ are defined as follows.

$$\mathcal{W}_{q_1} = \{\text{car, person, automobile, woman, vehicle}\}$$
$$\mathcal{W}_{q_2} = \{\text{office, building, school}\}.$$

The constructed characteristic graph G with the obtained optimal codeword assignment is depicted in Figure 2.1, illustrating $\mathcal{W}_{q_1}$ and $\mathcal{W}_{q_2}$. While both "car" and "automobile" are included in $\mathcal{W}_{q_1}$, no edge is assigned, as they are synonyms violating the second condition for establishing edges. Since the chromatic number of the characteristic graph in Figure 2.1 is 4, codeword lengths become $n = \lceil \log_2(|C|) \rceil = 2$, whereas without graph coloring it would be $\lceil \log_2(|\mathcal{W}|) \rceil = 3$. This trivial example highlights the importance of integrating semantics into the communication framework, benefiting both resource efficiency and error resilience.

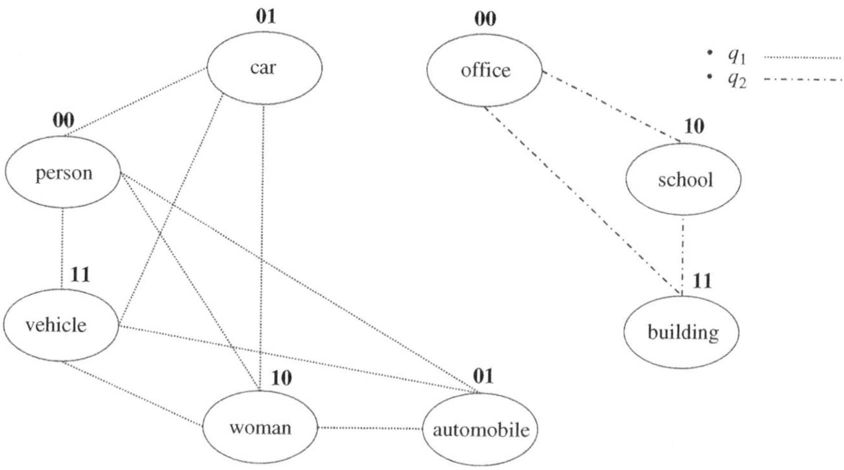

Figure 2.1 Characteristic graph with vertices $\mathcal{W}$ and edges constructed with $\mathcal{Q}$. Optimal codeword assignment is demonstrated in bold for joint graph coloring and index assignment. Source: Guler and Yener [2014]/With permission of IEEE.

2.3 The Rise of Machine Intelligence in Perception

The work in Section 2.2 is the earliest work highlighting the need for semantic metrics and their integration into communication. However, the absence of powerful tools available today, primarily provided by DL structures, necessitated the use cases to remain small groups of words. For a comprehensive understanding of meaning, one must consider the given text as a whole rather than isolated fragments. In particular, in cases involving polysemous words, i.e., words with multiple meanings, comprehending the intended meaning is not feasible without context. For instance, when Alice transmits the sentence "I love rock music," and Bob reconstructs it as "I love rock climbing," a single-word error results in a completely different meaning due to the polysemous nature of the word "rock." This difference cannot be accurately quantified if we calculate semantic distortion at the word level.

Word-level mechanisms, without the consideration of context, have also been considered as a bottleneck in the machine learning community, particularly for tasks that demand a comprehensive understanding of the message. Bahdanau et al. [2014] demonstrates a significant performance improvement in machine translation by introducing context awareness, recently known as the *attention mechanism*. The first attention mechanism relies on context vectors calculated for each word through bidirectional recurrent networks (B-RNNs), providing information about both preceding and succeeding words, i.e., context. While innovative, the structure exhibits a bias towards nearby contexts due to the iterative nature of B-RNNs. This bias arises because the iterations through words increase the complexity of establishing correlations for distant words, making it challenging to capture dependencies effectively [Vaswani et al., 2017]. To illustrate, in the initial attention model, the network must iterate through the entire sentence to establish relationships between the first and last words.

Recent attention models build upon the initial implementation by eliminating the B-RNN structure to effectively calculate long-range dependencies within the input message. Instead, learnable weights are employed to calculate correlations, resulting in a constant number of required computations for obtaining dependencies across all words, regardless of their positions. Specifically, the dot product attention mechanism is employed by utilizing the *query*, *key*, and *value* calculated for each word within the input. These calculations involve the dot product of continuous representations of words, encompassing both their individual semantic and positional information, with the set of learnable parameters associated with each expression, i.e., W^Q, W^K, and W^V [Vaswani et al., 2017]. First, the dot product between the *query* of word i and the *key* of word j determines the attention weight, indicating the extent to which word j influences the context of word i. Subsequently, the attention weights calculated for word i are scaled using the softmax function and then employed to calculate the weighted average of the *values* of words in the input message, thereby providing the context vector for word i. This structure, with a single context vector for each word calculated using a single set of learnable parameters W^Q, W^K, and W^V, forms the single-head attention mechanism. Recently, powerful neural models, e.g., *transformers*, stack multiple heads to grasp the underlying meaning more comprehensively by having several contextual vectors for each word through multiple sets of learnable parameters, i.e., W_i^Q, W_i^K, and W_i^V, where i denotes the ith attention head.

The transformer architecture has initiated a new paradigm in the machine learning community and has also become the primary building block in recent DL-enabled semantic communication models. To showcase its capabilities, the bidirectional encoder representations from transformers (BERT) model have achieved new state-of-the-art results in 11 natural language processing tasks immediately upon its introduction [Devlin et al., 2019]. In addition to its performance, the BERT model also offers flexibility by enabling its utilization for different tasks simply by modifying the last layer according to the task, without requiring substantial architecture modifications. The training of the BERT model involves two stages: unsupervised pretraining and fine-tuning, as outlined below [Devlin et al., 2019].

- Unsupervised pretraining involves two objectives. The first centers on masked language modeling, where some percentage of input words are randomly masked for the model to predict. The second objective is next-sentence

prediction, which employs binary classification to determine whether the given sentence follows the current one. This pretraining stage enables the model to learn how to extract underlying meaning from natural language without the need for human-annotated data, thereby facilitating training with larger datasets. Consequently, the model is exposed to various contexts and learns how to grasp the general context, rather than a specific one. This enables a single unsupervised pretrained model to be applicable for numerous tasks with associated fine-tuning, providing the aforementioned flexibility.
- The fine-tuning stage involves training the model for a specific task following the initial stage, utilizing the associated output layer and loss function. As the model has already undergone extensive training on a large dataset, it possesses the capability to understand the context underlying a given message proficiently, albeit not necessarily how to apply it in a specific manner. In this regard, pre-training enables the model to achieve high task performances with relatively inexpensive training during this stage.

Semantic communication necessitates a similarity metric to evaluate the performance in conveying the meaning. In Section 2.2, the similarity is considered at the word level and is calculated using lexical taxonomies. To extend this notion to the sentence level, BERT emerges as a powerful candidate due to its capability to extract semantic information. However, when dealing with a set of sentences, utilizing BERT to calculate similarities requires feeding all sentence combinations into the network, making it computationally unfeasible. To address this challenge, Sentence-BERT (SBERT) proposes a fine-tuning approach for BERT to calculate contextual embeddings for given sentences containing their semantic information, denoted as $Q \in \mathbb{R}^p$ [Reimers and Gurevych, 2019]. Unlike BERT, SBERT requires each sentence to be fed into the network only once to be assigned an embedding. These embeddings are structured in a way that ensures semantically similar sentences have nearby embeddings, while dissimilar ones are distant. This structure enables vector operations in $\mathbb{R}^p$ to accurately reflect the semantic relations of sentences.

SBERT fine-tunes BERT using three objectives: classification, regression, and triplet objective functions. In all these objectives, mean pooling is employed over the BERT outputs to obtain a fixed-size embedding for each message independent of its length. The classification objective utilizes cross-entropy loss, requiring the model to calculate embeddings for a sentence tuple to predict relative meaning, such as whether the sentences entail, contradict, or are neutral. The regression objective follows a similar structure, but in this case, the model uses mean squared error loss to estimate cosine similarity. The triplet objective function enables the model to position embeddings that reflect their semantic similarity.

In this scenario, each sentence has a positive and negative instance, representing semantically similar and different sentences, respectively. The model aims to minimize the $L2$ distance between the embeddings of the current sentence and its positive instance while simultaneously maximizing the distance between the negative instance. For a sentence s with a positive sample p and a negative sample n, with embeddings Q_s, Q_p, and Q_n, this operation is performed using the following objective function for minimization [Reimers and Gurevych, 2019].

$$\max\left(\|Q_s - Q_p\|_2 - \|Q_s - Q_n\|_2 + \epsilon, 0\right). \tag{2.10}$$

In Eq. (2.10), ϵ denotes the margin parameter controlling the relative distances. The training procedure of SBERT enables its utilization as a semantic encoding block, as it is not tailored to specific contexts or tasks. Moreover, a similar structure, incorporating the OpenAI Contrastive Language-Image Pre-training (CLIP) transformer instead of BERT, yields embeddings for both images and text by effectively embedding different modalities into the same semantic space [Radford et al., 2021].

2.4 Semantic Compression for Multimodal Sources

This section revisits the problem of semantic codebook construction in a task-oriented communication setup by utilizing contemporary tools enabled by DL. In particular, for text messages, the concept of semantic similarity is extended to the sentence level through the utilization of SBERT. As the calculated embeddings enable vector operations to reflect semantic relations, the semantic distortion is re-defined as follows [Kutay and Yener, 2023].

$$\delta(Q, \hat{Q}) = \|Q - \hat{Q}\|_2. \tag{2.11}$$

In Eq. (2.11), Q and $\hat{Q}$ denote the embeddings of the source message $\mathbf{s}$ and the reconstructed message $\hat{\mathbf{s}}$, respectively, i.e., $Q = SBERT(\mathbf{s})$ and $\hat{Q} = SBERT(\hat{\mathbf{s}})$. Since SBERT with OpenAI CLIP model provides embeddings for both image and text modalities, the source message $\mathbf{s}$ can take the form of either an image or text, highlighting the multimodal nature of the source.

2.4.1 System Model

In this section, the focus is on task-oriented semantic communication, where transmitted image and text messages are classified at the receiver's end. The primary objective is to efficiently communicate source messages while achieving high classification accuracies over additive white Gaussian noise (AWGN) and Rayleigh fading channels. The received signal $\mathbf{Y} \in \mathbb{C}^B$ is

$$\mathbf{Y} = h\mathbf{X} + \mathbf{Z}, \tag{2.12}$$

where $\mathbf{X} \in \mathbb{C}^B$ is the channel input, $\mathbf{Z}$ is Gaussian noise, $\mathcal{N}(0, \sigma_Z^2 \mathbf{I}_{B\times B})$, and h represents the channel gain. For AWGN channels $h = 1$, whereas for Rayleigh fading channels $h \sim \mathcal{N}(0, \sigma_h^2)$, with $\sigma_h^2 = 1$. A unit power constraint is assumed for the channel input.

The destination node is equipped with a classifier pair, one for images and one for text. These classifiers take the form of shallow neural networks (NNs), each taking reconstructed embeddings, denoted as $\hat{Q} \in \mathbb{R}^p$, as input and classifying them into labels $\hat{y}$. These models share the same NN structure consisting of three layers with the following numbers of hidden units: 128, 32, and n_{class}, respectively. Here, n_{class} represents the number of classes in the given datasets for text and images. The training involves cross-entropy loss and utilizes a dataset available exclusively at the destination node, consisting of training embeddings Q_{train} with associated labels y_{train}.

Three different separation-based semantic communication models are proposed for the given scenario. These models employ different procedures to construct a codebook for fixed length coding in semantic source coding block. In all models, the output of this block is transmitted using Reed Solomon (RS) channel coding (rate 25/31) and quadrature phase-shift keying (QPSK). For simplicity, demodulation and channel decoding operations are denoted as QPSK^{-1} and RS^{-1}, respectively. To design the codebook, an available dataset Q_{pre}^N is assumed, storing embeddings of previous source messages in an $N \times p$ matrix, where N and p denote the number and dimension of embeddings, respectively. This constraint, influenced by N, governs the amount of available data for design purposes. Having the same available dataset Q_{pre}^N with the same transmission scheme provides a basis for a fair comparison among introduced models.

2.4.1.1 Semantic Quantization

This section introduces the semantic quantization model which relies on the fact that source semantic structure can be discerned through prior realizations [Kutay and Yener, 2023]. To this end, the available dataset Q_{pre}^N is directly utilized as a codebook, having codewords of length $n = \lceil \log_2(N) \rceil$ bits. However, as communication takes place over these codewords only, a level of semantic distortion is introduced to the system. This distortion is quantified using the semantic distortion in Eq. (2.11). The optimal index assignment, denoted as $\mathbf{\Pi}_{Q_{pre}^N}(\cdot)$, aims to minimize this distortion by assigning indices as follows.

$$\mathbf{\Pi}_{Q_{pre}^N}(Q) = \underset{i' \text{ s.t. } 0 \leq i' < N}{\operatorname{argmin}}\ \delta\left(Q, Q_{pre}^N(i')\right). \tag{2.13}$$

Here, $Q_{pre}^N(i')$ denotes the ith embedding in Q_{pre}^N. As illustrated in Figure 2.2, embedding $\tilde{Q} \in Q_{pre}^N$ is selected using Eq. (2.13), its index $\mathbf{I}$ is transformed into binary, and transmitted through RS coding with QPSK modulation as stated. On the receiver side, the corresponding embedding $\hat{Q} \in Q_{pre}^N$ is selected using the reconstructed index $\hat{\mathbf{I}}$.

Semantic quantization preserves the modularity of conventional communication models while incorporating the meaning of the message. This is facilitated by not requiring any training related to the source or channel characteristics, thanks to the context-independent function of SBERT and the direct utilization of available data. Hence, the resulting semantic source coding block can be seamlessly integrated into practical systems without necessitating a substantial change. This property is lacking in recent DL-enabled semantic communication systems due to their purely data-driven nature. As explained in Section 2.1, these models follow an end-to-end training procedure that limits generalizability by customizing models for specific contexts and channels, inherently linked to the characteristics of the task and training dataset. In addition to the generalizable nature of semantic quantization, accessing message contents corresponding to the embeddings in Q_{pre}^N establishes an interpretable system. This enables observation of the human-understandable content of assigned codewords in the form of text or image. Such observation is impossible in pure data-driven models where assigned codewords lack understandable representation.

2.4.1.2 Semantic Compression

Semantic compression builds upon the semantic quantization model, improving its resource efficiency. In semantic quantization, larger values of N offer more points in the semantic space, enhancing the resolution of the operation. However, excessively large values lead to a sizable matrix Q_{pre}^N, increasing storage and communication costs significantly and demanding more resources. To tackle this issue, semantic compression clusters the embeddings in Q_{pre}^N, communicating via cluster labels instead of individual elements. This maintains resolution in the semantic space while eliminating redundant points for quantization, resulting in a smaller codebook [Kutay and Yener, 2023]. This process reduces codeword length to $n = \lceil \log_2(|\{AP_{Q_{pre}^N}\}|) \rceil$ bits, where $\{AP_{Q_{pre}^N}\}$ denotes the set of calculated centroids. Since $\lceil \log_2(|\{AP_{Q_{pre}^N}\}|) \rceil \leq \lceil \log_2(|N|) \rceil$, the length of codewords decreases in parallel with our objective.

Semantic compression employs the affinity propagation (AP) algorithm for clustering, known for its superior performance in minimizing Euclidean distances within clusters. This aligns with the goal of minimizing semantic distortion, described in Eq. (2.11) [Frey and Dueck, 2007]. Notably, this algorithm determines the number of clusters through its message-passing algorithm rather than assuming a fixed cluster number for the source. Hence, the modularity of

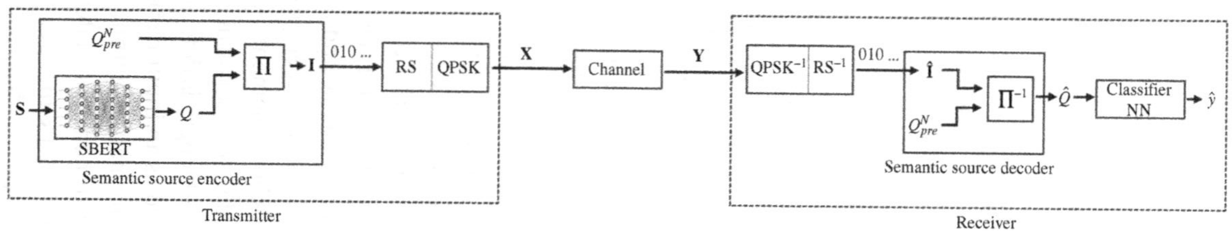

Figure 2.2 The model of introduced semantic quantization.

semantic quantization is preserved as the model does not impose any structure regarding the source or require hyperparameter tuning to obtain the number of clusters, e.g., investigation of cluster variances. In addition, since centroids are chosen from input data points by AP, the explainability of semantic quantization is retained.

$$\mathbf{L}(\mathbf{s}) = \mathbf{AP}_{Q_{pre}^N}(Q), \tag{2.14}$$

After clustering the Q_{pre}^N, $\{AP_{Q_{pre}^N}\}$ serve as the codebook. $\mathbf{AP}_{Q_{pre}^N}(Q)$ assigns a message $\mathbf{s}$ with embedding Q to a cluster as described in Eq. (2.14). The cluster label, denoted as $\mathbf{L}$, is then converted to binary, and transmitted through RS coding and QPSK, as illustrated in Figure 2.3. On the receiver side, the corresponding embedding $\hat{Q} \in \{AP_{Q_{pre}^N}\}$ is selected using the reconstructed cluster label $\hat{\mathbf{L}}$.

2.4.1.3 Semantic Vector Quantized Autoencoder

The semantic vector quantized autoencoder model (VQ AE) follows a different approach from the models introduced in Sections 2.4.1.1 and 2.4.1.2. Instead of directly using Q_{pre}^N as a codebook, the VQ AE employs it to learn a codebook from scratch, denoted as CD [Kutay and Yener, 2024]. This methodology is inspired by the concept of vector quantized variational autoencoders (VQ VAE) [van den Oord et al., 2017] and their recent applications in semantic image communication [Hu et al., 2023]. Notably, in this context, the model inherits the pure data-driven characteristics observed in existing DL-based semantic communication models.

During the communication phase, the model takes the SBERT embedding of the source message as input, similar to previous models. However, these embeddings are then fed into its encoder layers, denoted as AE Encoder, where subsequent vector quantization occurs in its latent space. The vector quantization is achieved through the minimization of the $L2$ norm, performed in parallel with Eq. (2.11). The resulting CD index $\mathbf{P}$ is converted to binary and transmitted with RS coding and QPSK, as depicted in Figure 2.4. On the receiver side, the received indices, denoted as $\hat{\mathbf{P}}$, are used to retrieve codebook elements, which are then fed into the decoder layers, denoted as AE Decoder, to reconstruct the semantic embeddings [Kutay and Yener, 2024]. The codeword length is determined by $n = \lceil \log_2(|CD|) \rceil$, where the cardinality $|CD|$ is specified as an input parameter.

The encoder aims to preserve information within the provided embeddings while projecting them into a lower-dimensional latent space for quantization. For a given dimension p, the embeddings are mapped to an α^j-dimensional space in the first FC layer, where j is the largest number such that $\alpha^j < p$. Subsequently, the dimension is iteratively scaled down by α^{-1} in each layer until $\alpha^{\tilde{j}}$, where $\tilde{j}$ is the smallest number such that $K < \alpha^{\tilde{j}}$. The last FC layer maps to $\mathbb{R}^K$, where K denotes the specified latent space dimension. The parameter $\alpha > 1$ determines

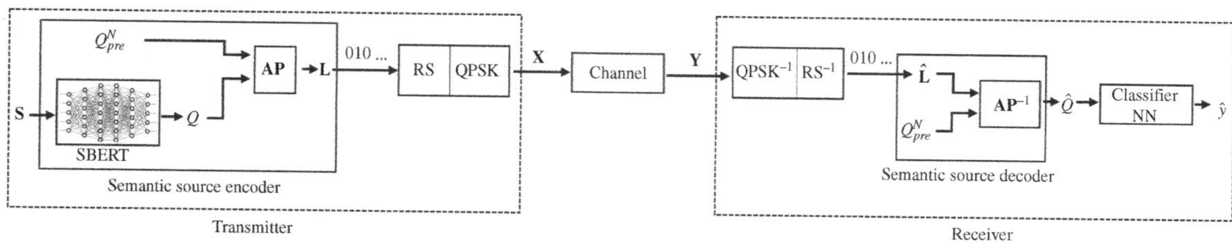

Figure 2.3 The model of introduced semantic compression.

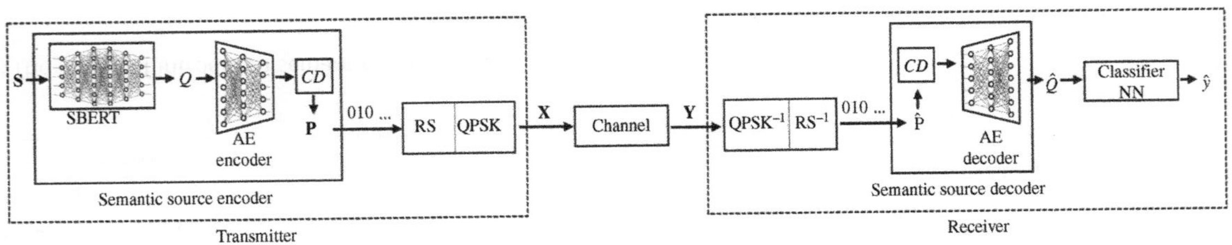

Figure 2.4 The model of introduced semantic VQ AE.

the size of the autoencoder, with larger α values resulting in smaller models. On the decoder side, the layers in the encoder are transposed to reconstruct the original embeddings given the codebook elements. PReLU is utilized as the activation function in the design. To train the model, the algorithm of the VQ VAE is adopted, setting the objective function as follows [van den Oord et al., 2017].

$$L = ||Q - \tilde{Q}||_2^2 + ||sg[z(Q)] - e||_2^2 + \beta||z(Q) - sg[e]||_2^2. \tag{2.15}$$

Here, $\tilde{Q}$ is the output of the decoder, $z(Q)$ denotes the output of the encoder, e represents the codebook element, β is the commitment parameter, and $sg[.]$ is the stop gradient operation, ensuring that its elements are not updated by having zero partial derivatives. The first term in Eq. (2.15) is the reconstruction loss of the semantic embeddings and is used to train the encoder and decoder. However, for the encoder model, the last term is also employed to ensure that the latent signals commit to the codebook elements. Finally, the codebook is trained through the middle term, which penalizes codebook elements as they diverge from the received latent signals [van den Oord et al., 2017].

2.4.2 Results

Simulations are conducted using two classification datasets: (i) AG's News, which encompasses articles from four news topics [Zhang et al., 2015] and (ii) STL 10, consisting of 96×96 RGB images belonging to 10 object classes [Coates et al., 2011]. For AG's News, 2500 samples from the training set and 2000 samples from the test set are reserved to form Q_{pre}^N and for testing, respectively. The remaining portion of the training set is utilized to train the classifier NN. Conversely, for STL 10, due to the small number of labeled samples, the training and test sets are merged, resulting in a set of 13000 samples. Subsequently, 2500 samples for Q_{pre}^N, and another 2000 samples for testing are reserved, while the remaining samples are employed for training the classifier. During these processes, it is ensured that embeddings in Q_{pre}^N exhibit uniformity in terms of class labels, thus ensuring a sufficient representation of embeddings from each class. Similarly, the test set is balanced in terms of labels, enabling the use of accuracy without requiring further metrics such as precision and recall. The result of this data processing provides a Q_{pre}^N with $N = 5000$ and a test set of size 4000, both equally distributed for image and text.

The semantic compression model utilizes negative Euclidean distance as a similarity measure and the median of the sample similarities as the preference for the AP algorithm, in conjunction with Eq. (2.11). Semantic VQ AE and classification models have hyperparameters given in Table 2.2, which are optimized through grid search. While a single set of hyperparameters is stated for the classifier NN, it should be noted that a pair of classifiers is trained, one for each modality. To

Table 2.2 Training parameters used for semantic VQ AE and classifier NNs.

Parameter	Classifier NN	Semantic VQ AE
Batch Size	128	128
Number of Epoch	15	50
LR Scheduler	Exponential LR	
$\gamma_{Scheduler}$	0.75	0.985
LR_0	0.001	0.01
Optimizer	Adam ($\beta_{1,2} = 0.9, 0.999$)	

ensure a fair comparison, the codebook size in the semantic VQ AE is set to the number of clusters generated by the semantic compression algorithm, as stated in Table 2.4. Additionally, the trade-off between training time and accuracy of the semantic VQ AE is addressed by minimizing the number of epochs and increasing the learning rate to the limit that allows proper convergence. In terms of the structure, the parameter α is set to 4. The optimal latent space dimension is tuned through grid search, which is found to be 16-dimensional space for the combined dataset.

To evaluate the resource efficiency introduced by the semantic approaches, semantic-agnostic conventional baselines are employed. Specifically, Huffman coding and JPEG2000 are used for text and image compression, respectively. Block coding is applied in Huffman coding, considering a group of three characters as a single symbol. Conversely, the quality layer parameter in JPEG2000 is configured to 20 using the Python Pillow package. Since these algorithms provide exact reconstruction, their task accuracies in source compression are calculated based on the exact message embeddings obtained from SBERT.

The structure of SBERT embeddings is investigated by projecting text embeddings into $\mathbb{R}^2$ space using the uniform manifold approximation and projection for dimension reduction (UMAP) [McInnes et al., 2018]. The embeddings of images are excluded due to their large number of classes, which makes observation with respect to classes unfeasible. The visualization of embeddings in $\mathbb{R}^2$, as depicted in Figure 2.5, supports the semantic compression model, as embeddings form clusters with respect to their labels.

To investigate explainability, Table 2.3 presents a random sample from the AG's News test set, denoted as $\mathbf{s}$, along with the messages corresponding to the index and cluster label assigned by the proposed semantic index assignment $\mathbf{I}(\mathbf{s})$ and semantic clustering $\mathbf{L}(\mathbf{s})$, respectively. The observation reveals that the assigned

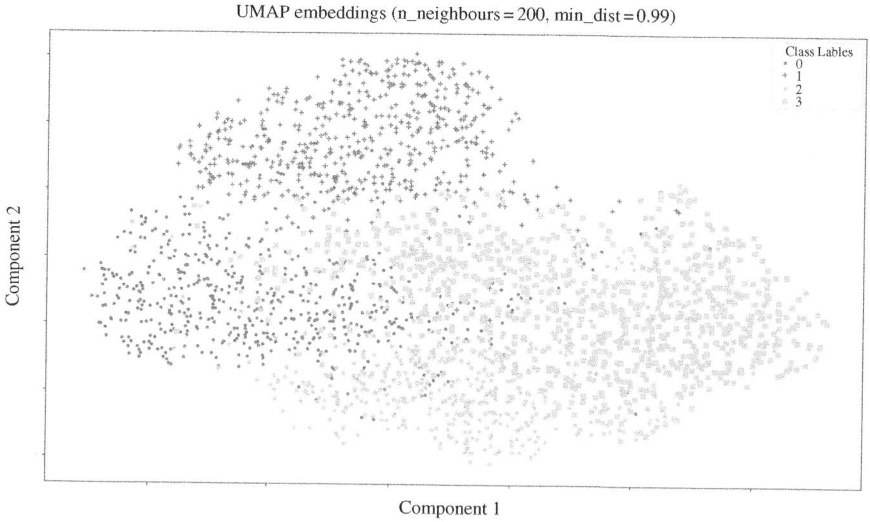

Figure 2.5 Projecting SBERT embeddings of text messages into $\mathbb{R}^2$ and visualizing them based on their class labels.

Table 2.3 A message **s** randomly selected from the AG's News test set and the sentences corresponding to the codewords assigned through semantic quantization **I**(**s**) and semantic compression **L**(**s**).

Model	Message
s	Southern California's smog-fighting agency went after emissions of the bovine variety Friday, adopting the nation's first rules to reduce air pollution from dairy cow manure.
I(**s**)	California will become hotter and drier by the end of century menacing valuable wine and dairy industries, even if dramatic steps are taken to curb global warming, researchers said on Monday.
L(**s**)	Global warming could cause dramatically hotter summers and a depleted snow pack in California, leading to a sharp increase in eat-related deaths and jeopardizing the water supply, according to a study released on Monday.

messages are consistent with human understanding, effectively conveying the semantics required for the classification task.

Table 2.4 presents the source compression results of the given models. Specifically, the number of required bits for compressing 4000 messages is provided

Table 2.4 Source compression results presented for 2000 text and 2000 image messages.

Model	Number of Bits	Classification Accuracies (%)
Conventional	23659964	95.82
Semantic Quantization	52000	92.17
Semantic Compression	32000	89.39
Semantic VQ AE	32000	86.62

The conventional baseline includes JPEG2000 and Huffman coding. The affinity propagation (AP) algorithm calculates 221 clusters for the constructed Q^N_{pre} with $N = 5000$. To ensure a fair comparison, $|CD|$ in semantic VQ AE is also set to 221.

alongside the classification accuracies. Notably, all our semantic approaches exhibit an order of magnitude improvement in resource efficiency compared to semantic-agnostic conventional models. However, semantic quantization and compression models maintain a higher level of accuracy compared to semantic VQ AE, demonstrating their resilience against scarce data. The data-greedy nature of pure data-driven models, inherent in semantic VQ AE, makes it prone to scenarios with limited data.

Extending the analysis to wireless channels, the system time efficiency metric is utilized instead of plain classification accuracies to address the cost of the pre-communication phase, e.g., training Kutay and Yener [2024]. This metric is denoted as $\eta_T(\tilde{T})$ and aims to evaluate the number of correct task executions (classifications) that proposed models can achieve within a given time budget $\tilde{T}$. For a model that achieves U correct classifications per second during the communication phase and requires T_{train} seconds for the training phase, $\eta_T(\tilde{T})$ is calculated as follows [Kutay and Yener, 2024].

$$\eta_T\left(\tilde{T}\right) = \left(\tilde{T} - T_{train}\right) * U. \tag{2.16}$$

Figure 2.6 demonstrates that semantic quantization and compression models achieve higher time efficiency in both channel models. This highlights the reduced computational demand and shorter training durations, enabling models to initiate the communication phase more quickly. Consequently, the elimination of the training phase holds merit, especially when assessing performance on noisy channels within a limited time frame.

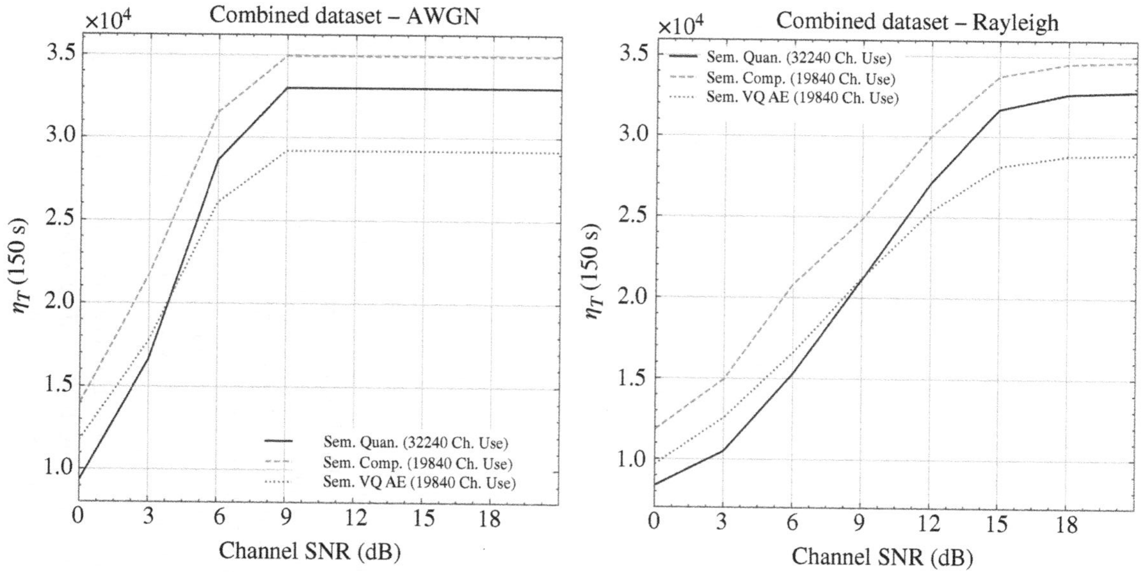

Figure 2.6 System time efficiency results over AWGN and Rayleigh Fading channels with a budget of 150 seconds.

2.5 Conclusion

In this chapter, we have described the genesis and the basics of semantic communications, i.e., reliably conveying the meaning of messages between a transmitter and a receiver. We have followed a chronological approach focusing on source compression while preserving the semantics. We have explained in detail the codebook construction problem by incorporating semantic distortion. In early work, the distortion calculation has relied on lexical taxonomies. With the arrival of DL systems, considering context and enabling similarity calculations by treating a text as a whole rather than individual words has become feasible. To this end, we have explained the codebook construction problem in a task-oriented communication scenario, utilizing the tools provided by pre-trained models. Results reveal that incorporating semantics offers both resource efficiency and error resilience compared to semantic-agnostic models. Semantic communications is a field yet in its nascence, especially in non-point-to-point settings. By drawing from contemporary AI tools and Natural Language Processing (NLP), and building on Shannon's information theory, this field is bound to impact next generation (6G) and beyond wireless communications system design.

Bibliography

Luigi Atzori, Antonio Iera, and Giacomo Morabito. The Internet of Things: A survey. *Computer Networks*, 54(15):2787–2805, Oct. 2010. ISSN 1389-1286. doi: 10.1016/j.comnet.2010.05.010.

Dzmitry Bahdanau, Kyunghyun Cho, and Yoshua Bengio. Neural machine translation by jointly learning to align and translate. *CoRR*, abs/1409.0473, 2014.

Eirina Bourtsoulatze, David Burth Kurka, and Deniz Gunduz. Deep joint source-channel coding for wireless image transmission. *IEEE Transactions on Cognitive Communications and Networking*, 5(3):567–579, 2019. doi: 10.1109/TCCN.2019.2919300.

R. Carnap and Y. Bar-Hillel. An Outline of a Theory of Semantic Information. Massachusetts Institute of Technology. *Technical Report No. 247*. Research Laboratory of Electronics, MIT, 1952.

Adam Coates, Andrew Ng, and Honglak Lee. An analysis of single-layer networks in unsupervised feature learning. In *Proceedings of the 14th International Conference on Artificial Intelligence and Statistics*, volume 15 of *Proceedings of MLR*, pages 215–223. PMLR, 2011.

Thomas M. Cover and Joy A. Thomas. *Elements of Information Theory (Wiley Series in Telecommunications and Signal Processing)*. Wiley-Interscience, USA, 2006. ISBN 0471241954.

Jacob Devlin, Ming-Wei Chang, Kenton Lee, and Kristina Toutanova. BERT: Pre-training of deep bidirectional transformers for language understanding. In *Proceedings of the 2019 Conference of the North American Chapter of the Association for Computational Linguistics*, pages 4171–4186, Minneapolis, Minnesota, Jun. 2019. Association for Computational Linguistics. doi: 10.18653/v1/N19-1423.

Brendan J. Frey and Delbert Dueck. Clustering by passing messages between data points. *Science*, 315(5814):972–976, 2007. doi: 10.1126/science.1136800.

Basak Guler and Aylin Yener. Semantic index assignment. In *2014 IEEE International Conference on Pervasive Computing and Communication Workshops (PERCOM WORKSHOPS)*, pages 431–436, 2014. doi: 10.1109/PerComW.2014.6815245.

Basak Guler, Aylin Yener, and Prithwish Basu. A study of semantic data compression. In *2013 IEEE Global Conference on Signal and Information Processing*, pages 887–890, 2013. doi: 10.1109/GlobalSIP.2013.6737034.

Basak Guler, Aylin Yener, and Ananthram Swami. The semantic communication game. *IEEE Transactions on Cognitive Communications and Networking*, 4(4):787–802, 2018. doi: 10.1109/TCCN.2018.2872596.

Deniz Gunduz, David Burth Kurka, Mikolaj Jankowski, Mohammad Mohammadi Amiri, Emre Ozfatura, and Sreejith Sreekumar. Communicate to learn at the edge. *IEEE Communications Magazine*, 58(12):14–19, 2020. doi: 10.1109/MCOM.001.2000394.

Deniz Gunduz, Zhijin Qin, Inaki Estella Aguerri, Harpreet S. Dhillon, Zhaohui Yang, Aylin Yener, Kai Kit Wong, and Chan-Byoung Chae. Beyond transmitting bits: Context, semantics, and task-oriented communications. *IEEE Journal on Selected Areas in Communications*, 41(1):5–41, 2023. doi: 10.1109/JSAC.2022.3223408.

Kaiming He, Xiangyu Zhang, Shaoqing Ren, and Jian Sun. Delving deep into rectifiers: Surpassing human-level performance on imagenet classification. In *2015 IEEE International Conference on Computer Vision (ICCV)*, pages 1026–1034, 2015. doi: 10.1109/ICCV.2015.123.

Kurt Hornik. Approximation capabilities of multilayer feedforward networks. *Neural Networks*, 4(2):251–257, 1991. ISSN 0893-6080. doi: 10.1016/0893-6080(91)90009-T.

Qiyu Hu, Guangyi Zhang, Zhijin Qin, Yunlong Cai, Guanding Yu, and Geoffrey Ye Li. Robust semantic communications with masked VQ-VAE enabled codebook. *IEEE Transactions on Wireless Communications*, 22(12):8707–8722, 2023. doi: 10.1109/TWC.2023.3265201.

Furqan Jameel, Zheng Chang, Jun Huang, and Tapani Ristaniemi. Internet of autonomous vehicles: Architecture, features, and socio-technological challenges. *IEEE Wireless Communications*, 26(4):21–29, 2019. doi: 10.1109/MWC.2019.1800522.

Jay J. Jiang and David W. Conrath. Semantic similarity based on corpus statistics and lexical taxonomy. In Keh-Jiann Chen, Chu-Ren Huang, and Richard Sproat, editors, *Proceedings of the 10th Research on Computational Linguistics International*

Conference, pages 19–33, Taipei, Taiwan, Aug. 1997. The Association for Computational Linguistics and Chinese Language Processing (ACLCLP).

Emrecan Kutay and Aylin Yener. Semantic text compression for classification. In *2023 IEEE International Conference on Communications Workshops (ICC Workshops)*, pages 1368–1373, 2023. doi: 10.1109/ICCWorkshops57953.2023.10283705.

Emrecan Kutay and Aylin Yener. Classification-oriented semantic wireless communications. In *ICASSP 2024 - 2024 IEEE International Conference on Acoustics, Speech and Signal Processing (ICASSP)*, 2024.

Wing Fei Lo, Nitish Mital, Haotian Wu, and Deniz Gunduz. Collaborative semantic communication for edge inference. *IEEE Wireless Communications Letters*, 12(7):1125–1129, 2023. doi: 10.1109/LWC.2023.3256006.

L. McInnes, J. Healy, and J. Melville. UMAP: Uniform manifold approximation and projection for dimension reduction. *arXiv preprint arXiv:1802.03426*, Feb. 2018.

Aaron van den Oord, Oriol Vinyals, and Koray Kavukcuoglu. Neural discrete representation learning. In I. Guyon, U. Von Luxburg, S. Bengio, H. Wallach, R. Fergus, S. Vishwanathan, and R. Garnett, editors, *Advances in Neural Information Processing Systems*, volume 30. Curran Associates, Inc., 2017.

Timothy O'Shea and Jakob Hoydis. An introduction to deep learning for the physical layer. *IEEE Transactions on Cognitive Communications and Networking*, 3(4):563–575, 2017. doi: 10.1109/TCCN.2017.2758370.

J. G. Proakis and M. Salehi. *Fundamentals of Communication Systems*. Pearson Prentice Hall, 2005. ISBN 9780131471351.

R. Rada, H. Mili, E. Bicknell, and M. Blettner. Development and application of a metric on semantic nets. *IEEE Transactions on Systems, Man, and Cybernetics*, 19(1):17–30, 1989. doi: 10.1109/21.24528.

Alec Radford, Jong Wook Kim, Chris Hallacy, Aditya Ramesh, Gabriel Goh, Sandhini Agarwal, Girish Sastry, Amanda Askell, Pamela Mishkin, Jack Clark, Gretchen Krueger, and Ilya Sutskever. Learning transferable visual models from natural language supervision. *arXiv preprint arXiv:2103.00020*, 2021.

Nils Reimers and Iryna Gurevych. Sentence-bert: Sentence embeddings using siamese bert-networks. In *Proceedings of the 2019 Conference on Empirical Methods in NLP*. Association for Computational Linguistics, Nov. 2019.

P. Resnik. Using information content to evaluate semantic similarity in a taxonomy. In *Proceedings of the 14th International Joint Conference on Artificial Intelligence*, volume 1, pages 448–453, 1995.

Yalin E. Sagduyu, Tugba Erpek, Sennur Ulukus, and Aylin Yener. Is semantic communication secure? A tale of multi-domain adversarial attacks. *IEEE Communications Magazine*, 61(11):50–55, 2023. doi: 10.1109/MCOM.006.2200878.

C. E. Shannon. A mathematical theory of communication. *The Bell System Technical Journal*, 27:379–423, 625–656, Jul. 1948.

Claude E. Shannon and Warren Weaver. *The Mathematical Theory of Communication*. University of Illinois Press, Urbana, IL, 1949. ISBN 978-0-252-72548-7.

Guangming Shi, Yong Xiao, Yingyu Li, and Xuemei Xie. From semantic communication to semantic-aware networking: Model, architecture, and open problems. *IEEE Communications Magazine*, 59(8):44–50, 2021. doi: 10.1109/MCOM.001.2001239.

Ashish Vaswani, Noam Shazeer, Niki Parmar, Jakob Uszkoreit, Llion Jones, Aidan N. Gomez, Łukasz Kaiser, and Illia Polosukhin. Attention is all you need. In I. Guyon, U. Von Luxburg, S. Bengio, H. Wallach, R. Fergus, S. Vishwanathan, and R. Garnett, editors, *Advances in Neural Information Processing Systems*, volume 30. Curran Associates, Inc., 2017.

Dylan Wheeler and Balasubramaniam Natarajan. Engineering semantic communication: A survey. *IEEE Access*, 11:13965–13995, 2023. doi: 10.1109/ACCESS.2023.3243065.

Huiqiang Xie, Zhijin Qin, Geoffrey Ye Li, and Biing-Hwang Juang. Deep learning enabled semantic communication systems. *IEEE Transactions on Signal Processing*, 69:2663–2675, 2021. doi: 10.1109/TSP.2021.3071210.

Wanting Yang, Hongyang Du, Zi Qin Liew, Wei Yang Bryan Lim, Zehui Xiong, Dusit Niyato, Xuefen Chi, Xuemin Shen, and Chunyan Miao. Semantic communications for future internet: Fundamentals, applications, and challenges. *IEEE Communications Surveys and Tutorials*, 25(1):213–250, 2023. doi: 10.1109/COMST.2022.3223224.

Xiang Zhang, Junbo Zhao, and Yann LeCun. Character-level convolutional networks for text classification. In *Advances in Neural Information Processing Systems*, volume 28. Curran Associates, Inc., 2015.

3

Toward a Theory of Semantic Information

Jean-Claude Belfiore[1] *and Daniel Bennequin*[1,2]

[1]*Advanced Wireless Technology Lab., Huawei Technologies, Boulogne Billancourt, France*
[2]*UFR de Mathématiques, Université Paris Cité, Paris, France*

3.1 Introduction

Humans have an intuitive understanding of the notion of information and can decide if a message/image/... contains more or less information, and if it is useful information or not. However, as for the notions of time or space, defining information, in general, is difficult. Convincing notions have been developed, as Shannon's theory, which is of statistical nature, or sensory spaces [Shepard, 1987] (colors, tastes, odors, sounds, textures, touch), which help to discriminate objects. In mathematics as well, Galois theory is seen as a theory of information, about the solutions of algebraic equations, then generalized to other problems under the name of ambiguity theory.

In this chapter, we present a notion of *semantic information*, which is related to artificial languages, but gives insight into natural languages. It is based on some notions of invariance, and has values that are topological spaces with additional structure. The fact that a semantic information theory cannot be represented by numbers, but, at least, needs sets to be expressed, was first noticed in Carnap and Bar-Hillel [1952]. However, they eventually proposed numerical measures deduced from Shannon's theory. The fact that information must correspond to *forms*, was strongly asserted by René Thom [1989]; he gave examples coming from dynamical systems and singularity theory. In this study, we will constantly be inspired by these authors: Shannon, Thom, Shepard, Carnap, and Bar-Hillel.

That Shannon's information has a form appeared in the interpretation of entropy in Baudot and Bennequin [2015]; (functions of) probability laws on a coherent set of random variables define an object (a presheaf) in a ringed topos in the sense of

Foundations of Semantic Communication Networks, First Edition.
Edited by Walid Saad, Christina Chaccour, Christo Kurisummoottil Thomas, and Merouane Debbah.

Alexander Grothendieck [Artin et al., 1963], where the *elements of forms* can be defined by cohomology classes (analogs of holes for ordinary manifolds). Entropy appears as the only universal class of this probabilistic object.

We will start this chapter by reminding what that means. We then propose to search for a similar class in a purely semantic context of languages. An obstacle appears, the same one as the one that was observed in Baudot and Bennequin [2015] in the quantum probability case, which can be solved by the introduction of specific fundamental symmetries, generalizing the unitary invariance. That suggestion was already considered in Belfiore and Bennequin [2021]. The axioms of the theory are summarized in Section 3.3. The most difficult part is rather technical and is summarized in Section 3.5 on a specific example: the construction of information spaces. The main clue that we are following the right direction is explained in Section 3.6 where we show that the notion of **semantic entropy** is naturally connected to psychological experiments of Shepard et al. [1961] about the different ways of learning concepts. In fact, the semantic entropy measures the difficulty to learn *classifications* when a hidden structure underlines their constructions. Experiments on artificial neural networks were performed in order to test the applications of our constructions, using various structures of (pre-)semantic invariance.

Reading this chapter requires some basic knowledge in topology and algebra that can be found, for instance, in Bennequin and Belfiore [2022].

3.2 Cohomological Nature of Information

It was shown, in Baudot and Bennequin [2015], that Shannon information is a theory of topological invariants of Bayes networks, i.e., networks made by a collection of random variables. Such a network defines a category, the arrows being the marginalizations. The *joint* operation on random variables Y, coarser than any given variable X, defines local commutative algebras. Conditioning defines an action of the presheaf of these algebras on the presheaf of functions $F_X(\mathbb{P})$ of local probability laws. This is an interpretation of Shannon's formula for average conditioning. It leads us to the general framework in order to define a *cohomology*, found by Grothendieck in the 1960s: topos theory.

The main result of Baudot and Bennequin [2015] is that the only universal class of degree one is Shannon's entropy H. They notice that nothing can happen without considering networks of several variables and localizing the invariant quantities (in a topos).

Especially, it means that H is not simply a real number; it is a function $H_X(Y; \pi_* \mathbb{P})$, where $\pi : X \to Y$, and the existence of H and its properties depend on the relations between random variables.

To define semantic information measures, we decided to follow the same path. According to Belfiore and Bennequin [2021], we start with an analogy between Logics and Statistics. Logical propositions are viewed as the analogs of random variables (they can have many values of truth). Theories are the analogs of probability laws, constraining the variables. Theories are generated by axioms, telling that some propositions are assumed to be true. The fact that R is true in the theory $\mathbb{T}$ is written $\vdash_{\mathbb{T}} R$. Inference rules are used to deduce other statements from the set of axioms.

We choose to work with typed languages in classical intuitionist type theory [Lambek and Scott, 1981; Bell, 2008]. As asserted by Lambek, a semantics is given by the embedding of the language in a topos. We restrict ourselves to Grothendieck toposes, made by sheaves over a site, because this gives both a limitation of size, and a geometrical interpretation. For instance, we can obtain a notion of sections [Artin et al., 1963; Bennequin and Belfiore, 2022].

The types give objects of the topos. A *proposition* corresponds to a subobject of an object in the topos. A *theory* is a set of propositions, which is closed by deduction.

The partial order on propositions is the inclusion, written $\leq$. The partial order on theories is denoted by this same symbol: $\mathbb{T} \leq \mathbb{T}'$, when the axioms of $\mathbb{T}'$ can be deduced from the axioms of $\mathbb{T}$. This partial order plays a role analog to the marginalization in the probabilistic case.

On propositions, we have the usual logical operators: $\leq, \top, \bot, \wedge, \vee, \neg, \wedge$, and $\Rightarrow$.

The local algebras of statistical variables are here replaced by Heyting algebras.

To define the action of propositions on the theories, we use the *logical conditioning*: the axioms of $Q.\mathbb{T} = \mathbb{T}|Q$ are the assertions ($\vdash Q \Rightarrow R$) with $\vdash R$ axiom of $\mathbb{T}$, and where the implication $\Rightarrow$ is the internal one, the right adjoint of $\wedge$ (AND). For instance, $Q \Rightarrow R$ denotes $R \vee \neg Q$ in the Boolean case.

In Belfiore and Bennequin [2021], we made more precise the locality by considering, for any propositions P, the set Θ_P, of theories implying $\neg P = (P \Rightarrow \bot)$ and the subalgebra of propositions Q such that $P \leq Q$, which preserve Θ_P.

The computations made in Belfiore and Bennequin [2021] showed that the analogy is stronger with quantum information theory than with the classical one [Witten, 2020].

In the semantic context, for every function $H_P^Q(\mathbb{T})$ that could play the role of the entropy, we prove that there exists a function $F_P(\mathbb{T})$, analog to the Von Neumann entropy, such that, when $P \leq Q, \mathbb{T} \leq \neg P$,

$$H_P^Q(\mathbb{T}) = F_P(\mathbb{T}|Q) - F_P(\mathbb{T}). \tag{3.1}$$

At first glance, this result looks deceptive. However, by analogy with unitary invariance of the Von Neumann entropy, we can use a notion of invariance to characterize the function F.

In the semantic case, there is no universal invariance replacing the unitary one. However, in most interesting cases (if not all), natural notions of invariance appear.

In the context of topos theory, the natural assumption is to require that some category C acts on a category of languages $\mathcal{H}$ through a contravariant functor $\mathcal{L} : C^{op} \to \mathcal{H}$.

The simplest case is to choose, for the category C, a group G. This is the main case we will encounter in this chapter, but we will also consider posets, which naturally arise in many applications (choice of hierarchy). C is a *presemantic* structure of cognitive origin according to the linguist Antoine Culioli [Belfiore and Bennequin, 2021, Chap. 4].

The function $F_P(\mathbb{T})$ has to be natural (i.e., equivariant) under the action of C, and to satisfy the two properties of *concavity* (as the Von Neumann entropy [Witten, 2020]) and *increasing monotonicity* as the logarithm.

Moreover, in all applications that we will consider, information has more than one dimension. It is not a number, it is at least a set [Carnap and Bar-Hillel, 1952]. Here, following Thom's intuition [Thom, 1989], we will prefer topological spaces to ordinary sets, for instance CW–complexes [Whitehead, 1949].

3.3 Axioms for Information Spaces

For simplicity reason, we do not consider other categories of spaces than CW-complexes, but some refined structures on these spaces, like *stratifications* and *textures* (which are notions of equivalence specific to a stratum). For instance, we frequently use marked points, up to isotopy of strata.

The analogs of growth and concavity on spaces refer to special notions of embeddings and differences between spaces.

A notion of special injections between spaces is given by *cofibration* in every closed model category of Quillen [1967]. For sets, they are injective maps; for groupoids, they are morphisms injective on the objects; for CW-complexes, they are the subspaces of which the complement is obtained by inductively gluing cells along continuous maps of spheres [Quillen, 1967].

In what follows, we use a combination of these notions: cofibrations of CW-complexes injective on the marked points, and we denote them,

$$\boxed{f : A \rightarrowtail B.}$$

The symbol $F_P(\mathbb{T})$ denotes a space associated to any theory $\mathbb{T}$ that excludes a given proposition P. In addition to (i) (the invariance under the action of C), its main properties are (ii) increasing monotonicity and (iii) concavity.

Property (ii) ensures that, if $\mathbb{T} \leq \mathbb{T}'$ (i.e., the axioms of $\mathbb{T}$ imply all the axioms of $\mathbb{T}'$), then there exists a cofibration

$$F_P(\mathbb{T}) \rightarrowtail F_P(\mathbb{T}'). \tag{3.2}$$

Property (iii) tells us that, if $\mathbb{T}_0 \leq \mathbb{T}_1 \leq \mathbb{T}_3$ and $\mathbb{T}_0 \leq \mathbb{T}_2 \leq \mathbb{T}_3$, we have

$$F_P(\mathbb{T}_3)\backslash F_P(\mathbb{T}_1) \rightarrowtail F_P(\mathbb{T}_2)\backslash F_P(\mathbb{T}_0). \tag{3.3}$$

Definition 3.1 If A and B are closed CW-complexes, with $A \rightarrowtail B$, the difference $B\backslash A$ is the closure of the complementary set of cells of A in B.

Definition 3.2 ***(semantic entropy)***

$$H_P^Q(\mathbb{T}) = F_P(\mathbb{T}|Q)\backslash F_P(\mathbb{T}), \tag{3.4}$$

when $P \leq Q$ and $\mathbb{T} \leq \neg P$

The concavity gives the two main geometrical properties of the function H_P,

(i) If $\mathbb{T} \leq \mathbb{T}'$, we have $H_P^Q(\mathbb{T}') \rightarrowtail H_P^Q(\mathbb{T})$,
(ii) If $Q \leq Q'$, then $H_P^{Q'}(\mathbb{T}) \rightarrowtail H_P^Q(\mathbb{T})$.

By convention (analogous to the choice of the basis of the log in Shannon's case) $F_P(\perp) = \emptyset$.

Definition 3.3 The *basic information spaces* are defined by

$$H^Q = H_\perp^Q = F_\perp(\neg Q)\,. \tag{3.5}$$

Axioms

For the basic semantic spaces H^Q they are,

(i) If there exists an isomorphism $\alpha \in Hom_C$, such that $\mathcal{L}(\alpha)(Q') = Q$, then the space H^Q is naturally isomorphic to the space $H^{Q'}$.
(ii) the existence of natural embeddings (or cofibrations) when $Q \leq Q'$,

$$H^{Q'} \rightarrowtail H^Q; \tag{3.6}$$

(iii) For any pair of propositions Q, Q',

$$j : H^{Q\wedge Q'}\backslash H^{Q'} \rightarrowtail H^Q, \tag{3.7}$$

which expresses the concavity of F.

Definition 3.4 ***(semantic independence)*** When the closed difference $H^{Q'}\backslash\left(H^{Q\wedge Q'}\backslash H^Q\right)$ (or the difference obtained by exchanging Q and Q') is empty, we say that Q and Q' are *semantically independent.*

Definition 3.5 ***(semantic mutual information)*** The semantic mutual information (in the general case) is defined as,

$$I(Q; Q'|\neg P) = H^Q(\neg P)\backslash H^Q(Q' \Rightarrow \neg P). \tag{3.8}$$

In the Boolean case, it can be rewritten as,

$$I(Q;Q'|\neg P) = \left(H^{Q\wedge P}\backslash H^{Q}\right)\backslash\left(H^{Q\wedge Q'\wedge P}\backslash H^{Q}\right). \tag{3.9}$$

3.4 Comparison with Other Propositions of Semantic Information Measures

The mathematical notion of semantic information, which is used today, comes from Carnap and Bar-Hillel [1952], and is defined in Bao et al. [2011]. A given set W models the *theories of the world*. State of knowledge about them is represented by a probability measure μ on W, and is expressed in a code X, through a conditional probability $P(x|w)$, then $P(x,w) = P(x|w)\mu$ and the marginalization m on X is viewed as a kind of *logical probability*. Its ambiguity is evaluated by Shannon's entropy $H(m)$. In this model, the structure of the language is hidden, especially the role of inference and deduction.

Finally, as we will see in Section 3.5, one of the main interests of our information spaces is that they underline few propositions, of the order of $2^{d \cdot n}$, for some fixed integer d. In Bao et al. [2011], if W has n elements, we get 2^n propositions, then more than 2^{2^n} degrees of freedom.

3.5 Carnap and Bar-Hillel Languages

We consider the language $\mathcal{L}_m^n$ generated by expressions sA, where s belongs to a finite set $\{s_1, \dots, s_n\}$ of "subjects" and A belongs to the set $\{A_1, A_1', \dots, A_m, A_m'\}$ made by the attributes with their two possible values.

The main ingredient for information spaces is the symmetry group $G = \mathfrak{S}_n \times \mathfrak{D}_m$, cartesian product of the permutation group of the subjects with the group generated by permutations of the attributes and the changes of their values.

Particular Case

The toy language $\mathcal{L}_2^3$, studied in Carnap and Bar-Hillel [1952], has three subjects a, b, c and two binary attributes $\mathcal{A}, \mathcal{G}$ (age, gender) with respective values O, Y (old, young) and F, M (female, male).

Let us describe the steps of construction of information spaces:

(1) We consider the natural groupoid of each orbit in the set of atoms (an atom is a full population of subjects), as in Figure 3.1, where the values of all attributes are given (see Figure 3.2). We then construct the corresponding Eilenberg–McLane space with marked points [Lewis Jr., 1992]. This is the information space H^T, where T is the type/orbit.

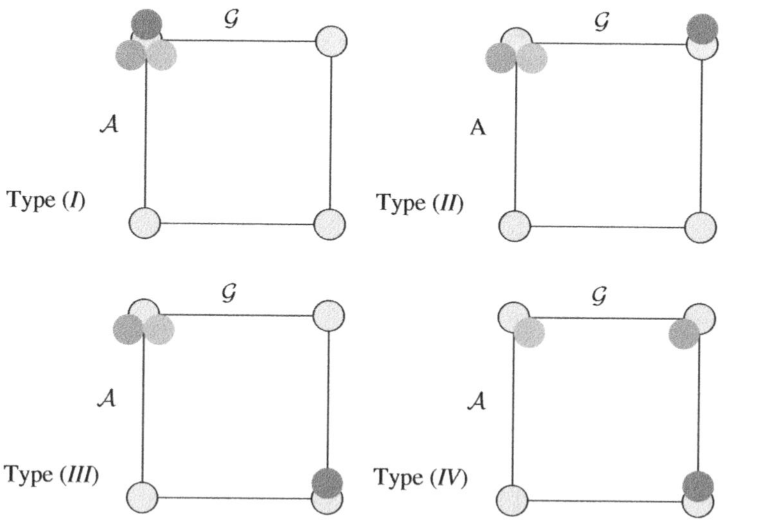

Figure 3.1 One representative (atom) for each of the four orbits (types) of G acting on $\mathcal{L}_2^3$. Colored disks are the subjects.

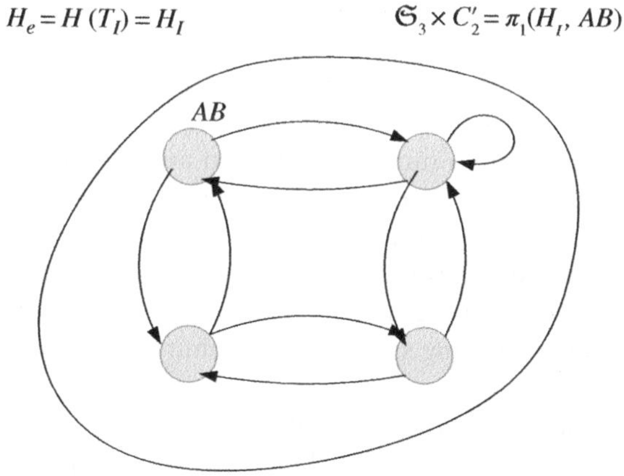

Figure 3.2 Groupoid representing the orbit of type I.

(2) For simple propositions sA (one subject with one attribute taking one value), we construct H^{sA} as the Eilenberg–McLane space, with one marked point, corresponding to the subgroup of G fixing s and A.

(3) Let $e \in X$ be any atom. We now want to construct H^e (maximal information space). Suppose e is in the orbit T. We choose a collection of minimal systems of simple propositions able to characterize e with respect to its orbit T. We form a groupoid associated to this collection of simple propositions, which

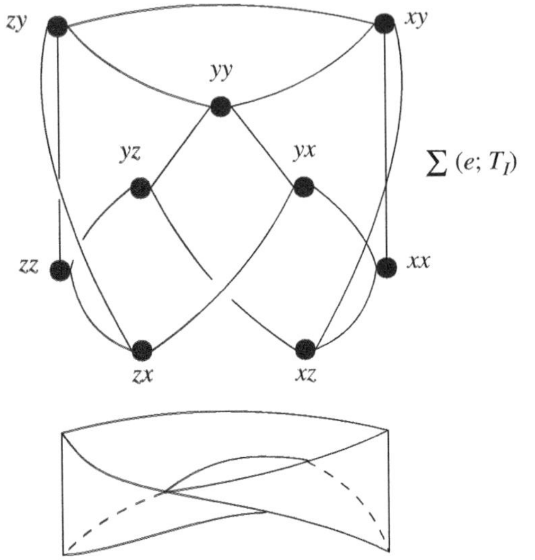

Figure 3.3 Changes of frames for the space of information H^e, for e of type I in $\mathcal{L}_2^3$.

Eilenberg–McLane space is denoted H_e'. The space H^e itself is obtained by grafting H_e' along H^T, using the mapping cylinders of natural morphisms $G^{sA} \to G^e$ from the collection of G^{sA} to the isotropy subgroup G^e of e (the fundamental group of T at point e). An important difficulty comes with the fact that systems that are both generators and independent (named basis) can have different cardinalities and that not every independent system belongs to a basis. However, the passage from a minimal system to another one is governed by a weak analog of the Steinitz lemma, which holds true in a finite dimensional linear space [van der Waerden, 1959]. Then we mark H^e by any system. However, its complete "texture" is given by the graph of exchanges, $\Sigma\left(e; T_I\right)$ in the Steinitz lemma. Even in the simplest case (type I), the graph (illustrated in Figure 3.3) has the amazing structure of a semiregular graph on a Mœbius band.

(4) To construct the space of a union $Q = \vee_i e_i$ of populations in the same orbit, we graft on the space H^T an analog of H_e' and a new kind space Y_Q taking into account the common symmetries and isotropy structures of the e_i.

(5) For unions over different types, the construction is similar, but uses a core space associated to the union of those orbits.

Remark 3.1 The complete construction is delicate (30+30 pages in the arXiv version Belfiore and Bennequin [2024]).

Theorem 3.1 There exists a collection of spaces satisfying axioms (*i*), (*ii*), and (*iii*), extending both the natural spaces of orbits and the spaces of simples.

Each space is made of three parts, of different textures, (i) orbital quotient with marked points corresponding to classes of subpopulations, (ii) joints of simple spaces, (iii) spaces for recording symmetries, which marked points are classes of systems sharing isotropy structures.

Theorem 3.2 With respect to the total number of propositions, $2^{2^{nm}}$, the number of non-trivial information spaces is negligible, i.e., polynomial in 2^{nm}.

An important variant is given by hierarchies in the classifications, inducing another semantics. Replace the group G by a poset, for instance a tree, given by an ordering of the attributes. Then information spaces come from the nerve of the poset, a simplicial complex.

In this case, the information spaces describe the sources of propositions with respect to a hierarchical structure. Using a fibered category in groups over the poset, we obtain a theory of information spaces that mixes both theories, Galois and ordering.

3.6 Shepard's Experiment

In a remarkable and influential study, Shepard et al. [1961] reported experiments with human subjects, showing that the relative difficulty of learning and memorization of classifications depends on the underlying principle (named *type*) of the classification. These types can be expressed in geometrical terms (see Figure 3.4), describing the structure of the distribution of attribute values among the stimuli. In fact, the example chosen in Shepard et al. [1961] is accounted by the binary language $\mathcal{L}_4^3$ of Carnap and Bar-Hillel, although not cited. And what they named type corresponds to the definition we used in this chapter (orbit). A difference is that Shepard et al. [1961] consider the part of the language where different subjects cannot all have the same attributes, like a "Fermionic" constraint. A consequence is that not all possible types, thus semantic spaces of $\mathcal{L}_4^3$ can appear in this study.

Amazingly, we can prove that our space of entropy of a population with respect to its orbit, taking into account both hierarchies and symmetries, explains fairly well the results of the experiments of Shepard et al. [1961].

This entropy is the space measuring the difficulty of discriminating between data, conditioned by a hypothesis on the concept underlying their structure.

By definition of our semantic entropy $H^T(\neg e) = H^{e\wedge T}\backslash H^T = H^e\backslash H^T$, we get, by only considering the **group semantics**, the following *complexities* for spaces of entropy for each type: for type I, dimension 3; for type II, dimension 4; for type III, dimension 9; for type IV, dimension 7; for type V, dimension 9; and for type VI, dimension 3.

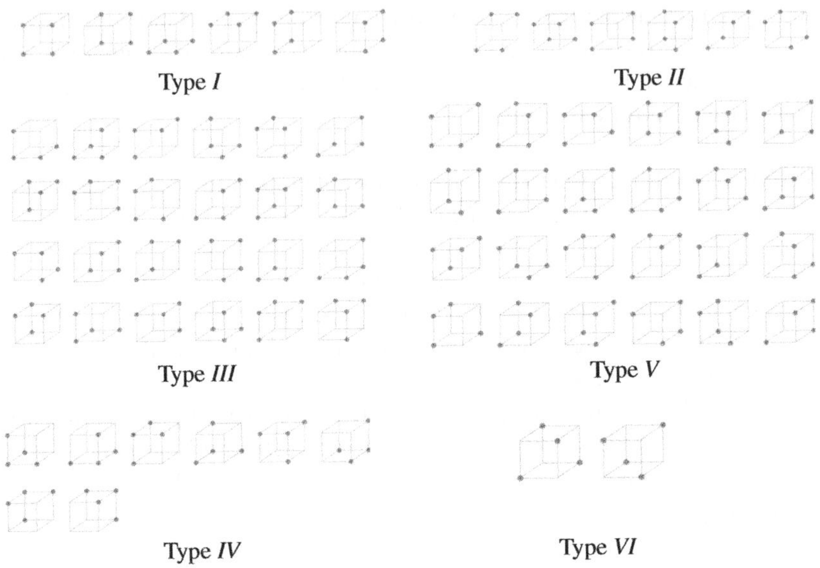

Figure 3.4 Types in Shepard's experiment.

And with a Tree Semantics

For type *I*, two optimal trees, pruning $p = 7$, dimension 1; type *II*, $p = 6$, only two optimal trees; two 2-cells; type *III*, $p = 5$, three optimal trees, one 2-cells, two 3-cells; type *IV*, $p = 6$, all six trees are equal, one 2-cell, two 3-cells; type *V*, $p = 5$, like type *III*; type *VI*, $p = 4$, all equal trees, four 3-cells, glued along two 1-cells.

Together, these computations explain the ranking results in Shepard et al. [1961], plus the transfer for *VI* (becoming easier after repetition), and the competition between *II* and *IV*, depending on mention of rules.

To our knowledge, this is the only theory that explains the observed evolution in time of the difficulty when repeated problems are solved, with some types but not all of them.

Especially, every known measure, which is based of purely logical complexity, fails to explain these results, even the basic ranking: **semantic information is not pure logic**.

For us, this application is fundamental, because it shows that our top-down mathematical method gives rise to a notion of *semantic entropy*, which corresponds to practical difficulties for humans to learn data.

Our next hope is to show that *semantic mutual information* can be used to define a *semantic capacity* in order to develop a theory of semantic communication.

Bibliography

Michael Artin, Alexander Grothendieck, and Jean-Louis Verdier. Théorie des topos et cohomologie étale des schémas. *SGA4, IHES*, 1963.

Jie Bao, Prithwish Basu, Mike Dean, Craig Partridge, Ananthram Swami, Will Leland, and James A. Hendler. Towards a theory of semantic communication. In *2011 IEEE Network Science Workshop*, pages 110–117, 2011.

Pierre Baudot and Daniel Bennequin. The homological nature of entropy. *Entropy*, 17(5):3253–3318, 2015.

Jean-Claude Belfiore and Daniel Bennequin. Topos and stacks of deep neural networks. arXiv:2106.14587, 2021.

Jean-Claude Belfiore and Daniel Bennequin. Spaces of semantic information. *arXiv*, 2024.

John L. Bell. *Toposes and Local Set Theories*. Dover, 2008.

Daniel Bennequin and Jean-Claude Belfiore. Mathematics for AI: Categories, toposes, types. In Liao Heng and Bill McColl editors, *Mathematics for Future Computing and Communications*. Cambridge University Press, 2022.

Rudolf Carnap and Jehoshua Bar-Hillel. An Outline of a Theory of Semantic Information. *Technical Report 247*, Research Laboratory of Electronics, MIT, 1952.

Joachim Lambek and Philip J. Scott. Intuitionist type theory and foundations. *Journal of Philosophical Logic*, 10:101–115, 1981.

L. Gaunce Lewis Jr. Equivariant Eilenberg-Maclane spaces and the equivariant Seifert-van Kampen and suspension theorems. *Topology and its Applications*, 48(1):25–61, 1992.

Daniel G. Quillen. *Homotopical Algebra (Lecture Notes in Mathematics)*. Springer-Verlag, 1967.

Roger N. Shepard. Toward a universal law of generalization for psychological science. *Science*, 237(4820):1317–1323, 1987.

Roger N. Shepard, Carl I. Hovland, and Herbert M. Jenkins. Learning and memorization of classifications. *Psychological Monographs: General and Applied*, 75:1–42, 1961.

René Thom. *Structural Stability and Morphogenesis*. CRC Press, 1989.

B. L. van der Waerden. *Algebra (Grundlehren der mathematischen wissenschaften)*. Julius Springer, 1959.

J. H. C. Whitehead. Combinatorial homotopy. I. *Bulletin of the American Mathematical Society*, 55(3.P1):213–245, 1949.

Edward Witten. A mini-introduction to information theory. *Nuovo Cimento Rivista Serie*, 43(4):187–227, Mar. 2020.

4

Deep Joint Source and Channel Coding

Haotian Wu[1], Chenghong Bian[1], Yulin, Shao[2], and Deniz Gündüz[1]

[1] *Department of Electrical and Electronic Engineering, Imperial College London, London, UK*
[2] *State Key Laboratory of Internet of Things for Smart City and Department of Electrical and Computer Engineering, University of Macau, Macau, China*

4.1 Introduction

In the evolving landscape of digital communications, the traditional adherence to Shannon's separation theorem [Shannon, 1948] – stating that source and channel coding can be optimized independently without loss of efficacy – faces practical challenges. This foundational principle, while elegant in theory, encounters suboptimality in the face of real-world constraints such as finite block lengths and dynamically changing channels, typical of modern wireless systems [Goldsmith, 1995; Vembu et al., 1995]. These limitations prompt a reevaluation of conventional methodologies, suggesting that an integrated approach to source and channel coding might better suit contemporary needs.

The advent of joint source-channel coding (JSCC) and its further innovation through deep learning (DeepJSCC) [Farsad et al., 2018; Bourtsoulatze et al., 2019] exemplify this shift. By intelligently combining the tasks of compression and error correction, these approaches aim to mitigate the inefficiencies brought about by the theorem's assumptions, addressing the critical issues of the cliff and leveling effects that can significantly impact system performance under varying channel conditions. This introduction sets the stage for a detailed exploration of how modern communication systems can transcend traditional limitations, leveraging the synergy of source and channel coding to achieve enhanced reliability and efficiency in data transmission.

DeepJSCC is a data-driven technique that uses deep neural networks (DNNs) to learn an optimal representation of the source data that is both compact and robust to channel impairments, effectively merging the compression and error-correction functionalities. The process typically involves:

Foundations of Semantic Communication Networks, First Edition.
Edited by Walid Saad, Christina Chaccour, Christo Kurisummoottil Thomas, and Merouane Debbah.

- **An encoder network**: A DNN that takes the source data and encodes it into a compact representation tailored for transmission over a specific communication channel. The encoded representation is designed to be resilient to the impairments of the channel, such as noise, fading, or interference.
- **A decoder network**: A DNN takes the received (and possibly corrupted) encoded representation and attempts to reconstruct the original source data. The decoder is designed to correct errors and recover the original data as accurately as possible.
- **End-to-end optimization**: The encoder and decoder are jointly optimized to minimize the end-to-end reconstruction error, taking into account the specific characteristics of the channel. This can lead to more efficient use of bandwidth and better performance under certain conditions compared to traditional methods.

DeepJSCC can be adapted to various types of data (e.g., texts Farsad et al. [2018], images Bourtsoulatze et al. [2019], videos Tung and Gndz [2022], and point clouds Bian et al. [2023a]) and different communication channels (e.g., orthogonal frequency division multiplexing (OFDM) Wu et al. [2022], Yang et al. [2022], multiple-input and multiple-output (MIMO) Bian et al. [2023b], Wu et al. [2023a], relay Bian et al. [2024a], multiple-access Yilmaz et al. [2023], and broadcast and feedback channels Wu et al. [2023c]) by training the encoder and decoder on relevant data and channel models, making it an appealing technology for emerging semantic communication systems [Gündüz et al., 2020; Jankowski et al., 2020; Gndz et al., 2023; Lo et al., 2023].

In this chapter, we embark on a detailed journey through three canonical channel models, exploring the transformative impact of DeepJSCC techniques on communication performance. These models include the MIMO channel, the relay channel, and the feedback channel. Each model presents unique challenges that test the limits of traditional communication systems. By integrating DeepJSCC, we aim to uncover its potential for superior performance, showcasing its adaptability, efficiency, and resilience in diverse communication scenarios.

4.2 DeepJSCC for MIMO Channels

To start with, our focus is on exploring the intricacies of point-to-point communication within the context of a MIMO channel. Throughout this chapter, we consider the task of transmitting images as a case study to underscore the superiority of DeepJSCC.

4.2.1 System Model

Assume that the transmitter and the receiver are equipped with M antennas. The primary objective of communication is to facilitate the transmission of an image

$\mathbf{S} \in \mathbb{R}^{h\times w\times c}$ from the transmitter to the receiver, where h, w, and $c = 3$ denote the height, width, and color channels of the image, respectively. The transmitter encodes the image into a matrix of channel symbols $\mathbf{X} \in \mathbb{C}^{M\times k}$, where k denotes the number of channel uses dedicated to transmitting one image. The transmitted signal $\mathbf{X}$ is subject to a power constraint P_s: $\frac{1}{Mk}\|\mathbf{X}\|_F^2 \le P_s$, where $\|\cdot\|_F$ is the Frobenius norm. Without loss of generality, we set $P_s = 1$. In the context of JSCC, the *bandwidth ratio*, which quantifies the available channel uses per pixel (CPP), is defined as $R \triangleq \frac{k}{chw}$.

The received signal at the receiver can be written as $\mathbf{Y} = \mathbf{HX} + \mathbf{W}$, where $\mathbf{X} \in \mathbb{C}^{M\times k}$ and $\mathbf{Y} \in \mathbb{C}^{M\times k}$ denote the channel input and output matrices, respectively, and $\mathbf{W} \in \mathbb{C}^{M\times k}$ is the additive white Gaussian noise (AWGN) term, whose entries $W[i,j] \sim \mathcal{CN}(0, \sigma_w^2)$ follow an independent and identically distributed (i.i.d.) complex Gaussian distribution with zero mean and variance σ_w^2. The entries of the channel gain matrix $\mathbf{H} \in \mathbb{C}^{M\times M}$ follow a complex Gaussian distribution with zero mean and variance σ_h^2, i.e., $\mathbf{H}[i,j] \sim \mathcal{CN}(0, \sigma_h^2)$. We consider block-fading, in which the channel matrix $\mathbf{H}$ remains constant for k channel uses, corresponding to the transmission of one image, and takes an independent realization in the next block.

Given the channel output $\mathbf{Y}$, the receiver reconstructs the source image as $\hat{\mathbf{S}} \in \mathbb{R}^{h\times w\times 3}$. The reconstruction quality is quantified through the peak signal-to-noise ratio (PSNR), which serves as an indicator of image distortion at the per-pixel level, defined as: $PSNR \triangleq 10\log_{10}\frac{\|\mathbf{S}\|_\infty^2}{MSE(\mathbf{S},\hat{\mathbf{S}})}$ (dB), where $\|\cdot\|_\infty$ is the infinity norm and $MSE(\mathbf{S}, \hat{\mathbf{S}}) \triangleq \frac{1}{3hw}\|\mathbf{S} - \hat{\mathbf{S}}\|_2^2$ is the mean squared error (MSE) between $\mathbf{S}$ and $\hat{\mathbf{S}}$. Other metrics, e.g., the learned perceptual image patch similarity (LPIPS), are also be used in the DeepJSCC literature to assess the perceptual similarity between images [Wu et al., 2023b].

In this section, we consider two approaches, i.e., the traditional separate source and channel coding scheme (SSCC) and the DeepJSCC scheme, to solve the image transmission problem over MIMO channels outlined a shown in Figure 4.1. In particular, we shall consider both the CSIR (open-loop MIMO) and CSIT (closed-loop MIMO) scenarios for each transmission scheme.

4.2.1.1 Open-loop MIMO with CSIR

In an open-loop MIMO system, channel state information (CSI) is only available at the receiver, enabling it to equalize the transmitted signals and subsequently decode the image.

SSCC In an open-loop MIMO system with SSCC, the transmitter follows a sequential process, applying source coding, channel coding, and modulation in order. Given the received signal $\mathbf{Y}$ and the CSI, the receiver decodes the source signal as $\hat{\mathbf{S}}$ by sequentially performing the MIMO detection, demodulation, channel decoding, and source decoding operations.

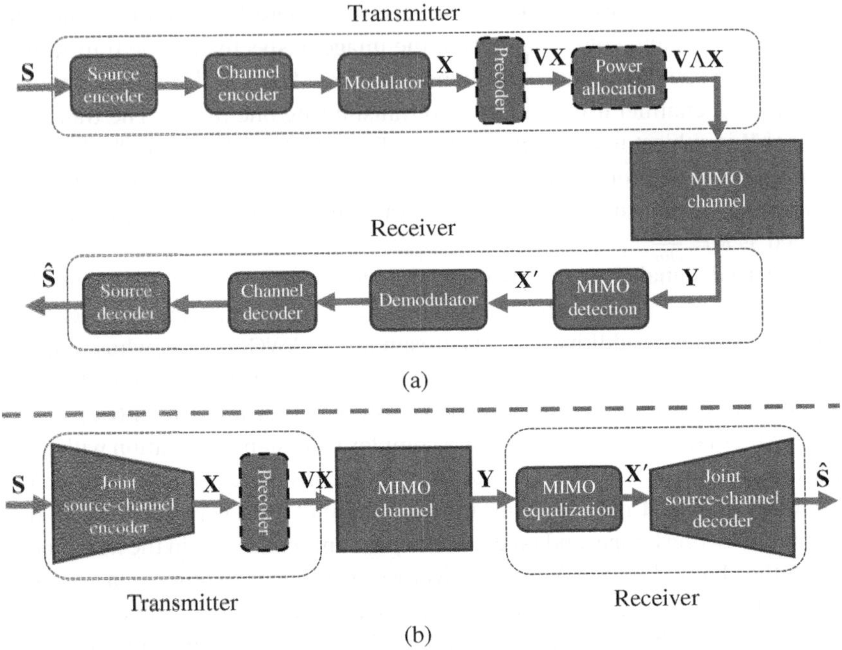

Figure 4.1 Block diagram of the MIMO image transmission system: (a) conventional separate source and channel coding scheme (SSCC) and (b) DeepJSCC scheme, where the gray blocks with dashed lines are the additional operations for the closed-loop MIMO system.

DeepJSCC DeepJSCC scheme uses a JSCC encoder at the transmitter to directly map the source signal $\mathbf{S}$ into the channel symbols $\mathbf{X}$. At the receiver, a MIMO equalization algorithm, such as zero-forcing (ZF) or minimum mean square error (MMSE), is employed first to exploit the CSI to decouple the entangled signal $\mathbf{Y}$ as $\mathbf{X}'$. Subsequently, a JSCC decoder is employed to reconstruct the source signal based on the available CSI represented by $(\mathbf{H}, \sigma_w^2)$, along with $\mathbf{X}'$.

4.2.1.2 Closed-loop MIMO with CSIT

Within a closed-loop MIMO system, the CSI is accessible to both the transmitter and receiver, allowing them to apply precoding and power allocation at the transmitter, as well as MIMO equalization at the receiver, thereby improving the image transmission quality.

SSCC The transmitter sequentially performs source coding, channel coding, and modulation to generate the channel input matrix $\mathbf{X}$. Additional operations at the transmitter with CSIT are precoding and power allocation, which can boost the communication rate. Specifically, given the CSI, we first decompose the channel matrix $\mathbf{H}$ by singular-value decomposition (SVD), yielding $\mathbf{H} = \mathbf{U}\boldsymbol{\Sigma}\mathbf{V}^H$, where $\mathbf{U} \in \mathbb{C}^{M\times M}$ and $\mathbf{V} \in \mathbb{C}^{M\times M}$ are unitary matrices, $\mathbf{V}^H$ is the conjugate transpose of $\mathbf{V}$, and $\boldsymbol{\Sigma} \triangleq \text{diag}(s_1, s_2, \dots, s_M)$ is a diagonal matrix whose singular values are in descending order, i.e., $s_1 \geq s_2 \geq \cdots \geq s_M$.

We denote the power allocation matrix by $\boldsymbol{\Lambda}$, where $\boldsymbol{\Lambda}$ is diagonal, with its diagonal elements being the power allocation weights for the signal streams of separate antennas. $\boldsymbol{\Lambda}$ can be derived using the standard water-filling algorithm [Scutari et al., 2009]. With power allocation and SVD precoding (precoding $\mathbf{X}$ into $\mathbf{VX}$), the channel model can be rewritten as: $\mathbf{Y} = \mathbf{HV}\boldsymbol{\Lambda}\mathbf{X} + \mathbf{W} = \mathbf{U}\boldsymbol{\Sigma}\boldsymbol{\Lambda}\mathbf{X} + \mathbf{W}$. Multiplying both sides of $\mathbf{Y}$ by $\mathbf{U}^H$ (MIMO detection) gives us: $\mathbf{X}' = \boldsymbol{\Sigma}\boldsymbol{\Lambda}\mathbf{X} + \mathbf{U}^H\mathbf{W}$. This SVD-based precoding converts the MIMO channel into a set of parallel subchannels with different SNR values. In particular, the SNR of the ith subchannel is determined by the ith singular value of s_i and the ith power allocation coefficient of $\boldsymbol{\Lambda}$. Given $\mathbf{X}' \in \mathbb{C}^{M\times k}$, the receiver performs demodulation, channel decoding, and source decoding sequentially to reconstruct the source image as $\hat{\mathbf{S}}$.

DeepJSCC Within the DeepJSCC scheme, we leverage the DeepJSCC encoder to perform feature extraction, channel symbol mapping, and power allocation all at once. Intuitively, the DNN is expected to transmit critical features over subchannels with higher SNRs, thereby improving the transmission performance.

In the above scenario, one approach could involve directly training the encoder/decoder networks with the expectation that they will autonomously learn to utilize the spatial degrees of freedom offered by the MIMO channel. However, we opt for a more structured, model-driven strategy, wherein we exploit the SVD and first convert the MIMO channel into subchannels. The received signal can then be written as $\mathbf{Y} = \mathbf{HVX} + \mathbf{W} = \mathbf{U}\boldsymbol{\Sigma}\mathbf{X} + \mathbf{W}$. To simplify the training, we apply MIMO equalization at the receiver by left multiplying both sides of $\mathbf{Y}$ by $\boldsymbol{\Sigma}^{\dagger}\mathbf{U}^H$ to obtain $\mathbf{X}' = \boldsymbol{\Sigma}^{\dagger}\mathbf{U}^H\mathbf{Y} = \mathbf{X} + \mathbf{W}'$, where $\mathbf{W}' \triangleq \boldsymbol{\Sigma}^{\dagger}\mathbf{U}^{\mathbf{H}}\mathbf{W} \in \mathbb{C}^{M\times k}$ is the equivalent noise term, $\boldsymbol{\Sigma}^{\dagger}$ is the Moore-Penrose inverse of matrix $\boldsymbol{\Sigma}$, and $\mathbf{U}^H$ is the conjugate transpose of matrix $\mathbf{U}$. Finally, we feed both $\mathbf{X}'$ and the CSI into the DeepJSCC decoder to recover the image.

4.2.2 A DeepJSCC-MIMO Solution

This section presents a new DeepJSCC architecture called DeepJSCC-MIMO, aiming to enable efficient image transmission over MIMO channels. The pipeline of our DeepJSCC-MIMO scheme is depicted in Figure 4.2, where a pair of vision transformer (ViT)-based encoder f_θ and decoder f_ϕ is employed to encode and decode channel symbols, utilizing the inherent self-attention mechanism over both source semantic features and channel conditions. Further

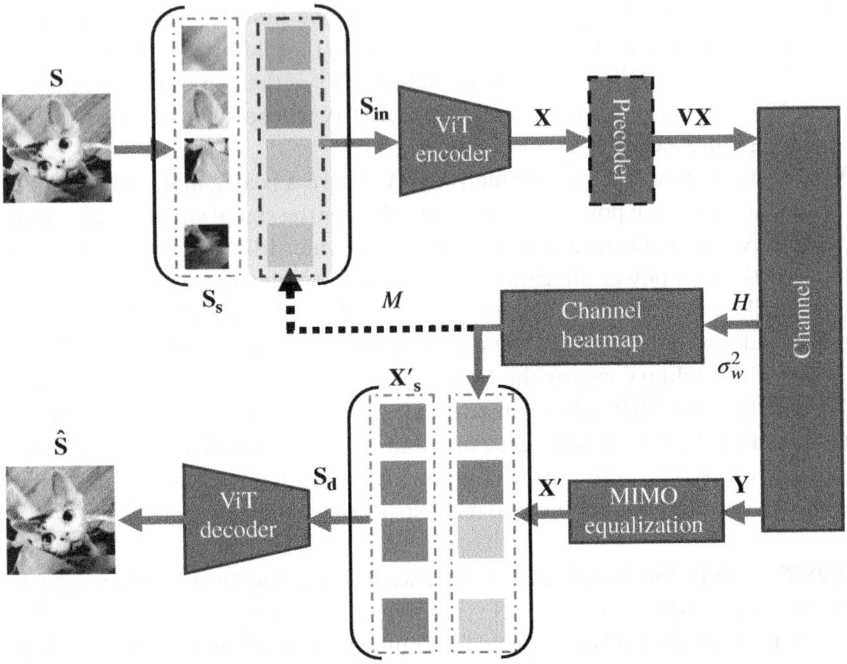

Figure 4.2 The pipeline of the DeepJSCC-MIMO scheme, where the source image **S** is encoded by a ViT encoder and reconstructed by a ViT decoder as $\hat{\mathbf{S}}$. Precoding operation in dashed line will be performed if the CSI is available at the transmitter, and the CSI $(\mathbf{H}, \sigma_w^2)$ is fed to the encoder, if available, and decoder in the form of a "heatmap" **M** to facilitate the JSCC encoding/decoding process. Source: Wu et al., 2023/With permission of IEEE.

elaborations on the inner structures of these ViTs are provided in Wu et al. [2023a]. In the following, we detail the pipeline of DeepJSCC-MIMO in five main steps: image-to-sequence transformation, channel heatmap construction, ViT encoding, ViT decoding, and the loss function.

4.2.2.1 Image-to-Sequence Transformation

We first convert the input image **S** into a sequence of vectors, denoted by $\mathbf{S_s} = Seq(\mathbf{S})$. Specifically, we divide **S** into a grid of $p \times p$ patches, and reshape each patch into a vector of dimension $\mathbb{R}^{\frac{3hw}{p^2}}$. In this way, **S** is converted to $\mathbf{S_s} \in \mathbb{R}^{l \times c}$, where $l = p^2$ and $c \triangleq \frac{3hw}{p^2}$.

4.2.2.2 Channel Heatmap Construction

To enable efficient training, we construct a channel heatmap from CSI, indicating the effective noise variance faced by each channel symbol generated by ViT. Let us define $\mathbf{P_n} \in \mathbb{R}^{M \times k}$ as the average power of the equivalent additive noise term

$\mathbf{W}'$. Specifically, for different scenarios, $\mathbf{P_n}$ is defined as:

$$\mathbf{P_n} \triangleq \begin{cases} \sigma_w^2(\mathbf{H}_w \odot \overline{\mathbf{H}}_w)\mathbf{J}_{M\times k}, & \text{MIMO with CSIR} \\ \sigma_w^2\boldsymbol{\Sigma}^\dagger\mathbf{U}^H\mathbf{J}_{M\times k}, & \text{MIMO with CSIT,} \end{cases} \tag{4.1}$$

where $\odot$ denotes the Hadamard product and $\overline{\mathbf{H}}_w$ is the element-wise conjugate of matrix $\mathbf{H}_w$, which is obtained from the ZF operation as: $\mathbf{H}_w \triangleq (\mathbf{H^H H})^{-1}\mathbf{H^H}$, and $\mathbf{J}$ is the matrix of ones.

Then the heatmap $\mathbf{M} \in \mathcal{R}^{l\times \frac{2Mk}{l}}$ is constructed as follows:

$$\mathbf{M} \triangleq \texttt{reshape}\left(\texttt{concat}\left(\frac{1}{2}\mathbf{P_n}, \frac{1}{2}\mathbf{P_n}\right)\right), \tag{4.2}$$

where $\texttt{concat}(\cdot)$ and $\texttt{reshape}(\cdot)$ denote the concatenation and reshape operations, respectively. We concatenate two matrices $\frac{1}{2}\mathbf{P_n}$ to get the shape of $\mathbb{R}^{M\times 2k}$, and reshape it into $\mathbf{M} \in \mathcal{R}^{l\times \frac{2Mk}{l}}$, which represents the equivalent noise term faced by each real encoding output element.

4.2.2.3 ViT Encoding

For the two scenarios, the input sequences to our ViT encoder $\mathbf{S_{in}} \in \mathbb{R}^{l\times c_{in}}$ are given by:

$$\mathbf{S_{in}} = \begin{cases} \mathbf{S_s}, & \text{MIMO with CSIR} \\ \texttt{concat}(\mathbf{S_s}, \mathbf{M}), & \text{MIMO with CSIT,} \end{cases} \tag{4.3}$$

where the dimensions of the vectors in these sequences are $c_{in} = c$ and $c_{in} = c + \frac{2Mk}{l}$ for the MIMO with CSIR and MIMO with CSIT scenarios, separately. The encoding process can then be represented as $\mathbf{X} = f_\theta(\mathbf{S_{in}}) \in \mathbb{C}^{M\times k}$, where the operations of our encoder are detailed in Wu et al. [2023b].

4.2.2.4 ViT Decoding

To simplify the decoding process, we perform an equalization operation on the channel output $\mathbf{Y}$ to decouple the transmitted data of each antenna as $\mathbf{X}' \in \mathbb{C}^{M\times k}$. We apply different equalization methods for the MIMO systems with CSIR and with CSIT, given as:

$$\mathbf{X}' = \begin{cases} \mathbf{H}_w\mathbf{Y} + f_{res}(\mathbf{Y}, \mathbf{H}), & \text{MIMO with CSIR} \\ \boldsymbol{\Sigma}^\dagger\mathbf{U^H Y}, & \text{MIMO with CSIT.} \end{cases} \tag{4.4}$$

Note that we can also apply other MIMO equalization methods. However, we have observed that a simple ZF estimation method [Bian et al., 2023b; Wu et al., 2023b], together with a learned compensation, is sufficiently effective.

Given the equalized signal $\mathbf{X}'$, which is reshaped into $\mathbf{X}'_\mathbf{s} \in \mathbb{R}^{l\times \frac{2Mk}{l}}$, and the noise heatmap $\mathbf{M}$, a ViT-based decoder f_ϕ is designed to recover the source image as $\hat{\mathbf{S}} = f_\phi(\mathbf{X}'_\mathbf{s}, \mathbf{M})$, as detailed in Wu et al. [2023b].

4.2.2.5 Loss Function

The encoder and decoder are optimized jointly to minimize the loss function: $\mathcal{L}(\theta, \phi) = MSE(\mathbf{S}, \hat{\mathbf{S}})$. We train the model to search for the optimal parameters (θ^*, ϕ^*) with a minimal $\mathcal{L}(\theta, \phi)$.

4.2.3 Training and Evaluation

We consider a 2×2 MIMO system and employ the practical separation-based codes and theoretical bounds (assuming capacity-achieving channel codes) as benchmarks. Unless stated otherwise, we consider the CIFAR10 dataset for evaluation. For the practical separation-based coding scheme, the BPG codec [Bellard, 2018], as well as the state-of-the-art neural compression algorithms are employed. For channel coding, we consider LDPC codes with various code lengths and constellation orders. For the theoretical bound, we assume capacity-achieving channel codes with the BPG codec and water-filling power allocation operations for transmission, denoted as the BPG-Capacity scheme.

Considering the model complexity and the performance, we set $p = 8$, $l = 64$, and $c = 48$. Each element of $\mathbf{H}$ is sampled as $H[i,j] \sim \mathcal{CN}(0,1)$. Results for a more general correlated MIMO channel model are presented in Wu et al. [2023b]. We define the channel SNR, denoted by μ, as $\mu \triangleq 10\log_{10}\frac{\mathbb{E}_{\mathbf{H},\mathbf{X}}[\|\mathbf{HX}\|_F^2]}{\mathbb{E}_{\mathbf{W}}[\|\mathbf{W}\|_F^2]}$ (dB) $= 10\log_{10}\frac{M}{\sigma_w^2}$.

4.2.3.1 Open-loop MIMO System with CSIR

We first evaluate DeepJSCC-MIMO in the open-loop MIMO system with CSIR, where we compare DeepJSCC-MIMO, trained at specific SNRs, with the BPG-Capacity[1] and BPG-LDPC schemes at different SNR values and bandwidth ratios. For the BPG-LDPC scheme, we apply the sphere decoding (SD) algorithm [Guo and Nilsson, 2006].

General Performance Figure 4.3a,b shows the PSNR performance of the DeepJSCC-MIMO, BPG-Capacity, and BPG-LDPC schemes over the SNR values from 0 to 20 dB when the bandwidth ratio is $R = 1/24$ and $R = 1/12$. We can observe that our DeepJSCC-MIMO can generally outperform the BPG-Capacity and BPG-LDPC schemes in all SNR values. We emphasize that the BPG-Capacity performance shown in these figures is not achievable by practical channel coding schemes and serves as an upper bound on any separation-based scheme that employs the BPG compression method. DeepJSCC-MIMO provides significant improvements

---

1 We assume that the transmitter knows the instantaneous capacity of each channel block without power allocation, and can select the best compression rate. Note that, in practice, the transmitter does not have access to the CSI, and would probably employ a conservative compression rate to avoid a potential outage.

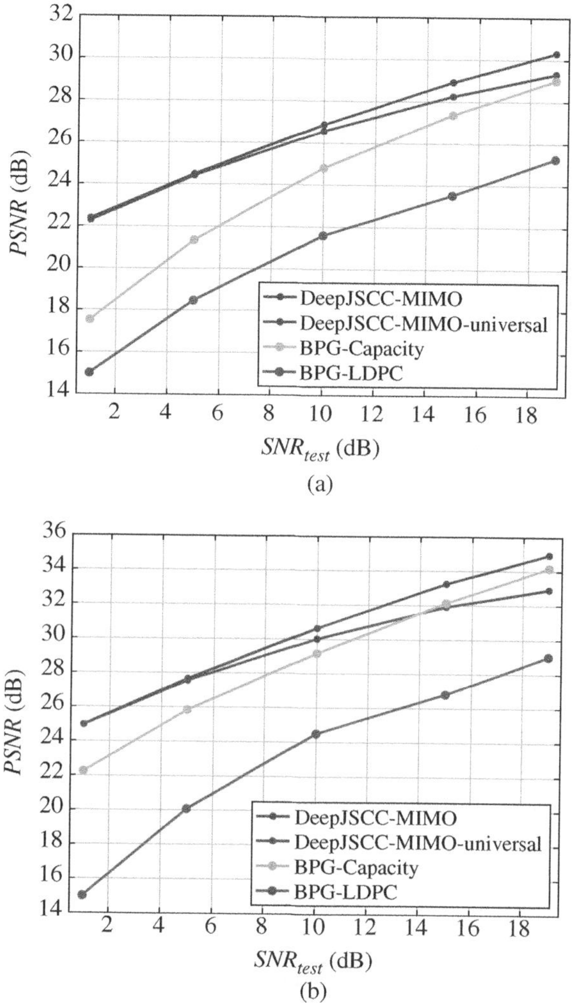

Figure 4.3 Performance comparisons between DeepJSCC-MIMO and BPG-Capacity over different SNR values and bandwidth ratios for the MIMO systems with CSIR. (a) $R = 1/24$ – *PSNR* and (b) $R = 1/12$ – *PSNR*.

(at least 1.3 and 5.05 dB) for $R = 1/24$, compared with the BPG-Capacity and BPG-LDPC schemes, respectively. For $R = 1/12$, our DeepJSCC-MIMO can still outperform the BPG-Capacity and BPG-LDPC schemes in all SNRs. We observe that the gap between DeepJSCC-MIMO and BPG-Capacity is more significant in the case of the lower bandwidth ratio.

SNR-adaptability Note that each point on the DeepJSCC-MIMO curve is obtained by training a separate encoder–decoder pair. We also consider a random SNR training strategy to evaluate the SNR-adaptability of the DeepJSCC-MIMO scheme, denoted by DeepJSCC-MIMO-universal. We train the model with random SNR values uniformly sampled from [0, 22] dB and test the well-trained model at different SNRs. The comparisons in Figure 4.3 show that there is a slight performance degradation compared to training a separate DeepJSCC-MIMO encoder–decoder pair for each SNR (up to 0.99 dB for $R = 1/24$ and 2.02 dB for $R = 1/12$); however, the DeepJSCC-MIMO-universal brings significant advantage in terms of training complexity and storage requirements. We can conclude that the DeepJSCC-MIMO-universal scheme learns to adapt to different SNRs from the random training strategy at the expense of a slight loss in PSNR, which tends to increase with SNR.

Robustness to Channel Estimation Errors We take the same approach as in Yoo and Goldsmith [2006] to evaluate the robustness of MIMO systems to channel estimation errors, where $\mathbf{H}$ is imperfectly estimated as $\hat{\mathbf{H}}$ at the receiver. Similar to the work in Hassibi and Hochwald [2003], Etkin and Tse [2006], and Yoo and Goldsmith [2006], let $\mathbf{E}_n \triangleq \mathbf{H} - \hat{\mathbf{H}}$, where the entries of $\mathbf{E}_n$ are zero-mean complex Gaussian with variance σ_e^2. Here, σ_e^2 is the "noise variance" term to capture the channel estimation quality, which can be appropriately selected depending on the channel conditions and channel estimation method.

We evaluate the DeepJSCC-MIMO-universal and BPG-LDPC schemes with different $\sigma_e^2 \in \{0, 0.2, 1\}$ values over different SNR values on the Kodak dataset with $R = 1/24$, as shown in Table 4.1. As expected, for both schemes, the performance degrades as the channel estimation error increases. Specifically, for $\sigma_e^2 = 0.2$, the performance loss of our DeepJSCC-MIMO scheme is up to 0.29 dB, compared with the performance loss up to 5.96 dB from the BPG-LDPC scheme. For $\sigma_e^2 = 1$, the BPG-LDPC scheme starts to fail to work. However, our DeepJSCC-MIMO scheme can still work with a performance loss of less than 0.84 dB. We want to emphasize that even with a significant channel estimation error ($\sigma_e^2 = 1$), DeepJSCC-MIMO

Table 4.1 Comparisons of DeepJSCC-MIMO-universal and BPG-LDPC schemes with sphere decoding operations under channel estimation errors over the Kodak dataset for $R = 1/24$.

Different SNR_{test} (dB)	**5**	**7**	**9**	**12**
DeepJSCC-MIMO ($\sigma_e^2 = 0$)	29.38	30.08	30.80	31.62
DeepJSCC-MIMO ($\sigma_e^2 = 0.2$)	29.18	29.86	30.53	31.33
DeepJSCC-MIMO ($\sigma_e^2 = 1$)	28.73	29.49	30.08	30.78
BPG-LDPC ($\sigma_e^2 = 0$)	23.79	27.23	28.92	30.27
BPG-LDPC ($\sigma_e^2 = 0.2$)	18.65	21.27	23.21	24.84
BPG-LDPC ($\sigma_e^2 = 1$)	15.01	15.37	15.72	16.03

continues to function, with only limited performance loss, compared with the significant performance loss with the traditional separation benchmark. We can conclude that DeepJSCC-MIMO is more robust in handling channel estimation errors, as it can learn to compensate for them during transmission.

4.2.3.2 Closed-loop MIMO System with CSIT

In this section, we consider the closed-loop MIMO system.

General Performance We first evaluate the DeepJSCC-MIMO model under the setup where the training SNR, denoted by SNR_{train}, matches the test SNR, denoted by SNR_{test}. In particular, we set $SNR_{train} = SNR_{test} \in \{1, 5, 10, 15, 19\}$ dB and the bandwidth ratio $R \in \{1/24, 1/12, 1/6\}$. The comparisons between DeepJSCC-MIMO and the BPG-Capacity benchmark is shown in Figure 4.4.

Regarding the PSNR performance, DeepJSCC-MIMO outperforms the SSCC benchmark in all SNR and bandwidth-ratio scenarios. Specifically, we can see that DeepJSCC-MIMO outperforms the benchmark by at least 1.98 and 1.78 dB for $R = 1/24$ and $R = 1/12$, respectively. When $R = 1/6$, improvements of up to 3.58 dB can be observed. These significant improvements demonstrate the superiority of the DeepJSCC-MIMO scheme and its capacity in extracting and mapping image features to the available channels in an adaptive fashion.

SNR-adaptability We also consider a random SNR training strategy to evaluate the channel adaptability of DeepJSCC-MIMO. We train the model with random SNR values uniformly sampled from $[0, 22]$ dB, denoted by DeepJSCC-MIMO-universal, and test it at different SNRs. The comparisons in Figure 4.4 show that there is a slight performance degradation compared to training a separate DeepJSCC-MIMO encoder–decoder pair for each channel SNR; however, DeepJSCC-MIMO-universal brings a significant advantage in terms of training complexity and storage requirements.

To evaluate the effectiveness of the SVD strategy in our scheme, we train the models of DeepJSCC-MIMO-universal without SVD decomposition-based precoding, denoted by DeepJSCC-MIMO-universal w/o SVD. Compared with the DeepJSCC-MIMO-universal, we can observe that the performance of DeepJSCC-MIMO-universal w/o SVD has a gap of up to 0.97, 1.91, and 4.21 dB for $R = 1/24$, $R = 1/12$, and $R = 1/6$, respectively. Although the networks can still learn to communicate over the MIMO channel, the SVD-based model-driven strategy can significantly simplify the training process and improve performance, illustrating the importance of exploiting domain knowledge in designing data-driven communication technologies.

High-Resolution Datasets To evaluate the model generalizability, we validate the DeepJSCC-MIMO on the CelebA dataset. Specifically, we train the

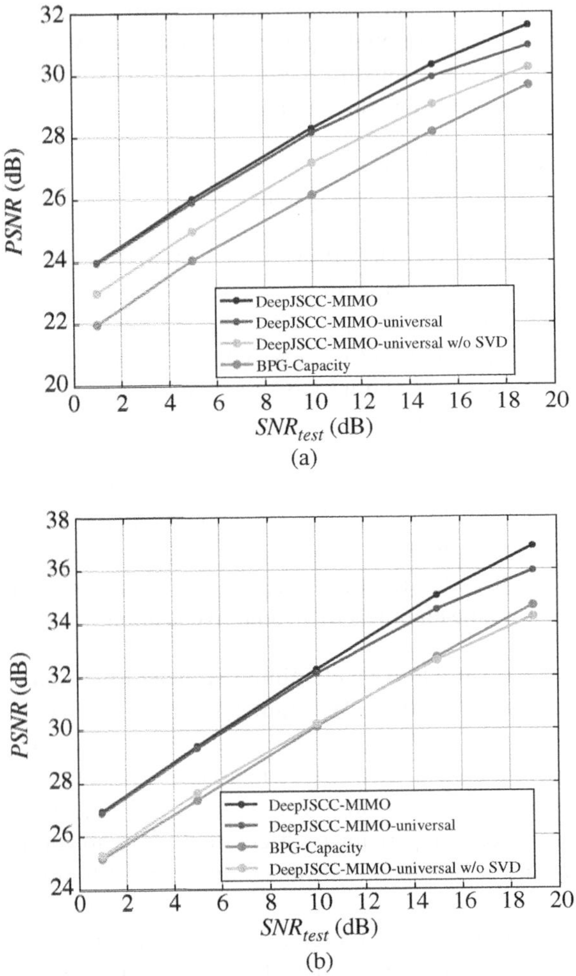

Figure 4.4 Performance of DeepJSCC-MIMO compared with the BPG-Capacity benchmark at different channel SNR and bandwidth-ratio scenarios for the closed-loop MIMO. (a) $R = 1/24 - PSNR$, (b) $R = 1/12 - PSNR$, and (c) $R = 1/6 - PSNR$.

DeepJSCC-MIMO and DeepJSCC-MIMO-universal schemes with randomly cropped 128×128 patches from the ImageNet dataset, and evaluate the well-trained models on the CelebA da.tasets. The detailed PSNR performance of the transmission, compared with BPG-Capacity, are shown in Figure 4.5a.

From Figure 4.5a, we can observe that our DeepJSCC-MIMO can outperform BPG-Capacity at all test SNRs and generally provide higher gains at lower SNRs (up to 2.1 dB) in the CelebA dataset. There is a slight performance degradation of the DeepJSCC-MIMO-universal scheme (up to 0.27 dB) compared with DeepJSCC-MIMO trained at specific SNRs; however, the DeepJSCC-MIMO-universal can still outperform the BPG-Capacity scheme at

Figure 4.4 (*Continued*)

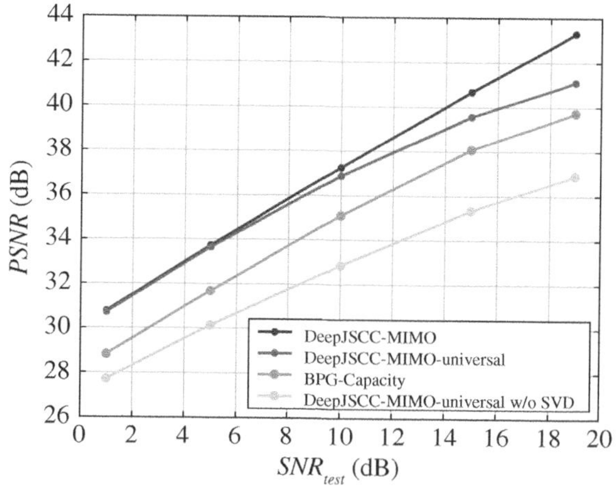

(c)

all SNRs. Additionally, we consider state-of-the-art learning-based compression algorithms along with capacity-achieving codes and water-filling algorithms, denoted as Cheng2020-Capacity [Cheng et al., 2020], MBT2018-Capacity [Minnen et al., 2018], and BMSHJ2018-Capacity schemes [Ballé et al., 2018], whose performance falls short of the DeepJSCC-MIMO. Considering that the BPG-Capacity performance shown in these figures is an upper bound on the performance of any separation-based scheme that employs the BPG codec, the comparison here clearly shows the benefits of the DeepJSCC-MIMO and DeepJSCC-MIMO-universal schemes in high-resolution datasets.

Visualization of the Channel Heatmap and Power Allocation We visualize the channel heatmap and the power allocation of the channel input matrix for a closed-loop MIMO system in Figure 4.5b. Within our channel heatmap, we visualize the equivalent noise power experienced by each channel symbol. We can observe that DeepJSCC-MIMO generally assigns more power to antennas with better channel conditions. Specifically, in the low SNR regime, DeepJSCC-MIMO concentrates most of the power on a few select antennas with exceptional quality. In certain extreme cases, DeepJSCC-MIMO refrains from allocating power to the subchannels characterized by the worst channel conditions. As the SNR increases, we note that DeepJSCC-MIMO tends to distribute power more evenly across a greater number of antennas, instead of concentrating on a limited set of antennas with the worst channel conditions.

Adaptability to the Antenna Numbers In this section, we investigate the generalizability of the DeepJSCC-MIMO architecture to different antenna numbers. In

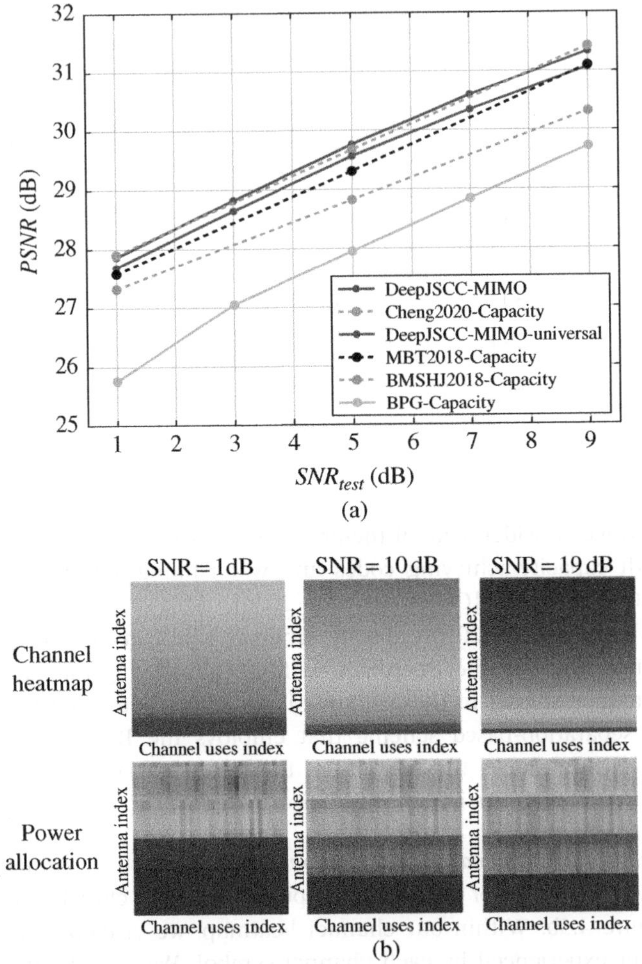

Figure 4.5 (a) Performance of DeepJSCC-MIMO and conventional separation-based scheme in the CelebA dataset when $R = 1/48$. (b) Visualization of the channel heatmap and power allocation across the channel symbols, where antenna number is set as $M = 32$ and $R = 1/6$.

particular, we repeat the experiment of DeepJSCC-MIMO in Figure 4.4b for different M values. The performance of our DeepJSCC-MIMO models with different antenna numbers $M \in \{2, 3, 4\}$ and channel SNRs $SNR_{test} \in \{1, 5, 10\}$ dB is shown in Figure 4.6. We can observe that DeepJSCC-MIMO can generally achieve better performance with the increased number of antennas, thanks to its more flexible power allocation. We also consider a random antenna number training strategy

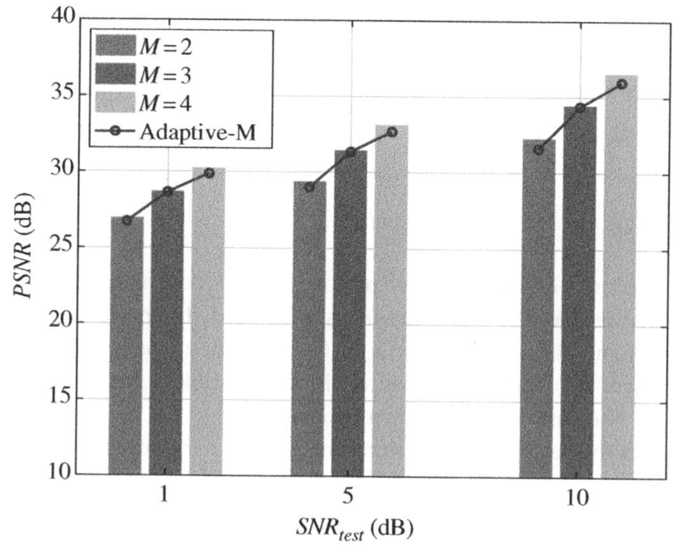

Figure 4.6 Performance of DeepJSCC-MIMO with different antennas number.

with a maximal antenna number setting M_{max}, denoted as "adaptive-M". From Figure 4.6, we observe that the adaptive-M strategy can achieve similar performance compared with models trained for specific M values: The performance loss is at most 0.6 dB. We can conclude that DeepJSCC-MIMO is highly flexible and adaptive and can be applied to different antenna numbers without retraining.

4.3 DeepJSCC for Relay Channels

In this section, our attention pivots to exploring the utilization of DeepJSCC in relay channels, with a keen interest in two specific types of relay channels: The classical three-node cooperative relay channel and the multihop relay channel. These channels are instrumental in facilitating the transmission of information from a source node to a destination node that is significantly distant.

4.3.1 Cooperative Relay

4.3.1.1 System Model

We consider classical three-node cooperative relay channel consisting of a source node S, a destination node D, and a relay node R, as shown in Figure 4.7. The goal is to deliver an image $\mathbf{S}$ from S to D with the help of R. We assume a half-duplex relay node that cannot receive and transmit at the same time.

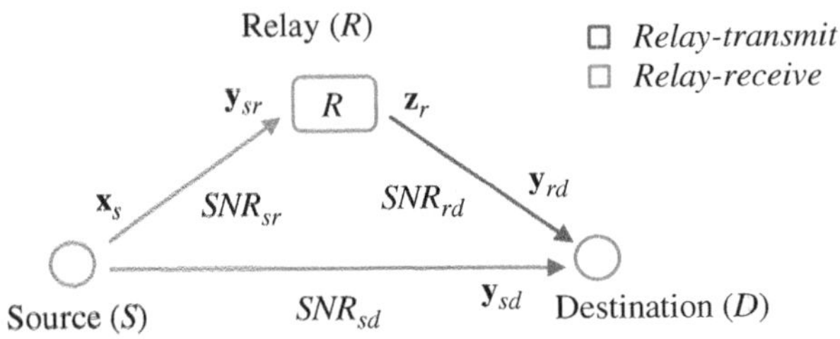

Figure 4.7 The half-duplex relay channel.

Therefore, the transmission is divided into two periods [Host-Madsen and Zhang, 2005]: The *relay-receive* period and the *relay-transmit* period. In the former, S encodes the image $\mathbf{S}$ into a channel codeword $\mathbf{x}_s \in \mathbb{C}^k$ using an encoder function $f_s : \mathbb{R}^{c\times h\times w} \to \mathbb{C}^k$, where $\mathbf{x}_s = f_s(\mathbf{S})$. The source input is subject to a power constraint: $\frac{1}{k}\|\mathbf{x}_s\|^2 \leq 1$. The bandwidth ratio is given as $R \triangleq \frac{k}{chw}$.

The received signal at the relay R and destination D at the *relay-receive* period can be written as

$$\mathbf{y}_{sr} = \alpha_{sr}\mathbf{x}_s + \mathbf{n}_r, \tag{4.5}$$

$$\mathbf{y}_{sd} = \alpha_{sd}\mathbf{x}_s + \mathbf{n}_d, \tag{4.6}$$

where α_{sr}, α_{sd} are real constants governed by the transmission distances of the S–R and S–D links, respectively; $\mathbf{n}_r \sim \mathcal{CN}(0, N_r)$, $\mathbf{n}_d \sim \mathcal{CN}(0, N_d)$ denote complex AWGN signals. The SNR of the S–R and S–D links are defined as $SNR_{sr} \triangleq \frac{\alpha_{sr}^2}{N_r}$ and $SNR_{sd} \triangleq \frac{\alpha_{sd}^2}{N_d}$.

Upon receiving $\mathbf{y}_{sr}$, the relay processes it by a function $f_r(\cdot)$ and forwards the processed signal to the destination in the relay-transmit period. Denoting the signal transmitted from the relay by $\mathbf{z}_r \in \mathbb{C}^k$, we have $\mathbf{z}_r = f_r(\mathbf{y}_{sr})$, where $\mathbf{z}_r$ is also subject to a power constraint: $\frac{1}{k}\|\mathbf{z}_r\|^2 \leq 1$.

The signal received at the destination in the *relay-transmit* period can be written as:

$$\mathbf{y}_{rd} = \alpha_{rd}\mathbf{z}_r + \mathbf{n}_{rd}, \tag{4.7}$$

where α_{rd}, $\mathbf{n}_{rd}$, and SNR_{rd} are defined similarly as in the S–R and S–D links.

Given the received signal from the sour .ce and relay in the two periods, the destination reconstructs the image using a decoding function $g = \mathbb{C}^k \times \mathbb{C}^k \to \mathbb{R}^{C\times H\times W}$. The reconstructed image is denoted by $\tilde{\mathbf{S}} \in \mathbb{R}^{C\times H\times W}$, where $\tilde{\mathbf{S}} = g(\mathbf{y}_{sd}, \mathbf{y}_{rd})$. The PSNR will be used to evaluate the reconstruction quality of $\tilde{\mathbf{S}}$.

4.3.1.2 DeepJSCC for Cooperative Relay

With DeepJSCC, we can parameterize the encoder f_s, decoder g, and the transformation f_r at the relay by DNNs. Function f_r at the relay terminal depends on the particular relaying scheme. In the case of DeepJSCC-AF, the relay simply amplifies its received signal, and no DNN is needed. The architectures used for the relay function in DeepJSCC-DF and DeepJSCC-PF schemes, denoted by f_{DF} and f_{PF}, respectively, are shown in figure 2 of Bian et al. [2024a]. Specifically, "ResNet" refers to a block of 2D convolutional neural networks with residual connections. We emphasize that residual connections are essential in our problem, especially at the relay. The upsampling blocks in g are realized by pixel shuffling. The parameter C_{feat} denotes the number of filters and C_{out} is determined by the bandwidth ratio R.

Channel attention (CA) modules are introduced to improve the robustness to different channel conditions. These CA modules take both the features and channel conditions as inputs, where the channel conditions are a collection of SNRs defined as $\mathbf{SNR} = (SNR_{sr}, SNR_{sd}, SNR_{rd})$, allowing DNNs to assign different weights to different input features according to the channel conditions. Based on the DNNs described above, we next explain the processing at the relay and destination nodes when operating under DeepJSCC-AF, -DF, and -PF protocols, respectively.

DeepJSCC-AF As shown in Figure 4.8, f_{AF} is simply a linear function, and the signal transmitted by the relay is a scaled version of its received signal: $\mathbf{z}_r = \beta \mathbf{y}_{sr}$, where $\beta \triangleq \sqrt{\frac{1}{\alpha_{sr}^2 + N_r}}$ ensures that the power constraint is satisfied. Substituting $\mathbf{z}_r$ into (4.7) yields

$$\mathbf{y}_{rd} = \beta \alpha_{rd} \alpha_{sr} \mathbf{z} + \tilde{\mathbf{n}}, \tag{4.8}$$

where $\tilde{\mathbf{n}}$ is AWGN with zero mean and variance $N_d + \frac{N_r \alpha_{rd}^2}{\alpha_{sr}^2 + N_r}$.

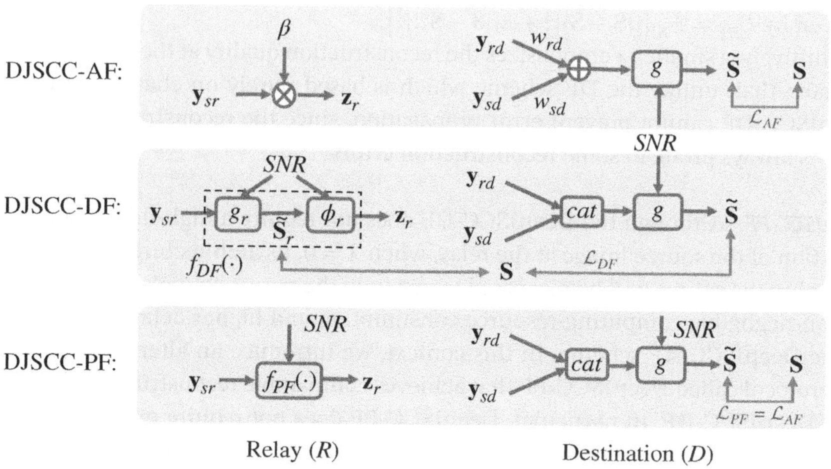

Figure 4.8 The processing of DeepJSCC-AF, -DF and -PF at the relay (R) and destination (D), where β is the scaling factor while w_{rd} and w_{sd} are the MRC coefficients in (4.9). **SNR** denotes the collection of channel qualities, consisting of SNR_{sr}, SNR_{rd}, and SNR_{sd}.

Given (4.6) and (4.8), the received signal at the destination from two transmission periods can be combined by maximum ratio combining (MRC) to generate an estimate of $\mathbf{z}$, expressed as:

$$\tilde{\mathbf{z}} = \frac{(\beta^2\alpha_{rd}^2 N_r + N_d)\alpha_{sd}\mathbf{y}_{sd} + N_d\beta\alpha_{rd}\alpha_{sr}\mathbf{y}_{rd}}{N_d\beta^2\alpha_{rd}^2\alpha_{sr}^2 + \alpha_{sd}^2(\beta^2\alpha_{rd}^2 N_r + N_d)}. \tag{4.9}$$

Then we convert $\tilde{\mathbf{z}}$ into a real vector, reshape it, and feed it to the DeepJSCC-AF decoder along with **SNR**. The whole system is trained in an end-to-end fashion with mean square error (MSE) as the loss function: $\mathcal{L}_{AF} = \mathbb{E}_{\mathbf{S}}\left[||\mathbf{S} - \tilde{\mathbf{S}}||_2^2\right]$.

DeepJSCC-DF In the conventional (digital) DF protocol, the relay decodes the original information bits, re-encodes the bits to create additional parity symbols, and forwards these parity symbols to the destination. We can apply a similar technique in conjunction with DeepJSCC: the relay tries to reconstruct the source image from $\mathbf{y}_{sr}$, then performs another DeepJSCC encoding on the recovered image into $\mathbf{z}_r$ for transmission. Specifically, we first feed $\mathbf{y}_{sr}$ to the relay decoder network g_r (along with the **SNR**), yielding an estimated image $\tilde{\mathbf{S}}_r = g_r(\mathbf{y}_{sr}, \mathbf{SNR})$ followed by the relay encoder $\phi_r(\cdot)$ to generate $\mathbf{z}_r = \phi_r(\tilde{\mathbf{S}}_r, \mathbf{SNR})$. Note that $\mathbf{z}_r$ will be power normalized before transmission. As shown in Figure 4.8, we use $f_{DF} = \phi_r(g_r(\cdot))$ to represent the operations at the relay in the DF mode. At the destination, we concatenate the received signals $\mathbf{y}_{sd}$ and $\mathbf{y}_{rd}$ to reconstruct $\tilde{\mathbf{S}}$.

The loss function of DeepJSCC-DF is comprised of two parts, representing the image reconstruction quality at the relay and the destination, respectively. To balance the two, we introduce a hyper-parameter $\lambda \geq 0$, and the overall loss function is given by $\mathcal{L}_{DF} = \mathbb{E}_{\mathbf{S}}\left[||\mathbf{S} - \tilde{\mathbf{S}}||_2^2 + \lambda||\mathbf{S} - \tilde{\mathbf{S}}_r||_2^2\right]$.

Intuitively, a smaller λ emphasizes the reconstruction quality at the destination. We note that, unlike the DF scheme which is based purely on channel coding, DeepJSCC-DF cannot prevent error propagation, since the reconstruction at the relay is always prone to some reconstruction errors.

DeepJSCC-PF Although the DeepJSCC-DF does not require a high-fidelity reconstruction of the source image at the relay, when $\lambda = 0$, its architecture is designed assuming the source will be recovered to generate the parity symbols. This results in non-negligible computing-resource consumption and higher delays compared to the DeepJSCC-AF scheme. In this context, we introduce an alternative relaying protocol called DeepJSCC-PF that achieves comparable reconstruction results with DeepJSCC-DF. In particular, DeepJSCC-PF does not require explicit reconstruction at the relay, employing a lightweight neural network to model relay processing.

We have $\mathbf{z}_r = f_{PF}(\mathbf{y}_{sr})$, which is normalized before transmission. The decoder for DeepJSCC-PF at the destination is identical to that of DeepJSCC-DF: The received signals are concatenated, reshaped, and fed into the decoder. The training loss $\mathcal{L}_{PF}$ is the same as that of DeepJSCC-AF. The relaying and decoding processes for DeepJSCC-AF, -DF, and -PF are illustrated in Figure 4.8.

4.3.1.3 Numerical Experiments

Parameter settings and training details Throughout this section, we assume that the S–D and R–D links have the same quality, while the S–R link has better quality, i.e., $SNR_{sr} > SNR_{sd} = SNR_{rd} \triangleq \gamma$.

We evaluate DeepJSCC-AF, -DF, and -PF architectures considering the transmission of images from the CIFAR-10 dataset. The Adam optimizer is adopted to train the model with a varying learning rate. Specifically, the learning rate is initialized to 10^{-4} and will be dropped by a factor of 0.8 if the validation loss does not improve in four consecutive training epochs. Throughout this section, we fix $R = 0.125$, and hence $C_{out} = 12$.

Comparison of different relaying protocols Next, we compare the performances of DeepJSCC-AF and DeepJSCC-DF, benchmarked against noncooperative transmission, which refers to transmission from S to D in the absence of R Bourtsoulatze et al. [2019]. The non-cooperative scheme uses the same neural network structure while the source utilizes k symbols to transmit the image. The PSNR comparisons of these schemes are shown in Figure 4.9, where we train and test different DeepJSCC-AF and -DF pairs with $SNR_{sr} \in \{0, 12, 24, \infty\}$ dB. Note that DeepJSCC-DF curves shown in this simulation are obtained with the optimal λ for each SNR_{sr} value, which will be detailed later.

In cooperative channel coding, the condition of the S–R link determines the relative performances of AF, DF, and noncooperative transmission in terms of the achievable rates [Kramer et al., 2005]. In particular, when SNR_{sr} is low, it is no longer possible to decode at the relay; hence, DF becomes worse than AF and noncooperative transmission in this regime. Our results in Figure 4.9 tell a different story: The PSNR of our DeepJSCC-DF scheme is strictly better than that of the non-cooperative transmission, even when $SNR_{sr} = 0$ dB. This is because, even though we require the relay to decode the image, we allow lossy reconstruction in DeepJSCC. Therefore, in DeepJSCC-DF, the reconstruction quality at the destination is not limited by that of the relay, unlike in the digital DF scheme.

The performance gain of DeepJSCC-DF with respect to DeepJSCC-AF increases with SNR_{sr}. In particular, when $SNR_{sr} = \infty$, DeepJSCC-DF outperforms DeepJSCC-AF by approximately 1 dB. This can be understood as a coding gain: when $SNR_{sr} = \infty$, DeepJSCC-AF is essentially a repetition code. Notice that we have set $SNR_{sd} = SNR_{rd}$, as a result, the PSNR performance of DeepJSCC-AF is 3 dB better than that of the noncooperative transmission. For DeepJSCC-DF, the relay generates "parity" symbols from the received signal, providing an additional

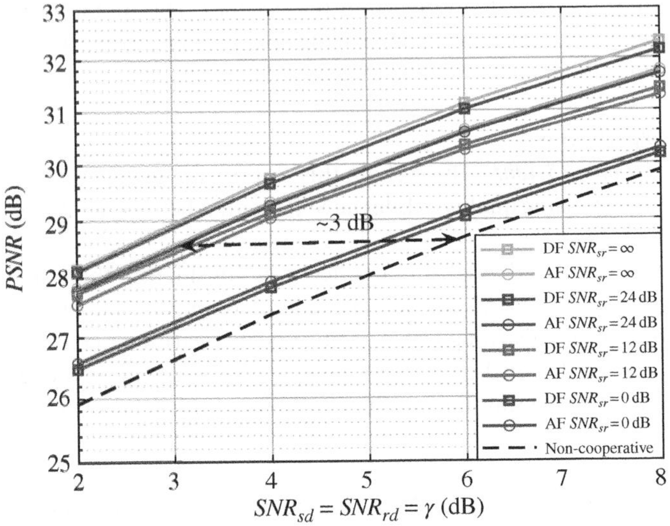

Figure 4.9 Comparison between DeepJSCC-AF and DeepJSCC-DF with $SNR_{sr} \in \{0, 12, 24, \infty\}$ dB. We also include the noncooperative scheme [Bourtsoulatze et al., 2019] as a benchmark. Source: Bourtsoulatze et al. [2019]/With permission of IEEE.

coding gain compared to DeepJSCC-AF. However, under lower SNR_{sr}, the noise in $\mathbf{y}_{sr}$ will hinder the parity generation process and the coding gain vanishes.

We then compare the DeepJSCC-PF with the DeepJSCC-DF, which can be seen as a special case of DeepJSCC-DF with $\lambda = 0$, but employs a simplified DNN at the relay. We observe from our simulation results that the performance of DeepJSCC-DF is robust with respect to different λ values and DeepJSCC-PF can achieve comparable performance with DeepJSCC-DF, thus we will focus on the DeepJSCC-PF, thanks to its low computation complexity.

Comparison with the Digital Baseline Finally, we compare DeepJSCC-PF with a digital baseline in Figure 4.10. In this simulation, we keep $SNR_{sr} = 12$ dB and vary γ from 0 to 8 dB. For the baseline, we use the state-of-the-art BPG compression algorithm with the 3GPP polar code. To exploit the full potential of the baseline, we assume that the relay can always decode $\mathbf{y}_{sr}$ perfectly. The processing of the digital baseline is as follows: the source node first compresses the image via BPG and encodes it using a rate $R \in (0, 1)$ polar code followed by a given modulation scheme. The relay perfectly decodes the information bits, generates parity bits, and forwards them to the destination in the second half. The received signal at the destination is a noisy version of the rate $R/2$ polar code with the same modulation. We also provide an upper bound on the performance of the separation scheme by calculating the capacity of the system described above (with

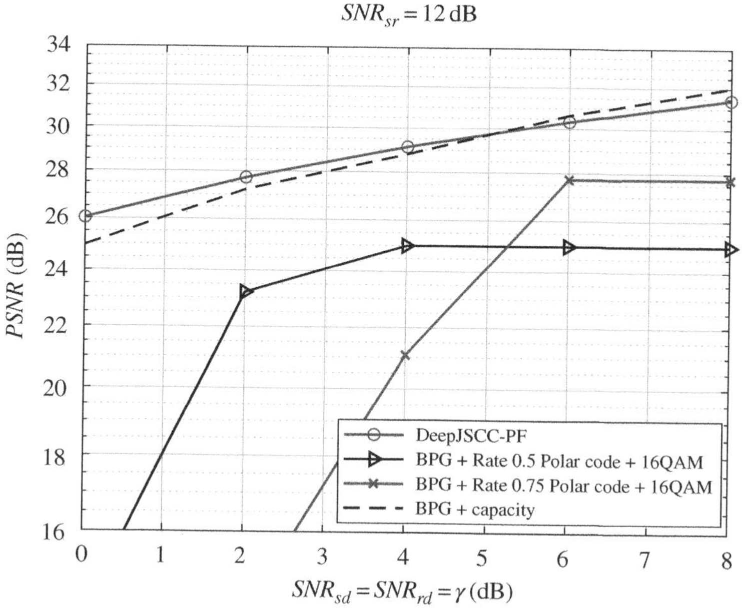

Figure 4.10 The PSNR and SSIM performance of the DeepJSCC-PF scheme compared with separation-based schemes employing BPG algorithm using either Polar codes or assuming capacity-achieving channel codes for $SNR_{sr} = 12$ dB.

perfect S–R link), given by $\log_2(1 + 10^{\gamma/10})$. It can be seen that DeepJSCC-PF outperforms the separation-based approach at all noise levels in terms of both PSNR and SSIM. We note that DeepJSCC-PF has comparable performance even with the ideal separation-based scheme with capacity-achieving channel codes, which also assumes an ideal S–R link.

4.3.2 Multihop Relay

In this section, we delve into the challenge of ensuring robust image transmission across multi-hop networks. The scenario under consideration involves a mobile user's intention to transmit an image to a designated receiver via a mobile cellular network. We consider a practical setup by assuming that the connections among the nodes within the mobile core network are stable and of superior quality. In contrast, the link between the mobile user and the initial node (for instance, an access point) is characterized by variability and potentially lower quality.

To address the intricacies of this channel model, we introduce a hybrid solution. This approach employs DeepJSCC for the initial hop to adeptly navigate the challenges of the fluctuating quality. Subsequently, the signal received at the first relay undergoes digital compression before being forwarded through the stable, high-quality links of the mobile core network. Through numerical simulations, we demonstrate the efficacy of such a hybrid method. The results indicate that the hybrid solution not only navigates the complexities of mixed-quality links but also surpasses the performance of both entirely analog and entirely digital transmission schemes. This finding underscores the potential of integrating DeepJSCC with legacy digital communication techniques to enhance the reliability and efficiency of data transmission in dynamically varying network environments.

4.3.2.1 System Model

We consider a multihop network, where a mobile edge user S aims to deliver an image to a destination node D with the aid of $(n-1)$ relay nodes $\{R_1, \dots, R_{n-1}\}$ residing in the mobile core network, as depicted in Figure 4.11. In our framework, we consider the channels between adjacent nodes to be AWGN channels. In particular, since the relay nodes and the destination node lie within the mobile core network, the links among them are more reliable than that between S and R_1 in the edge network. This framework can be easily extended to slow-fading channels when the CSI is known to all nodes in the network.

As in the cooperative relay settings in Section 4.3.1.1, the edge user S encodes an image $\mathbf{S} \in \mathbb{R}$ to a latent vector $\mathbf{x}_s \in \mathbb{C}^k$ using an encoding function $f_s(\cdot) : \mathbb{R}^{c \times h \times w} \to \mathbb{C}^k$. The latent vector is subject to a power constraint: $\frac{1}{k}\|\mathbf{x}_s\|^2 \leq 1$.

The signal received at the first relay node R_1 (i.e., the access point of the edge network) can be expressed as $\mathbf{y}_1 = \mathbf{x}_s + \boldsymbol{w}_s$, where $\boldsymbol{w}_s \sim \mathcal{CN}(0, \sigma_s^2 \mathbf{I}_k)$. The SNR of the first hop is defined as $SNR_s \triangleq \frac{1}{\sigma_s^2}$.

In the mobile core network, the relays can adopt different processing strategies, such as a fully digital scheme, a fully analog scheme, and the hybrid JSCC scheme detailed later. In general, we denote by $f_{R_i}(\cdot)$ the operations at R_i, $\mathbf{x}_i$ the signal transmitted from the ith relay to the $(i+1)$th relay, and $\mathbf{y}_i$ the signal received by the ith relay from the $(i-1)$th relay. Thus, we have

$$\mathbf{x}_i = f_{R_i}(\mathbf{y}_i) \in \mathbb{C}^{k_i}, \tag{4.10}$$

$$\mathbf{y}_{i+1} = \mathbf{x}_i + \boldsymbol{w}_i. \tag{4.11}$$

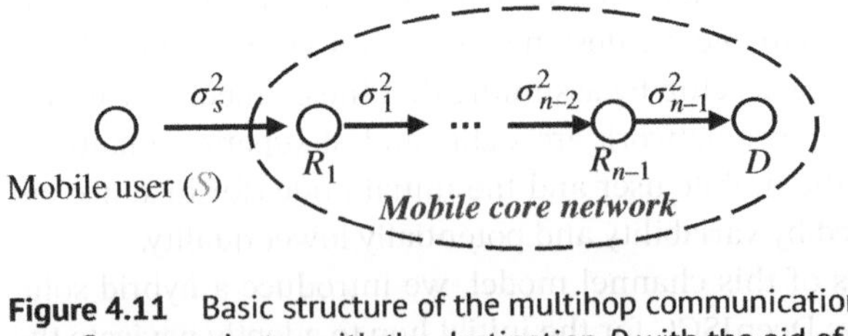

Figure 4.11 Basic structure of the multihop communication network where the mobile user S communicates with the destination D with the aid of the relays $\{R_1, \dots, R_{n-1}\}$ located in a mobile core network with high-capacity wireless connections.

where $i = 1, 2, \ldots, n-1$ and $\boldsymbol{w}_i \sim \mathcal{CN}(0, \sigma_i^2 \mathbf{I}_{k_i})$. In particular,

- $\mathbf{x}_i$, $\forall i$, are subject to a power constraint as $\frac{1}{k_i}\|\mathbf{x}_i\|^2 \leq 1$.
- $k_1, \ldots, k_{n-1}$ are determined by the latency requirement of the destination, and may not be equal.
- All relays work in a half-duplex mode: R_i starts to transmit $\mathbf{x}_i$ after it receives and processes the entire $\mathbf{y}_i$.
- The noise variance in the edge network σ_s^2 is generally larger than that within the mobile core network $(\sigma_1^2, \ldots, \sigma_{n-1}^2)$.
- To simplify the presentation, we assume that all the link qualities in the mobile core network are identical, i.e., $\sigma_{1:(n-1)}^2 = \sigma^2$, and the SNR is denoted by $SNR_N = \frac{1}{\sigma^2}$.

Finally, the received signal at the destination D can be written as $\mathbf{y}_d \triangleq \mathbf{y}_n = \mathbf{x}_{n-1} + \boldsymbol{w}_{n-1}$. Upon receiving $\mathbf{y}_d$, the destination reconstructs the original image by a decoding function $f_D(\cdot)$ as $\hat{\mathbf{S}} = f_D(\mathbf{y}_d)$. To evaluate the reconstruction performance of different processing strategies (specified by f_s, $\{f_{R_i}, \forall i\}, f_D$), we adopt the PSNR of the image as the metric.

4.3.2.2 Existing Methods

We first review existing methods for communication over multi-hop wireless networks, namely, the fully analog and fully digital transmissions.

Fully Analog Transmission Transmitted signal at each hop is produced by a DeepJSCC encoder. Since the relay cannot recover the message exactly, the signal gradually becomes noisier with each hop. Two protocols are considered, namely, the DeepJSCC-AF and DeepJSCC-PF corresponding to the conventional amplify-and-forward and decode-and-forward schemes which have been illustrated in Section 4.3.1.2. In particular, the encoder and decoder functions, $f_s(\cdot), f_d(\cdot)$ are parameterized using 2D CNN.

In the DeepJSCC-AF protocol, since the relay nodes simply amplify their received signals, we have $k_1 = \cdots = k_{n-1} = k'$. Note that the number of complex channel uses k for the first hop may be different. We employ neural networks comprising 2D CNNs at R_1 to change the number of channel uses (for $k \neq k'$ case) by controlling the number of output channels. The processing at the i-th relay node, R_i can be expressed as $\mathbf{x}_i = \alpha \mathbf{y}_i$, $i = 2, \ldots, n-1$ with $\alpha = \sqrt{\frac{1}{1+\sigma^2}}$. The loss function is simply: $\mathcal{L}_{AF} = \mathbb{E}_{\mathbf{S} \sim P_\mathbf{S}}\left[\|\mathbf{S} - \hat{\mathbf{S}}\|_2^2\right]$.

In DeepJSCC-PF, we instead employ neural networks to generate the signals transmitted by the relays, instead of simple amplification. At R_1, DeepJSCC-PF is identical to that in DeepJSCC-AF (except for the case when $k = k'$, where R_1 of the DeepJSCC-AF protocol simply amplifies the signal), while for the remaining hops, the input–output relationship for R_i can be written as: $\mathbf{x}_i = f_{R_i}(\mathbf{y}_i)$, $2 \leq i \leq (n-1)$, where the forwarding functions $f_{R_i}(\cdot)$ are parameterized by neural networks. DeepJSCC-PF employs the same loss function as in AF.

Fully Digital Transmission The encoder at S first compresses the input image into a bit sequence, denoted by $\mathbf{b}_s$, which is then channel coded and modulated to form the channel codeword $\mathbf{x}_s$. Since the state-of-the-art source encoders (e.g., BPG) output variable-length codewords, the number of complex channel uses k will refer to the expected length of $\mathbf{x}_s$. The processing operation $f_{R_1}(\cdot)$ at the first relay is comprised of a channel decoder, denoted as $D_1(\cdot)$, to decode the received signal $\mathbf{y}_1$ to the bit sequence $\mathbf{b}_s$ and re-encodes it to $\mathbf{x}_1$ using an appropriate coded modulation scheme, $M_1(\cdot)$. Since we assume identical channels over the mobile core network, $M_i(\cdot)$s and $D_i(\cdot)$s adopted by the relay nodes are the same.

To better illustrate the processing of the full digital transmission scheme, we provide a concrete example where $SNR_s = 2$ dB and $SNR_N = 10$ dB, with corresponding channel capacities $C_s = 1.37$ and $C_N = 3.46$, respectively. We adopt two coded modulation schemes with rates below these capacities (for reliable transmission). To be precise, we consider a rate-1/2 LDPC code with 4-QAM corresponding to a achievable rate $R_s = 1$ for the first link, and a rate-1/2 LDPC code with 16-QAM, which corresponds to $R_N = 2$, for the remaining links. It has been shown by simulations that the coded modulation schemes along with the corresponding belief propagation (BP) decoders are capable to achieve almost zero packet error rate (PER) in the considered scenario. We also note that $k = 2k'$ holds for this case.

A key weakness of the fully digital transmission is that, though it is plausible to assume SNR_N is stable, due to the link between the source and the first relay node is versatile, it will still suffer from the cliff and leveling effects.

4.3.2.3 A Hybrid JSCC Solution

Hybrid Solution We consider a hybrid scheme (see Figure 4.12), where the mobile user adopts DeepJSCC protocol in the first hop to avoid the cliff and leveling effects, whereas the first relay node, R_1, uses a DNN, $g_c(\cdot)$ to compress its received signal, $\mathbf{y}_1$ to a bit sequence $\mathbf{b}_1$. Then, each R_i adopts a coded modulation scheme, $M_i(\cdot)$, which is identical to the fully digital scheme in Section 4.3.2.2 for transmission over the $(n-1)$ hops in the mobile core network. All coded modulation schemes are carefully chosen to exhibit zero PER; thus, it can be guaranteed that D correctly decodes the bit sequence $\mathbf{b}_1$ regardless of the number of hops. This avoids performance degradation caused by noise accumulation as in the fully analog scheme when the number of hops increases. Then, the decoder at D, takes $\mathbf{b}_1$ as input to generate the final reconstruction, $\hat{\mathbf{S}}$. Note that the processing functions, $\{f_{R_2}(\cdot), \ldots, f_{R_{n-1}}(\cdot)\}$ at $\{R_2, \ldots, R_{n-1}\}$, only involve channel decoding $D_i(\cdot)$ and re-encoding $M_i(\cdot)$ without neural processing.

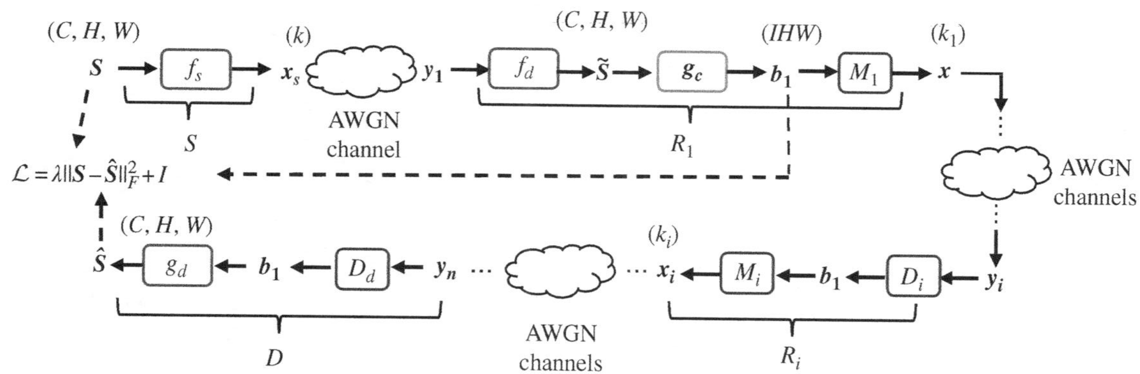

Figure 4.12 The hybrid JSCC framework, where $f_s(\cdot)$ denotes the DeepJSCC encoder, $f_d(\cdot)$ denotes the DeepJSCC decoder located at R_1 and D. $g_c(\cdot)$ and $g_d(\cdot)$ represent the compression and decompression modules, M_i, D_i refer to the coded modulation block and the channel decoding block at the ith relay, respectively. We show the dimensions on top of the tensors and vectors for clarity. The loss function, $\mathcal{L}$, is a summation of the distortion and the bpp (I) defined in (4.14) used to compress the image $\tilde{S}$.

Encoder at S The encoder function $f_s(\cdot)$ at S is shown in figure 2 of Bian et al. [2024b], and is identical to the DeepJSCC encoder introduced in Section 4.3.2.2.

Processing at R_1 A naive solution is to directly quantize each element of the received signal $\mathbf{y}_1$ to a given precision, e.g., m bits, and protect the bit sequence with length km using a coded modulation scheme. However, naively quantizing the received signal leads to sub-optimal solution, which motivates the employment of a neural compression model, as shown in Fig. 4.13.

Instead of directly compressing $\mathbf{y}_1$, we found it is more beneficial to first transform $\mathbf{y}_1$ into a tensor with the same dimension of the original image $\mathbf{S}$, denoted by $\tilde{\mathbf{S}}$, via a DeepJSCC decoder, $f_d(\cdot)$, introduced in Section 4.3.2.2, and then apply the neural image compression algorithm [Ballé et al., 2018], denoted as $g_c(\cdot)$, to compress $\tilde{\mathbf{S}}$ into a bit sequence $\mathbf{b}_1$. The processing of the neural compression model is summarized as follows (see also Figure 4.12): After obtaining the reconstructed image $\tilde{\mathbf{S}}$, we first use nonlinear transform blocks, denoted by $g_a(\cdot)$, which is comprised of 2D CNNs with C_z output channels to generate a latent tensor $\mathbf{z} \in \mathbb{R}^{C_z \times H/4 \times W/4}$, which is fed to the hyper-latent encoder, $h_a(\cdot)$, to generate the hyper-latent $\boldsymbol{v} \in \mathbb{R}^{C_v \times H/16 \times W/16}$. The hyper-latent $\boldsymbol{v}$ is introduced with the aim to remove the dependency between the elements in $\mathbf{z}$, such that given $\boldsymbol{v}$, the probability density function (pdf) of the elements in $\mathbf{z}$ will be in product form.

It is worth mentioning that the neural image compression model involves two phases, the training and deployment phases. In the deployment phase, we adopt arithmetic encoder to encode both latent vectors, where elements, z_i and v_i, are rounded to the nearest integers, $\hat{z}_i$ and $\hat{v}_i$, and then compressed according to their individual learned distributions. In the training phase, however, the rounding operation is infeasible as its gradient is one only at integer points, and zero otherwise. Thus, during training, we use $\tilde{\mathbf{z}}, \tilde{\boldsymbol{v}}$ to replace the rounded vectors $\hat{\mathbf{z}}, \hat{\boldsymbol{v}}$, where each element $\tilde{z}_i, \tilde{v}_i$ is obtained by adding uniform noise (to model the quantization noise) to z_i, v_i. A hyper-latent decoder, $h_s(\cdot)$, with two upsampling layers takes $\tilde{\boldsymbol{v}}$ as input, and outputs two tensors $\tilde{\mu}$ and $\tilde{\sigma}$ with the same dimensions

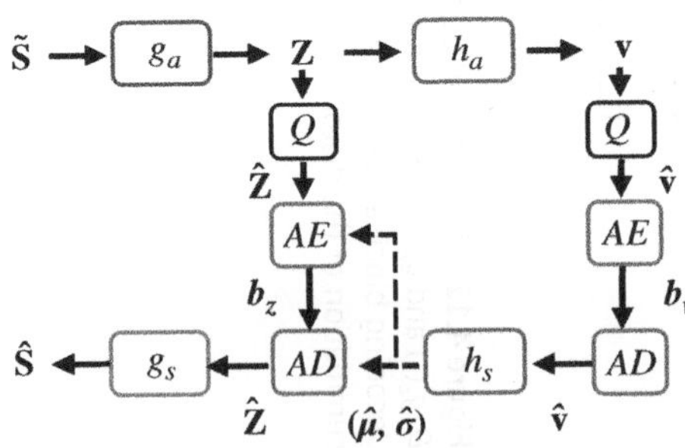

Figure 4.13 The compression and decompression modules, $g_c(\cdot)$ and $g_d(\cdot)$, respectively. In particular, $g_a(\cdot), h_a(\cdot)$ are the non-linear transform blocks to generate the latent vectors $\mathbf{z}, \boldsymbol{v}$, whereas $g_s(\cdot), h_s(\cdot)$ are adopted to facilitate the synthesis of $\hat{\mathbf{S}}, \hat{\mathbf{z}}$. The predicted tensors $\hat{\mu}, \hat{\sigma}$ are used for the arithmetic encoding and decoding of $\hat{\mathbf{z}}$ and $\mathbf{b}_z$.

as $\tilde{\mathbf{z}}$. Then, we model the pdf of $\tilde{\mathbf{z}}$ as:

$$p(\tilde{\mathbf{z}}|\tilde{\boldsymbol{v}}) = \prod_i \left(\mathcal{N}(\tilde{\mu}_i, \tilde{\sigma}_i^2) * \mathcal{U}(-\frac{1}{2}, \frac{1}{2}) \right) (\tilde{z}_i),$$
$$\tilde{\mu}, \tilde{\sigma} = h_s(\tilde{\boldsymbol{v}}, \theta), \tag{4.12}$$

where θ denotes the parameters for $h_s(\cdot)$ and $*$ represents convolution. For the hyper-latent $\tilde{\boldsymbol{v}}$, we consider modeling it also using a fully factorized model as:

$$p(\tilde{\boldsymbol{v}}|\phi) = \prod_i \left(p_{\tilde{v}_i|\phi_i}(\phi_i) * \mathcal{U}(-\frac{1}{2}, \frac{1}{2}) \right) (\tilde{v}_i), \tag{4.13}$$

where ϕ denotes a collection of vectors to parameterize the univariate distribution.

Once the models are trained and deployed, the latent and hyper-latent vectors $\mathbf{z}, \boldsymbol{v}$ can be obtained from the reconstructed image $\tilde{\mathbf{S}}$. We use an arithmetic encoder to encode the rounded $\hat{\mathbf{z}}, \hat{\boldsymbol{v}}$ to the bit sequence $\mathbf{b}_1 = (\mathbf{b}_z, \mathbf{b}_v)$ according to the optimized pdfs in (4.12) and (4.13) by simply replacing $\tilde{\mathbf{z}}, \tilde{\boldsymbol{v}}$ by $\hat{\mathbf{z}}, \hat{\boldsymbol{v}}$. The bit sequence will be further coded and modulated before forwarding to R_2. Due to the fact that different images contain different amount of information, the length of the bit sequence varies and we evaluate the expected number of complex channel uses as $k' = \frac{1}{R_N}\mathbb{E}_{\tilde{\mathbf{S}}\sim p_{\tilde{\mathbf{S}}}}(\#\mathbf{b}_1)$, where $\#\mathbf{b}_1$ denotes the number of bits in the sequence while R_N is defined in Section 4.3.2.2.

Neural Processing at D Since the links in the mobile core network are assumed static, and protected by channel codes exhibiting zero PER, the destination node can retrieve the bit sequence $\mathbf{b}_1$ correctly, and employs an arithmetic decoder (*AD*) to first generate the rounded latent vector $\hat{\boldsymbol{v}}$, which is then fed to $h_s(\cdot)$ to obtain $\hat{\mu}, \hat{\sigma}$. Then, another *AD* takes the mean and variance, $\hat{\mu}, \hat{\sigma}$ along with the sequence $\mathbf{b}_z$ to produce the rounded latent vector $\hat{\mathbf{z}}$. Finally, the DeepJSCC decoder, $f_d(\cdot)$ takes $\hat{\mathbf{z}}$ as input and outputs $\hat{\mathbf{S}}$.

Loss Function for the Hybrid JSCC Framework Similarly to the general source compression problem, there is a trade-off between the amount of bits used to compress $\mathbf{y}_1$ (or equivalently, $\tilde{\mathbf{S}}$) and the reconstruction performance at D. To achieve different trade-off points in the rate-distortion plane, we introduce a variable λ and the loss function is written as: $\mathcal{L} = \lambda\|\mathbf{S} - \hat{\mathbf{S}}\|_F^2 + I$, where $I = I_z + I_v$ is the summation of the bit per pixel (bpp) for compressing $\tilde{\mathbf{z}}$ and $\tilde{\boldsymbol{v}}$, which is defined by the number of bits divided by the height H and width W of the image. Note that the notation $\tilde{\mathbf{z}}, \tilde{\boldsymbol{v}}$ indicate the training phase[2] and the rate can be further expressed as:

$$I = \frac{1}{HW}\mathbb{E}_{\tilde{\mathbf{z}},\tilde{\boldsymbol{v}}\sim q}\left[-\log_2(p_{\tilde{\mathbf{z}}|\tilde{\boldsymbol{v}}}(\tilde{\mathbf{z}}|\tilde{\boldsymbol{v}})) - \log_2(p_{\tilde{\boldsymbol{v}}}(\tilde{\boldsymbol{v}}))\right], \tag{4.14}$$

2 Definition of I is the same for the deployment phase obtained by replacing $\tilde{\mathbf{z}}$ and $\tilde{\boldsymbol{v}}$ with $\hat{\mathbf{z}}$ and $\hat{\boldsymbol{v}}$.

where the first and second terms represent I_z and I_v, respectively, while q denotes the posterior of $\tilde{\mathbf{z}}, \tilde{\boldsymbol{v}}$ given reconstructed image $\tilde{\mathbf{S}}$ at R_1, which follows a uniform distribution centered at $g_a(\cdot)$ and $h_a(\cdot)$ output, $\mathbf{z}$ and $\boldsymbol{v}$:

$$q(\tilde{\mathbf{z}}, \tilde{\boldsymbol{v}}|\tilde{\mathbf{S}}) = \prod_i \mathcal{U}\left(\tilde{z}_i|z_i - \frac{1}{2}, z_i + \frac{1}{2}\right) \cdot \mathcal{U}\left(\tilde{v}_i|v_i - \frac{1}{2}, v_i + \frac{1}{2}\right). \tag{4.15}$$

4.3.2.4 Numerical Experiments

Parameter settings and training details Unless otherwise mentioned, the number of complex channel uses, k for the first hop is fixed to 768, which corresponds to $C_{out} = 24$. The number of channels, C_z and C_v, for $\mathbf{z}$ and $\boldsymbol{v}$ are set to 256 and 192, respectively. For the training settings, we note that it is identical to the cooperative ones, thus we skip it here.

We then show by experiments the gain of the hybrid JSCC scheme over the fully analog and fully digital counterparts measured by reconstruction PSNR values. We then illustrate that by training different models with different λ values, the hybrid JSCC framework can adapt to the user specific reconstruction requirement and latency constrain. Note that similar with the cooperative relay settings, the hybrid JSCC scheme can also avoid the cliff and leveling effects, which is not shown due to the page limit.

Comparison with Fully Analog and Fully Digital Schemes We consider the scenario presented in Section 4.3.2.2, where the SNR between the mobile user S and the first relay R_1 is $SNR_s = 2$ dB, while the link qualities within the core network are $SNR_N = 10$ dB.

We first compare the performance between the hybrid JSCC framework with fully analog transmission using DeepJSCC-AF and DeepJSCC-PF, respectively. We make simplification such that $k' = k$, thus, for DeepJSCC-AF, all relays only need to amplify the signal.

Since the coded modulation adopted in this case is a 1/2-rate LDPC with 16-QAM, for a fair comparison, the bpp, I, used to compress the received signal $\mathbf{y}_1$ (or equivalently, $\tilde{\mathbf{S}}$) should satisfy $I \leq \frac{2k'}{HW} = 1.5$ averaged over all images from the dataset. To achieve this, we carefully select the hyper-parameter, $\lambda = 3200$ leading to $I \approx 1.25$. Figure 4.14a shows the relative performance of the hybrid JSCC framework, the naive quantization solution, and the DeepJSCC-AF/PF protocols. When the number of hops is one, i.e., a point-to-point channel, all schemes yields the same PSNR performance around 28.5 dB. When there is at least one relay in the network, due to the lossy compression at R_1 for the hybrid JSCC scheme, its performance drops to $\approx$ 27.6 dB and then keeps constant since the noise introduced by each hop is removed by the channel code before being forwarded to the next relay; that is, there is no noise accumulation. The DeepJSCC-AF and -PF protocols, however, both suffer from noise accumulation,

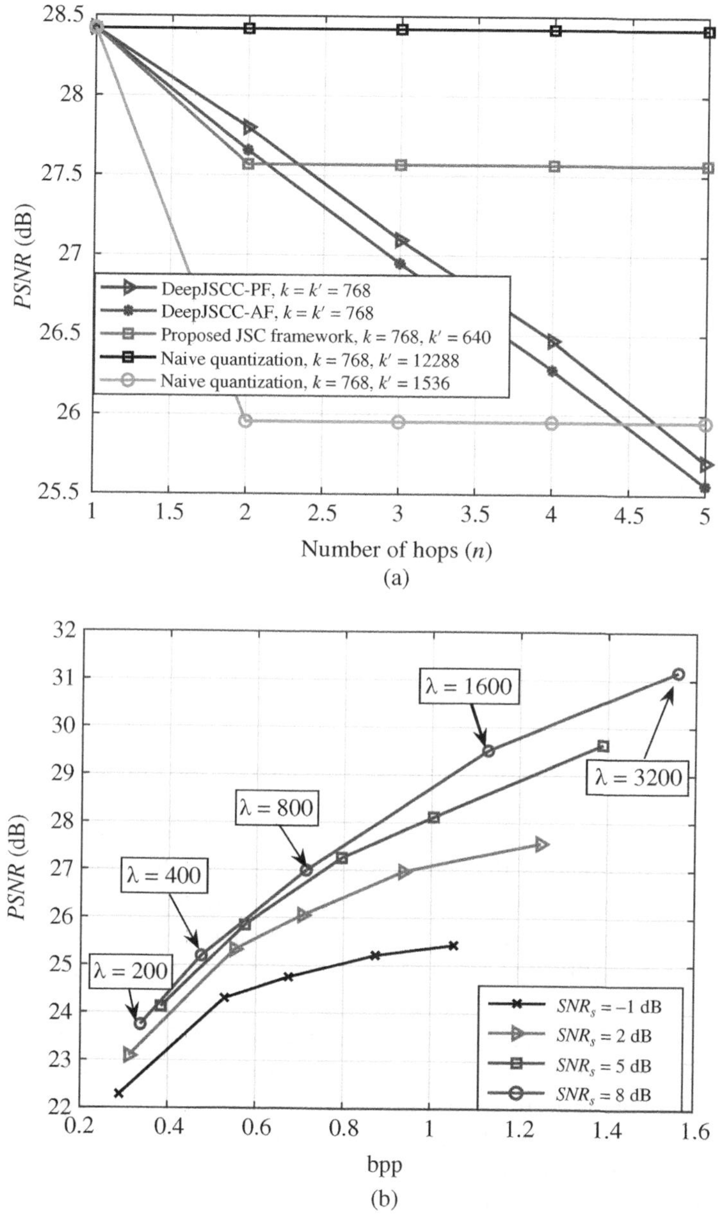

Figure 4.14 (a) The comparison of PSNR performance between the hybrid JSCC framework along with the DeepJSCC-PF and DeepJSCC-AF protocol. We set $k = k' = 768$ for the two benchmarks, while $\lambda = 3200, k' = 640$ for the hybrid JSCC framework. (b) Rate-distortion trade-off for the hybrid JSCC framework where 20 JSCC models are trained and evaluated corresponding to $SNR_s \in \{-1, 2, 5, 8\}$ dB and $\lambda \in \{200, 400, 800, 1600, 3200\}$.

and their performances drop as the number of hops increases. To be precise, when the number of hops becomes larger than 3, both of them are outperformed by the hybrid JSCC framework. We also observe that DeepJSCC-PF outperforms DeepJSCC-AF only by a small margin, although requiring the training of neural network processors at all the relays. The curves marked "Naive Quantization" represent the scheme, in which each element of $\mathbf{y}_1$ is quantized to 4 and 32 bits[3] and further protected by channel codes. It can be seen that the hybrid JSCC scheme outperforms the 4-bit quantization baseline with smaller k'. The 32-bit precision scheme, as expected, outperforms all other schemes, but it only serves as an upper bound with no loss over the hops, as the overhead, i.e., k', makes it unpractical.

We then compare the hybrid JSCC scheme with the fully digital scheme. In this case, the digital scheme uses a 1/2-rate LDPC with 4-QAM for the first hop leading to an expected number of compressed bits $\#\mathbf{b}_1 = k = 768$. The coded modulation scheme for the remaining relays has a rate $R_N = 2$, thus we have $k' = k/2$. For a fair comparison, the number of compressed bits for the hybrid JSCC framework should also be less or equal to k. We choose $\lambda = 800$ with a corresponding $\#\mathbf{b}_1 = 727$ and find that the hybrid JSCC scheme achieves $PSNR = 26.05$ dB, which is higher than 25.39 dB achieved by the fully digital scheme.

Rate-distortion Curves for the Hybrid JSCC Framework Next, we focus on the hybrid JSCC framework and show that by training models with different λs, the hybrid JSCC framework can achieve different points on the rate-distortion plane, enabling flexible adaptation to the user specific reconstruction requirement and latency constraint. In this simulation, we do not specify the coded modulation scheme used within the core network but assume the compressed bits generated by R_1 can be delivered to the destination without error. Different $SNR_s \in \{-1, 2, 5, 8\}$ dB are evaluated and for each of the SNR value, we train different models with respect to different $\lambda \in \{200, 400, 800, 1600, 3200\}$ values.

We plot the PSNR performance versus bpp I defined in (4.14) in Figure 4.14b. For all SNR_s values, different bpp values are obtained with different models. When the latency constraint of the destination is stringent, we adopt a model with small λ, when the destination requires a better reconstruction performance; however, a large λ is preferable. We also find that a better rate-distortion curve is obtained when SNR_s is larger, which is intuitive as the first hop becomes less noisy.

---

3 Note that we consider a nonuniform scalar quantization for the 4-bit quantization shown in the figure where the codebook is obtained via Lloyd algorithm. We also consider 32-bit quantization as a bound since the precision of float numbers in Pytorch is 32 bits.

4.4 DeepJSCC for Feedback Channels

This section investigates the DeepJSCC approach in the presence of feedback signals from the receiver to the transmitter. Feedback does not increase the capacity of a memoryless channel [Shannon, 1956; Dobrushin, 1958]. However, it can simplify the coding mechanism, as demonstrated in Schalkwijk [1966] and Schalkwijk and Bluestein [1967]. Notably, only a limited number of studies have addressed the JSCC problem with channel feedback in the literature. In most of the earlier existing JSCC works, a scalable compression scheme is considered for the source, and an unequal error protection scheme is employed for the channel code with their rates adjusted based on channel feedback [Chande et al., 1998; Kafedziski, 1998; Lu et al., 1999].

The first truly joint coding scheme for transmitting images over a wireless channel with feedback was proposed in Kurka and Gündüz [2020]. The authors presented a stacked autoencoder structure called DeepJSCC-f to transmit images in multiple blocks. As illustrated in Figure 4.15, DeepJSCC-f utilizes a set of encoders and decoders in the transmitter and receiver, respectively, and a different pair of encoder and decoder is used at each iteration. It is shown in Kurka and Gündüz [2020] that feedback can improve the performance of DeepJSCC, and the DeepJSCC-f scheme achieves state-of-the-art performance. To improve the DeepJSCC-f in terms of complexity, channel adaptability, and the performance, this section presents a new JSCC paradigm with feedback, called JSCCformer-f [Wu et al., 2023c].

4.4.1 System Model

We consider the problem of wireless transmission of images in the presence of noisy channel output feedback. The transmission of each image is done in m blocks, where $m = 1$ corresponds to the no-feedback scenario. After each block, we assume the channel output is fed back to the transmitter via a noisy or noiseless feedback link. At each block, the transmitter can exploit the feedback signals it has received up until that block.

Accordingly, for a fixed total bandwidth allocation, a JSCC code that transmits an image of n pixel intensities in m blocks, each consisting of k channel symbols, is denoted by the triplet (n, m, k), where $n = 3hw$ denotes the source bandwidth, and mk denotes the total channel bandwidth cost. The bandwidth ratio $R = mk/n$.

In the ith transmission block, the encoder function E_{θ_i}, parameterized by $\boldsymbol{\theta}_i$, maps $\mathbf{S}$ and all the available feedback signals into a complex channel vector $\mathbf{X}_i = E_{\theta_i}(\mathbf{S}, \hat{\mathbf{Y}}_{1:i-1}) \in \mathbb{C}^k$, where $\hat{\mathbf{Y}}_{1:i-1} \triangleq [\hat{\mathbf{Y}}_1, \dots, \hat{\mathbf{Y}}_{i-1}]$ denotes all the previous feedback signals available at the i-th transmission block, and $\hat{\mathbf{Y}}_i \in \mathbb{C}^k$ is the channel output feedback received after the i-th block. Note that the encoder, in general, can consist

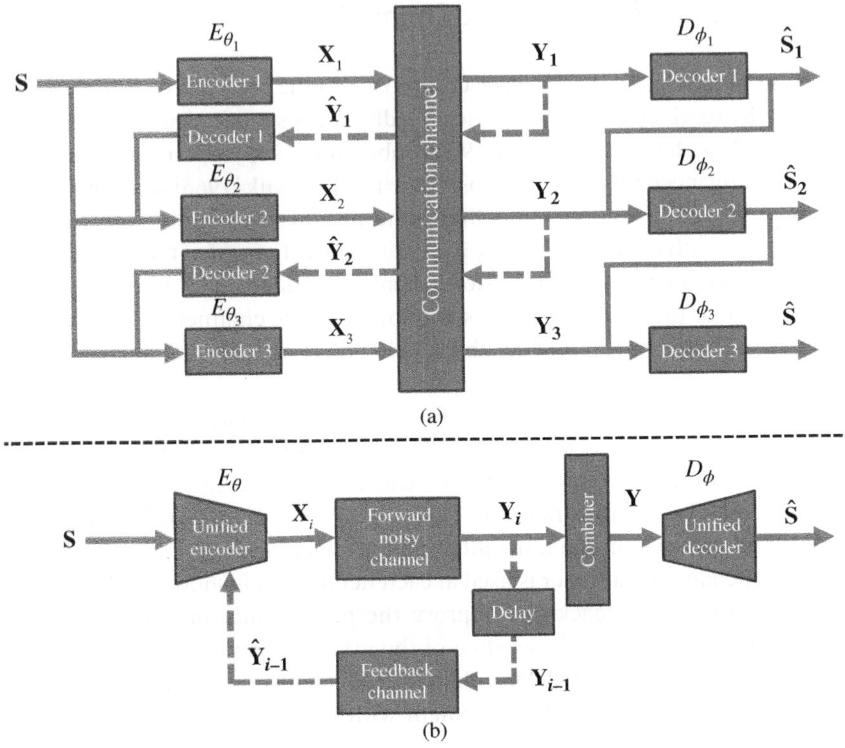

Figure 4.15 Alternative JSCC schemes for channels with feedback, where the solid lines represent forward channels, while the dotted lines represent feedback channels. (a) The DeepJSCC-f scheme with $m = 3$, where multiple independent encoders and decoders are trained as the channel codes in each block. (b) The JSCCformer-f scheme, where a single unified encoder–decoder pair is employed at each block, significantly reducing the training and memory requirements as well as the coding complexity.

of m encoding functions, one for each block. The transmitted signal $\mathbf{X_i}$ is subject to a power constraint as $\frac{1}{mk}\sum_{i=1}^{m}\mathbb{E}\left[\|\mathbf{X_i}\|_2^2\right] \leq 1$.

The forward channel in the i-th transmission block is modeled as an AWGN channel with fading as: $\mathbf{Y_i} = h\mathbf{X_i} + \mathbf{W_i}$, where $\mathbf{Y_i} \in \mathbb{C}^k$ is the channel output in ith forward transmission block, $h \in \mathbb{C}$ is the channel gain, and $\mathbf{W_i} \in \mathbb{C}^k$ is the AWGN term. Each element of $\mathbf{W_i}$ is sampled from an i.i.d. complex Gaussian noise as $W[i] \sim \mathcal{CN}(0, \sigma_w^2)$. We consider a slow fading scenario and assume that h remains constant for the whole duration of the transmission of one image. We assume that h is sampled from an i.i.d. complex Gaussian distribution, i.e., $h \sim \mathcal{CN}(0, \sigma_h^2)$, where $h = 1$ for the AWGN scenario. To measure the channel quality, we define the SNR as $10\log_{10}\frac{\sigma_h^2}{\sigma_w^2}$(dB).

After each block, the channel output vector $\mathbf{Y_i}$ is then fed back to the transmitter through a feedback link, which can be either perfect or noisy, modeled as: $\hat{\mathbf{Y}}_\mathbf{i} = \mathbf{Y_i} + \mathbf{W_f}$, where $\mathbf{W_f} \in \mathbb{C}^k$ is the AWGN term sampled from $W_f[i] \sim \mathcal{CN}(0, \sigma_f^2)$ and $\mathbf{W_f} = 0$ in the perfect feedback link.

The decoding function D_ϕ, parameterized by ϕ, reconstructs the transmitted image as $\hat{\mathbf{S}} = D_\phi(\mathbf{Y}) \in \mathbb{R}^{h\times w\times 3}$, where $\mathbf{Y} \triangleq [\mathbf{Y_1}, \dots, \mathbf{Y_m}]$. The reconstruction quality is quantified through the PSNR and the LPIPS metrics [Wu et al., 2023c]. Given the code parameters (n, m, k), we aim to optimize the encoder and decoder parameters $\theta_\mathbf{1}, \dots, \theta_\mathbf{m}$, and ϕ, respectively, to improve the average reconstruction performance at the receiver.

4.4.2 A JSCCFormer-f Solution

This section details the JSCCformer-f architecture, which consists of three key components: ViT encoder, ViT decoder, and the loss function, as depicted in Figure 4.16, and the transmission is partitioned into m blocks.

4.4.2.1 ViT Encoder

The encoding process entails sequentializing the image, embedding the feedback signal, and utilizing a ViT-based encoder E_θ to map the updated input sequence to channel symbols. The encoding process is elaborated in four steps: (i) image-to-sequence transformation, (ii) feedback embedding, (iii) input sequence update, and (iv) ViT-based encoding.

Image-to-Sequence Transformation We first convert the input image $\mathbf{S}$ into a sequence of vectors, denoted by $\mathbf{S_s}$, where we divide $\mathbf{S}$ into $p \times p$ patches and flatten each patch to form a sequence of vectors. In this way, $\mathbf{S}$ is converted to $\mathbf{S_s} \in \mathbb{R}^{l\times c}$, where $l = p^2$ and $c \triangleq \frac{3hw}{p^2}$.

Feedback Embedding Within each transmission block, a feedback embedding module is employed at the transmitter, parameterized as $f_e(\cdot) : \mathbb{R}^{l\times \frac{2mk}{l}} \to \mathbb{R}^{l\times z}$, to embed all the received feedback signals $\hat{\mathbf{Y}}$ into $\mathbf{Z_i}$ for the subsequent ViT-encoding step, where z is the output dimension for the feedback embedding block. We propose two different feedback embedding schemes, JSCCformer-f and JSCCformer-f (lite), respectively. Specifically, the embedding scheme of JSCCformer-f utilizes an additional ViT-based decoder at the transmitter (the weights of which are copied from the decoder at the receiver) to estimate the receiver's current belief from the received feedback, which is then combined with the raw received feedback as the input to the encoder. The second scheme, in contrast, directly feeds the received channel feedback to the encoder. For JSCCformer-f and JSCCformer-f (lite), we have $z = \frac{2k}{l} + c$ and $z = \frac{2k}{l}$, respectively. Thanks to the integration of

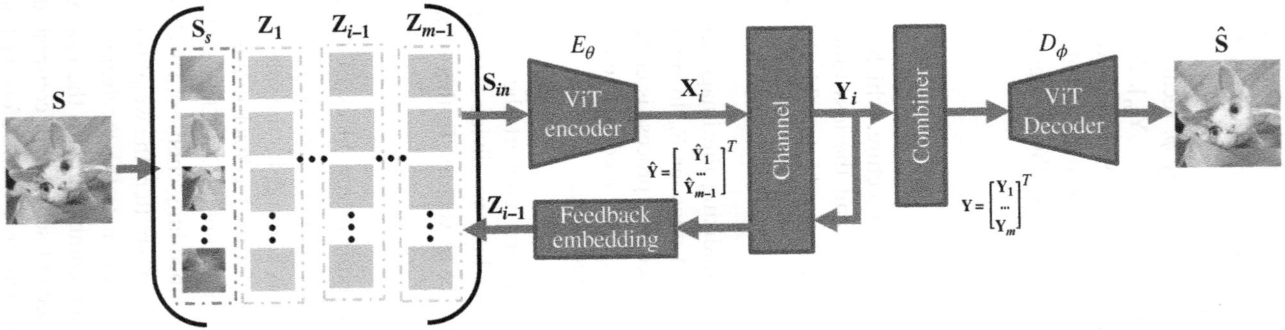

Figure 4.16 The pipeline of our JSCCformer-f scheme. In the ith transmission block, the ViT-based encoder E_θ encodes the channel symbols $\mathbf{X}_i$ based on the input image and the channel feedback signals received until that time, where $\hat{\mathbf{Y}}_{i-1}$ is the noisy channel feedback for the $i-1$th transmission and $\mathbf{Z}_{i-1}$ is its embedding. Note that $(\hat{\mathbf{Y}}_i, \dots, \hat{\mathbf{Y}}_m)$ and $(\mathbf{Z}_i, \dots, \mathbf{Z}_{m-1})$ are padded with zeros as they correspond to future channel feedback signals. Source: Wu et al., 2024/With permission of IEEE.

the receiver's belief, JSCCformer-f achieves a better performance at the cost of additional complexity.

The detailed formula of $\mathbf{Z_i} = f_e(\hat{\mathbf{Y}})$ is provided as below:

$$\mathbf{Z_i} = \begin{cases} concat(sg(D_\phi(\hat{\mathbf{Y}})), \hat{\mathbf{Y}}_\mathbf{i}), & \text{JSCCformer-f,} \\ \hat{\mathbf{Y}}_\mathbf{i}, & \text{JSCCformer-f (lite),} \end{cases} \tag{4.16}$$

where $\hat{\mathbf{Y}}_\mathbf{i} \in \mathbb{R}^{l \times \frac{2k}{l}}$ is the channel output feedback of the ith block, reshaped from $\mathbb{C}^k$, $\hat{\mathbf{Y}} = \hat{\mathbf{Y}}_{\mathbf{1:m}} \triangleq [\hat{\mathbf{Y}}_\mathbf{1}, \dots, \hat{\mathbf{Y}}_\mathbf{m}] \in \mathbb{R}^{l \times \frac{2mk}{l}}$, $sg(\cdot)$ means stop gradient operation, $concat(\cdot)$ denotes concatenation, and $\mathbf{Z_i} \in \mathbb{R}^{l \times z}$ is the embedded feedback matrix given $\hat{\mathbf{Y}}_\mathbf{i}$ for the $(i+1)$th block. We pad $\hat{\mathbf{Y}}_\mathbf{t}$ with a matrix of zeros when there is no knowledge of $\hat{\mathbf{Y}}_\mathbf{t}$. In the ith block of a total m block transmission, $\hat{\mathbf{Y}}_{\mathbf{1:i-2}}$ is given, $\hat{\mathbf{Y}}_{\mathbf{i-1}}$ is updated, while $\hat{\mathbf{Y}}_{\mathbf{i:m}}$ is set to zeros.

Input Sequence Update We concatenate $\mathbf{S_s}$ and the embedded feedback matrix $\mathbf{Z}_{\mathbf{1:m-1}} \triangleq [\mathbf{Z_1}, \dots, \mathbf{Z_{m-1}}]$ to form the input sequence $\mathbf{S_{in}}$ of the ViT-based encoder in the i-th block as: $\mathbf{S_{in}} = concat(\mathbf{S_s}, \mathbf{Z_{1:m-1}})$, where $\mathbf{Z_{1:m-1}} \in \mathbb{R}^{l \times (m-1)z}$ and $\mathbf{S_{in}} \in \mathbb{R}^{l \times (c+(m-1)z)}$.

In each transmission block (ith block), we update $\mathbf{Z_{1:m-1}}$ and $\mathbf{S_{in}}$, once $\mathbf{Z_{i-1}}$ generated from Eq. (4.16) with a new $\hat{\mathbf{Y}}_{\mathbf{i-1}}$ becomes available. Similarly, if there is no knowledge of feedback $\hat{\mathbf{Y}}_\mathbf{t}$ in the current ith transmission block (i.e., $t \geq i$), $\mathbf{Z_k}$ would be padded with zeros. In particular, for the ith block of a total m transmission blocks, $\mathbf{Z_{i-1}}$ is updated while $\mathbf{Z_{i:m-1}}$ is set to zeros.

ViT-based Encoding In each block, we use a ViT-based encoder $E_\theta : \mathbb{R}^{l \times (c+(m-1)z)} \to \mathbb{C}^k$, parameterized by θ, to generate the channel symbols based on the updated $\mathbf{S_{in}}$. The channel output for each block is given as $\mathbf{X_i} = E_\theta(\mathbf{S_{in}})$, where the architecture and operations of our ViT-based encoder are detailed in Wu et al. [2023c].

4.4.2.2 ViT Decoder

At the decoder, we first employ a combiner to form the input sequence for the ViT decoder by concatenating all the received signals as: $\mathbf{Y} \triangleq [\mathbf{Y_1}, \dots, \mathbf{Y_m}] \in \mathbb{R}^{l \times \frac{2mk}{l}}$. While we use the concatenation operation as a combiner for simplicity, we could also use a learning-based method, such as a CNN or a fully connected layer.

Given $\mathbf{Y}$, a symmetrical ViT-based decoder $D_\phi : \mathbb{R}^{l \times 2mk} \to \mathbb{R}^{l \times c}$, parameterized by ϕ is employed to reconstruct the source signal as $\hat{\mathbf{S}} = D_\phi(\mathbf{Y})$, where the detailed operations and model structure are given in Wu et al. [2023c].

Loss Function The encoder and decoder are optimized jointly by minimizing the mean square error between the input image $\mathbf{S}$ and its reconstruction $\hat{\mathbf{S}}$ as: $\mathcal{L}(\theta, \phi) = \mathbb{E}\left[MSE(\mathbf{x}, \hat{\mathbf{x}})\right] \triangleq \mathbb{E}\left[\|\mathbf{S} - \hat{\mathbf{S}}\|_2^2\right]$, where the expectation is taken over the randomness both in the source pixels and the channel state.

4.4.3 Training and Evaluation

This section presents experimental results to evaluate the performance of the JSCCformer-f model across various scenarios. DeepJSCC-f [Kurka and Gündüz, 2020] and the capacity bounds with BPG are employed as benchmarks. Notably, the BPG-Capacity scheme is practically unattainable and requires perfect CSI at both the transmitter and receiver ends. To achieve the optimal result, models of JSCCformer-f and DeepJSCC-f are all trained at a fixed average channel SNR value and tested at the same SNR value. We set the $p = 8$, $l = 64$, and $c = 48$ for the ViT.

4.4.3.1 Transmission Performance

Considering AWGN channels, Figure 4.17a,b presents the PSNR performance versus different forward channel SNR values from −2 to 13 dB, where the bandwidth ratio is set to $R = 1/6$ and $R = 1/3$, respectively. We assume the feedback link is noiseless for now, and the noisy feedback case will be considered later.

We first evaluate JSCCformer-f without channel feedback, i.e., $m = 1$. As shown in Figure 4.17a,b, JSCCformer-f with $m = 1$ significantly outperforms the BPG-Capacity scheme in all SNRs and bandwidth ratios. Again, we note that the BPG-Capacity performance is not achievable in the short block length regime. Next, we compare JSCCformer-f ($m = 1$) with DeepJSCC-f ($m = 2$). As shown in Figure 4.17a, JSCCformer-f ($m = 1$) can surprisingly outperform DeepJSCC-f ($m = 2$) (up to 0.97 dB) at all SNR values, even though DeepJSCC-f utilizes additional feedback from the receiver.

Then, we consider JSCCformer-f with $m > 1$. For a fair comparison with the simulation results in Kurka and Gündüz [2020], we use the same number of blocks for JSCCformer-f. Specifically, we set $m = 2$ for $R = 1/6$ in Figure 4.17a and $m = 4$ for $R = 1/3$ in Figure 4.17b. JSCCformer-f ($m > 1$) improves the performance of JSCCformer-f ($m = 1$) in all SNR values (up to 1.06 dB for $m = 2$ and 2.11 dB for $m = 4$). Further, it can be observed that a larger bandwidth ratio leads to a larger gain. We also observe that with the same number of blocks, JSCCformer-f significantly outperforms DeepJSCC-f at all test SNRs. The gains are up to 2.03 and 3.88 dB for $R = 1/6$ and $R = 1/3$, respectively, demonstrating that JSCCformer-f can exploit feedback better than DeepJSCC-f. Moreover, with the increase in the bandwidth ratio, the improvement from JSCCformer-f becomes even more significant.

Figure 4.17c shows the performance of JSCCformer-f and DeepJSCC-f in fading channels, where the fading gain is sampled as $h \sim \mathcal{CN}(0, 1)$. We observe that DeepJSCC-f outperforms the BPG-Capacity scheme only in the low SNR regime (−2 to 10 dB), although we should remind that this is a highly favorable bound for the separation-based scheme. However, JSCCformer-f consistently outperforms both DeepJSCC-f and BPG-Capacity schemes in all SNR regimes with a significant gain of up to 2.71 and 3.57 dB, respectively.

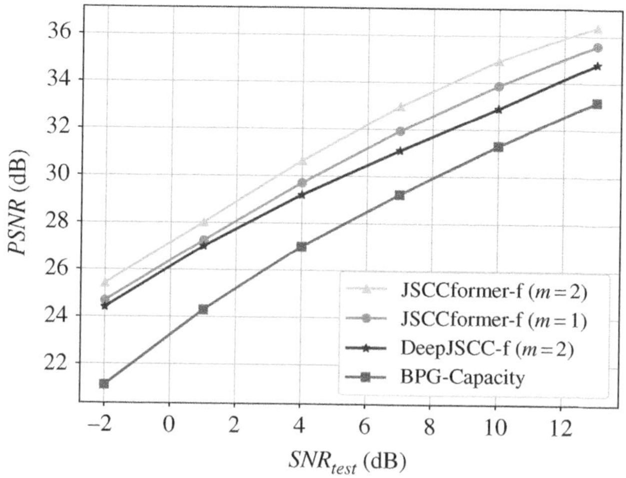

(a)

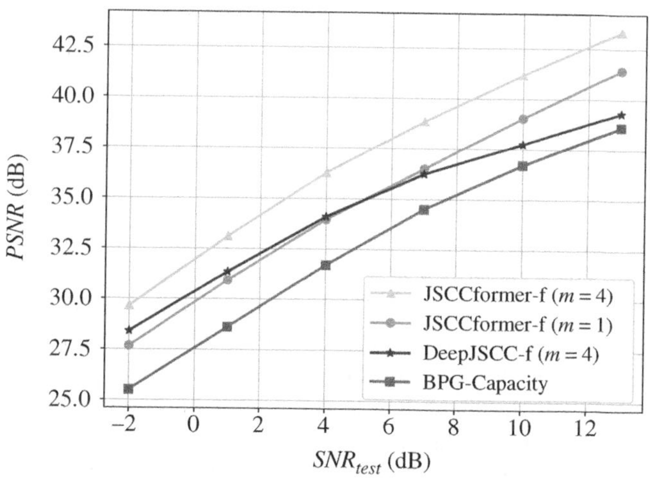

(b)

Figure 4.17 Performance of different schemes at various SNR values and bandwidth ratios with noiseless feedback, where models in subfigures (a) and (b) undergo assessment within the AWGN channel, and models in subfigure (c) are evaluated in a Rayleigh fading channel. (a) $R = 1/6$, (b) $R = 1/3$, and (c) $R = 1/6$.

4.4.3.2 Impacts of Bandwidth Ratio and Block Number

We evaluate the performance of JSCCformer-f over a wide range of bandwidth ratios in Figure 4.18a, where the forward SNR is fixed to 10 dB, the feedback is noiseless, and $m = 4$. JSCCformer-f outperforms DeepJSCC-f in all bandwidth ratio settings with a gain of up to 3.42 dB. Interestingly, even with no feedback, JSCCformer-f ($m = 1$) outperforms DeepJSCC-f ($m = 4$) in the low bandwidth-ratio regime ($R \leq 0.45$).

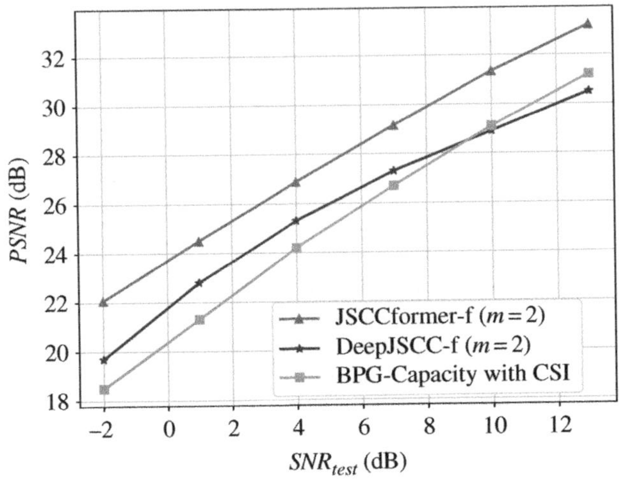

(c)

Figure 4.17 *(Continued)*

Next, we assess the impact of the block number on the PSNR of DeepJSCC-f and JSCCformer-f. To this end, we consider different feedforward SNR values (−2, 4, and 10 dB), noiseless feedback, $R = 1/2$, and vary the number of blocks m, from 1 to 16. We note that, in principle, increasing m means we benefit from more feedback at the encoder. However, it can be observed in Figure 4.18b that the performances of JSCCformer-f and DeepJSCC-f do not always improve with m. Specifically, increasing m brings some improvements at first for both DeepJSCC-f and JSCCformer-f; however, the PSNR declines beyond a certain threshold. One possible explanation for this is the increasing learning difficulty brought about by increasing m. We can observe that the optimal m values for JSCCformer-f are $m = 4, 6, 12$ corresponding to $SNR = 10, 4, -2$ dB, respectively. Notably, the optimal number of blocks tends to be larger under worse channel conditions. This can be explained by the fact that more feedback blocks can provide more information about the decoder's current beliefs and the channel conditions, which is more beneficial for the model in poor channel conditions.

Interestingly, both DeepJSCC-f and JSCCformer-f tend to converge toward a suboptimal performance as m becomes sufficiently large, thus revealing the existence of the trade-off between the training complexity and model performance. Our experimental results suggest that $m = 4$ provides a reasonable performance and complexity trade-off for JSCCformer-f in this channel condition, which can achieve competitive performance with an acceptable computation complexity.

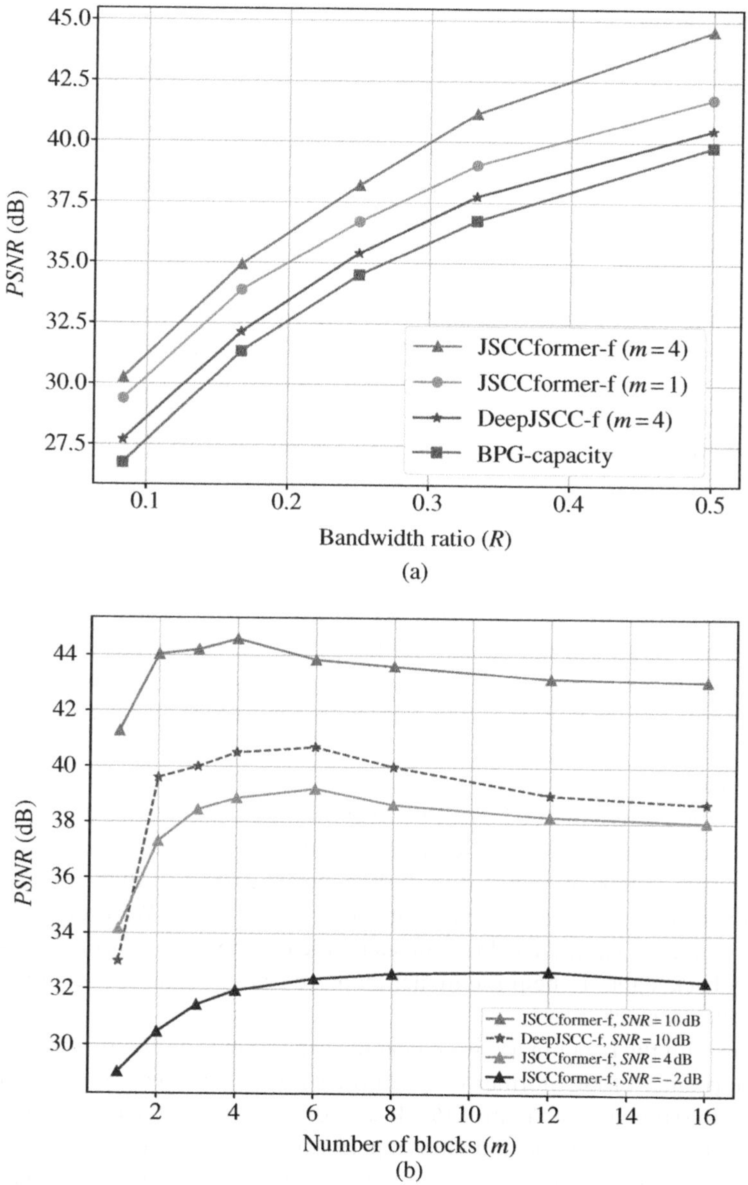

Figure 4.18 (a) Performance comparison of different models versus bandwidth ratio R in AWGN channel when $SNR = 10$ dB with noiseless feedback. (b) Performance of models over different block numbers and SNRs in AWGN channel with $R = 1/2$.

4.4.3.3 Noisy Feedback Channel

In this section, we investigate the impact of noisy feedback links for JSCCformer-f. Introducing noise in the feedback channel can lead to a drastic performance decrease [Kim et al., 2018] or even system breaking down [Kailath, 1967], as feedback can be dominated by noise. DeepJSCC-f is the first practical image transmission scheme that considers noisy feedback, which we shall use as a benchmark.

We repeat the experiments in Figure 4.17a by replacing the noiseless feedback channel with an AWGN channel. The feedback channel SNR is set to $SNR_{fb} = 0, 10,$ and 20 dB. Figure 4.19a shows the performance of the JSCCformer-f model with different SNR_{fb} values. When $SNR_{fb} = 20$ dB (high-quality feedback), the performance only degrades slightly (for at most 0.48 dB), when compared with the perfect feedback case. When the feedback channel quality gets worse, e.g., $SNR_{fb} = 10$ dB, the PSNR continues to decrease, but the gap is less than 0.88 dB compared with the noiseless case. This verifies the robustness of JSCCformer-f to feedback channel noise. In particular, it can be observed that the performances of JSCCformer-f ($m = 2$) with $SNR_{fb} = 20$ dB and $SNR_{fb} = 10$ dB are still better than JSCCformer-f ($m = 1$) without feedback, indicating that JSCCformer-f can still make good use of the noisy feedback. Even with larger feedback noise $SNR_{fb} = 0$ dB, our model can still outperform DeepJSCC-f with a perfect feedback link. It again demonstrates the superiority of our scheme in noisy feedback scenarios. Overall, we conclude that JSCCformer-f is robust to feedback noise.

4.4.3.4 Adaptability

This section validates the SNR-adaptability of JSCCformer-f. To verify the SNR-adaptability, we adopt a random SNR training method, where the SNRs in the training phase are uniformly sampled from the range $[-2, 15]$ dB. After training, the well-trained model is then evaluated at different test SNRs. We expect the JSCCformer-f and JSCCformer-f (lite) encoder to learn the channel conditions from the feedback signal to help the model map the source signal into channel symbols via the SA mechanism.

To compare with the benchmark, we repeat the experiments in Figure 4.17 and plot the performance of JSCCformer-f, JSCCformer-f (lite), and DeepJSCC-f with different training strategies in Figure 4.19b. It can be observed that the DeepJSCC-f scheme exhibits a significant degradation in performance when trained with the random SNR strategy, indicating its inability to adapt to diverse channel conditions. Then, we compare the performance of the JSCCformer-f (lite) model trained using random SNRs and the optimal DeepJSCC-f models trained at each individual SNR. As shown, JSCCformer-f (lite) outperforms DeepJSCC-f across all SNR scenarios, with a maximum gain of 1.64 dB. JSCCformer-f (lite) can adapt to different test SNRs. Importantly, this channel adaptability is attributed to the JSCCformer-f

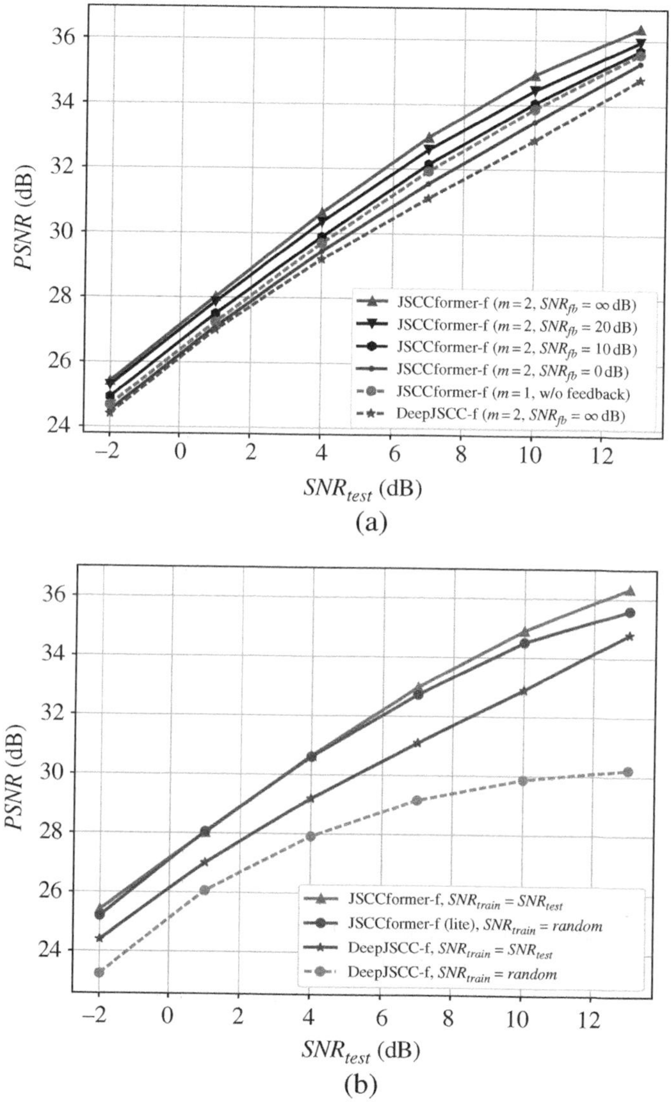

Figure 4.19 (a) Performance of JSCCformer-f over noisy feedback link in AWGN channel, where $R = 1/6$ and $m = 2$. (b) Performance of JSCCformer-f (lite) compared with other schemes in AWGN channel, where $R = 1/6$ and $m = 2$.

method's inherent attention mechanism rather than being solely an outcome of the random training strategy.

In a comparative analysis between JSCCformer-f (lite) and JSCCformer-f from Figure 4.19b, the results indicate that the performance of JSCCformer-f (lite) is only slightly worse than the optimal performance of JSCCformer-f at all SNRs. The performance gap is smaller than 0.74 dB. In summary, our JSCCformer-f (lite) is able to achieve close performance to JSCCformer-f in a wide range of SNRs and is significantly better than DeepJSCC-f. It is worth noting that we do not feed the channel SNRs explicitly to the encoder and decoder. Instead, they are implicitly learned by JSCCformer-f from the observed channel outputs.

4.4.3.5 High Resolution Dataset and Visualization

In order to evaluate the model generalizability, we validate the JSCCformer-f and JSCCformer-f (lite) with $m = 2$ on the Kodak dataset using $SNR_{test} \in [1, 10]$ dB and $R = 1/12$. Specifically, models are trained with randomly cropped 256×256 patches from the ImageNet dataset and subsequently evaluated on the Kodak dataset.

We visualize the comparisons of the recovered image patches by DeepJSCC-f, JSCCformer-f (lite), and JSCCformer-f in Figure 4.20. We can observe that JSCCformer-f performs the best with more detailed high-frequency features (e.g., hair and trees), particularly in the low SNR regime, e.g., at −2 and 1 dB. JSCCformer-f (lite) demonstrates comparable visual performance to JSCCformer-f and exhibits a noteworthy improvement over DeepJSCC-f.

Additionally, the effect of each interaction and the attention of the JSCCformer-f model in each transmission block are presented in Figure 4.21, where we trained the models from the ImageNet dataset and tested the models over the Kodak dataset. We can observe that the reconstruction performance is significantly improved after each interaction, particularly in some high-frequency details. Interestingly, the model tends to acquire an initial understanding of the overall semantic structure after the initial interaction, subsequently recovering a rough sketch. It then proceeds to enhance and refine its prior reconstructions through successive interactions. Furthermore, the attention map of the model in the final interaction revealed that our model could concentrate on some blurry areas and refine them in the next block, demonstrating how our method refines results in a coarse-to-fine manner based on the attention mechanism.

4.4.3.6 Variable Rate Transmission

The JSCC coding problem with channel feedback can be reformulated by imposing a particular transmission quality objective and minimizing the corresponding channel bandwidth, where more significant gains can be observed when considering variable-length coding [Kostina et al., 2017; Kurka and Gündüz, 2020]. Specifically, the transmitter can determine the stopping time by analyzing the decoder's current knowledge of the transmission with the help of perfect channel output feedback.

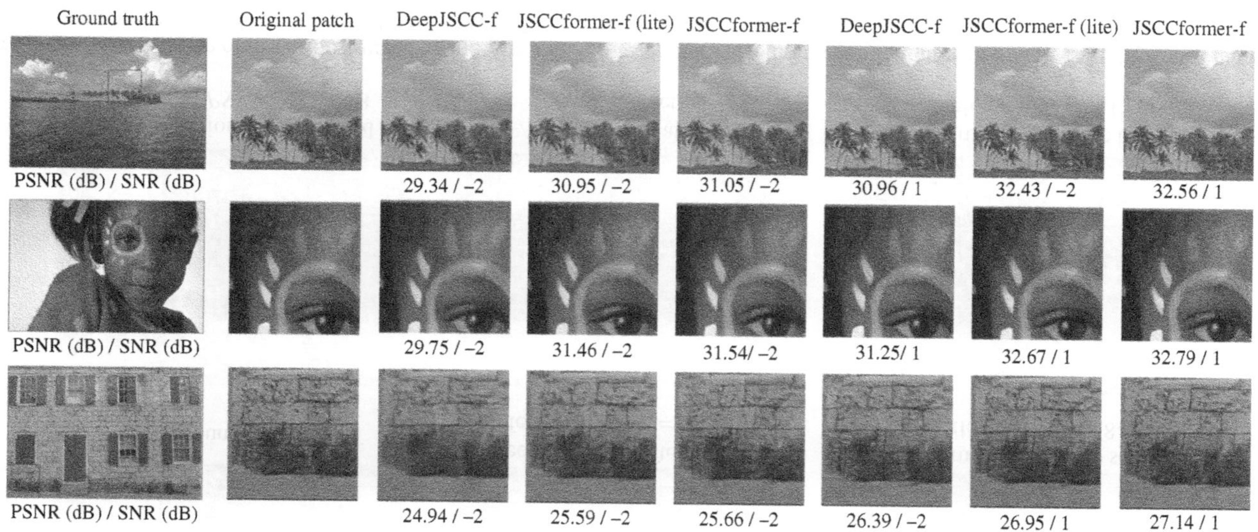

Figure 4.20 Visual comparisons of images transmitted by DeepJSCC-f, JSCCformer-f (lite), and JSCCformer-f over the AWGN channel, where the model is trained on the ImageNet dataset and validated on the Kodak dataset at various SNR values and $R = 1/12$. Source: Wu et al. [2024]/IEEE.

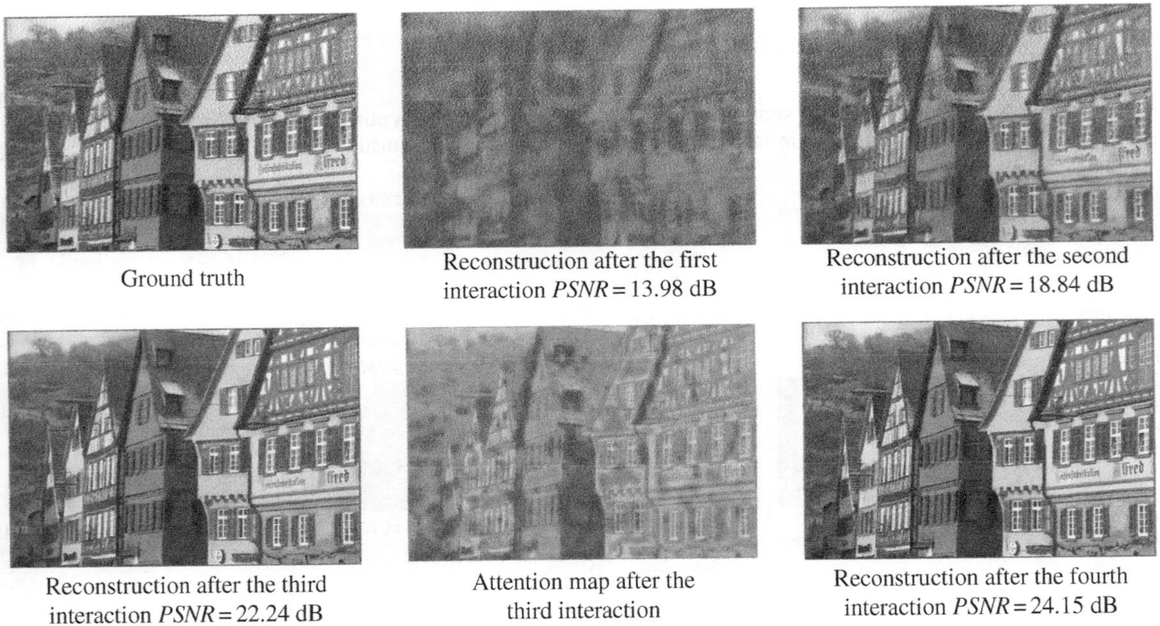

Figure 4.21 Visualization of the reconstructed image with JSCCformer-f after each interaction for an input image from the Kodak dataset, where the model is trained on the ImageNet dataset with $R = 1/12$ and $SNR = -2$ dB. Source: Wu et al. [2024]/IEEE.

We evaluate the performance of JSCCformer-f with variable transmission rate, where we compute the average bandwidth necessary to achieve a predefined PSNR target. For a fair comparison, we set $m = 8$, same with DeepJSCC-f. In the BPG-Capacity scheme, we compute the number of channel uses required to transmit the minimum number of compressed bits while satisfying the target

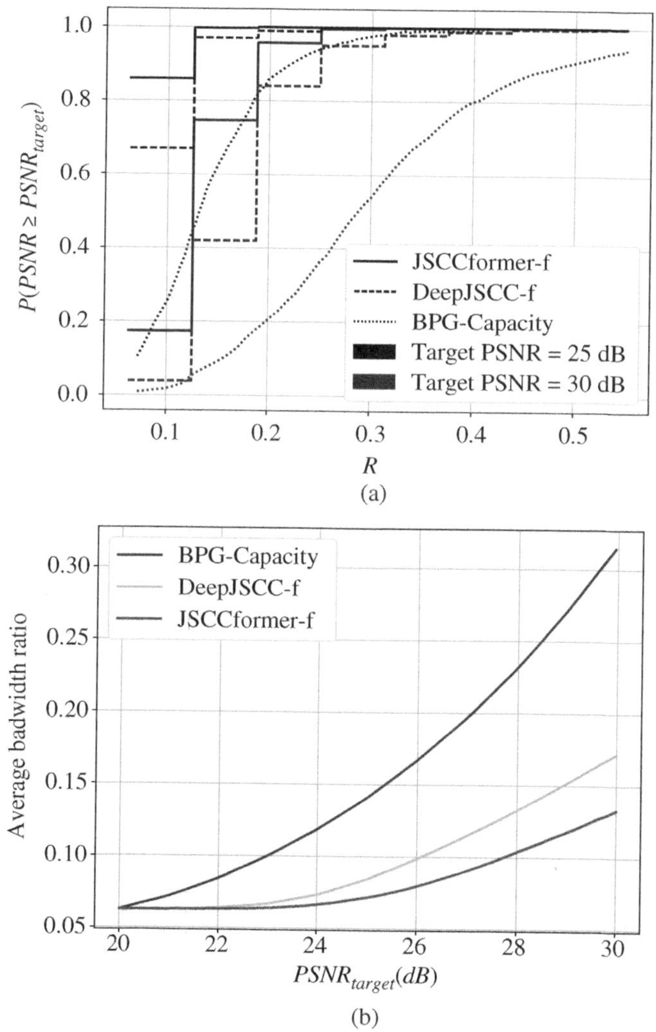

Figure 4.22 Target PSNR analysis of different schemes for variable rate transmission. (a) The cumulative distribution function over various bandwidths given a specific target PSNR. (b) The average bandwidth ratio to achieve a specific target PSNR.

PSNR value using a capacity-achieving channel code. In order to improve the efficacy of variable rate transmission, we fine-tune the well-trained JSCCformer-f while preserving the end-to-end transmission performance, which is achieved through the utilization of the subsequent loss function: $\mathcal{L}(\theta, \phi) = \sum_{i=1}^{8} \|\mathbf{S} - \hat{\mathbf{S}}_\mathbf{i}\|_2^2$, where $\hat{\mathbf{S}}_\mathbf{i} \triangleq D_\phi(\hat{\mathbf{Y}})$ is the intermediate reconstruction of the source signal at the transmitter after each block.

The cumulative distribution function of the required bandwidth to achieve two distinct PSNR values (25 and 30 dB) is presented in Figure 4.22a. We can observe that JSCCformer-f provides significant improvements compared to DeepJSCC-f and BPG-Capacity across the entire range, particularly when the target PSNR is higher. Figure 4.22b displays the average bandwidth ratio needed to achieve distinct target PSNR values. The observed significant improvements compared to the digital scheme confirm the theoretical results in this practical scenario. Furthermore, the JSCCformer-f model exhibits superior performance and establishes a new state-of-the-art performance, particularly in the high target PSNR value regimes. We can conclude that the JSCCformer-f maintains its superiority in the variable rate transmission scenarios and achieves the state-of-the-art performance.

4.5 Concluding Remarks

Throughout this chapter, we embarked on a comprehensive exploration of DeepJSCC, a pioneering technology that fundamentally challenges and extends beyond the traditional paradigms established by Shannon's separation theorem. By integrating the processes of source compression and error correction into a singular, unified framework facilitated by deep learning, DeepJSCC heralds a new era of communication systems characterized by enhanced efficiency, adaptability, and resilience.

The journey through the realms of MIMO channels, relay channels, and feedback channels has illuminated the versatility and transformative potential of DeepJSCC across diverse communication scenarios. Each of these canonical channel models presented unique challenges, serving as a testament to the limitations of conventional communication systems and underscoring the necessity for innovative approaches like DeepJSCC. We would like to highlight that, despite years of study on channel coding, there are no practical channel codes deployed for relay or feedback channels. We believe that the flexibility offered by DeepJSCC and its data-driven nature makes it a compelling alternative for 6G and future communication networks, as well as other nonstandardized communication scenarios.

Overall, DeepJSCC stands out as an emerging tool of innovation in semantic communications, extending the boundaries of what is possible with JSCC. Its significance lies not only in challenging theoretical limits but also in making JSCC practical and applicable across a broad spectrum of sources and channels. As we continue to push the frontiers, DeepJSCC offers a promising pathway toward realizing systems that are not only highly efficient and reliable but also adaptable to the ever-evolving landscape of wireless communications and many emerging applications that rely on it with ever increasing diversity of requirements and constraints. This chapter has laid the groundwork for further exploration and application of DeepJSCC, including its application in multiuser scenarios, setting the stage for future advancements that will undoubtedly continue to shape the field in profound ways.

Bibliography

Johannes Ballé, David Minnen, Saurabh Singh, Sung Jin Hwang, and Nick Johnston. Variational image compression with a scale hyperprior. In *International Conference on Learning Representations, ICLR*, 2018.

Fabrice Bellard. Better portable graphics 0.9.8. URL https://bellard.org/bpg/, 2018.

Chenghong Bian, Yulin Shao, and Deniz Gunduz. Wireless point cloud transmission. arXiv preprint arXiv:2306.08730, 2023a.

Chenghong Bian, Yulin Shao, Haotian Wu, and Deniz Gündüz. Space-time design for deep joint source channel coding of images over MIMO channels. In *2023 IEEE 24th International Workshop on Signal Processing Advances in Wireless Communications (SPAWC)*, pages 616–620. IEEE, 2023b.

C. Bian, Y. Shao, H. Wu, and D. Gündüz. Deep joint source-channel coding over cooperative relay networks. In *ICMLCN*, 2024a.

Chenghong Bian, Yulin Shao, and Deniz Gunduz. A hybrid joint source-channel coding scheme for mobile multi-hop networks. In *IEEE International Conference on Communications (ICC)*, 2024b. doi: 10.1109/ICC.2016.7511265.

Eirina Bourtsoulatze, David Burth Kurka, and Deniz Gündüz. Deep joint source-channel coding for wireless image transmission. *IEEE Transactions on Cognitive Communications and Networking*, 5(3):567–579, 2019.

Vinay Chande, Hamid Jafarkhani, and Nariman Farvardin. Joint source-channel coding of images for channels with feedback. *Proceedings of IEEE Information Theory Workshop*, Feb. 1998.

Zhengxue Cheng, Heming Sun, Masaru Takeuchi, and Jiro Katto. Learned image compression with discretized Gaussian mixture likelihoods and attention modules. In *Conference on Computer Vision and Pattern Recognition (CVPR)*, 2020.

Roland L'vovich Dobrushin. Information transmission in a channel with feedback. *Theory of Probability & Its Applications*, 3(4):367–383, 1958.

R. H. Etkin and D. N. C. Tse. Degrees of freedom in some underspread MIMO fading channels. *IEEE Transactions on Information Theory*, 52(4):1576–1608, 2006. doi: 10.1109/TIT.2006.871595.

Nariman Farsad, Milind Rao, and Andrea Goldsmith. Deep learning for joint source-channel coding of text. In *2018 IEEE International Conference on Acoustics, Speech and Signal Processing (ICASSP)*, pages 2326–2330. IEEE, 2018.

Deniz Gndz, Zhijin Qin, Inaki Estella Aguerri, Harpreet S. Dhillon, Zhaohui Yang, Aylin Yener, Kai Kit Wong, and Chan-Byoung Chae. Beyond transmitting bits: Context, semantics, and task-oriented communications. *IEEE Journal on Selected Areas in Communications*, 41(1):5–41, 2023. doi: 10.1109/JSAC.2022.3223408.

Andrea Goldsmith. Joint source/channel coding for wireless channels. In *1995 IEEE 45th Vehicular Technology Conference. Countdown to the Wireless Twenty-First Century*, volume 2, pages 614–618. IEEE, 1995.

Deniz Gündüz, David Burth Kurka, Mikolaj Jankowski, Mohammad Mohammadi Amiri, Emre Ozfatura, and Sreejith Sreekumar. Communicate to learn at the edge. *IEEE Communications Magazine*, 58(12):14–19, 2020.

Zhan Guo and Peter Nilsson. Algorithm and implementation of the K-best sphere decoding for MIMO detection. *IEEE Journal on Selected Areas in Communications*, 24(3):491–503, 2006.

Babak Hassibi and Bertrand M. Hochwald. How much training is needed in multiple-antenna wireless links? *IEEE Transactions on Information Theory*, 49(4):951–963, 2003.

A. Host-Madsen and J. Zhang. Capacity bounds and power allocation for wireless relay channels. *IEEE Transactions on Information Theory*, 51(6):2020–2040, 2005. doi: 10.1109/TIT.2005.847703.

Mikolaj Jankowski, Deniz Gündüz, and Krystian Mikolajczyk. Wireless image retrieval at the edge. *IEEE Journal on Selected Areas in Communications*, 39(1):89–100, 2020.

V. Kafedziski. Joint source channel coding of images over frequency selective fading channels with feedback using DCT and multicarrier block pulse amplitude modulation. In *Proceedings of Asilomar Conference on Signals, Systems and Computers*, volume 1, pages 37–41, Nov. 1998. doi: 10.1109/ACSSC.1998.750824.

T. Kailath. An application of Shannon's rate-distortion theory to analog communication over feedback channels. *Proceedings of the IEEE*, 55(6):1102–1103, 1967.

Hyeji Kim, Yihan Jiang, Sreeram Kannan, Sewoong Oh, and Pramod Viswanath. Deepcode: Feedback codes via deep learning. *Advances in Neural Information Processing Systems 31*, 2018.

Victoria Kostina, Yury Polyanskiy, and Sergio Verd. Joint source-channel coding with feedback. *IEEE Transactions on Information Theory*, 63(6):3502–3515, 2017.

G. Kramer, M. Gastpar, and P. Gupta. Cooperative strategies and capacity theorems for relay networks. *IEEE Transactions on Information Theory*, 51(9):3037–3063, 2005. doi: 10.1109/TIT.2005.853304.

David Burth Kurka and Deniz Gündüz. DeepJSCC-f: Deep joint source-channel coding of images with feedback. *IEEE Journal on Selected Areas in Information Theory*, 1(1):178–193, 2020.

Wing Fei Lo, Nitish Mital, Haotian Wu, and Deniz Gndz. Collaborative semantic communication for edge inference. *IEEE Wireless Communications Letters*, 12(7):1125–1129, 2023. doi: 10.1109/LWC.2023.3256006.

J. Lu, A. Nosratinia, and B. Aazhang. Progressive joint source-channel coding in feedback channels. In *Proceedings DCC'99 Data Compression Conference (DCC)*, pages 140–148, Mar. 1999. doi: 10.1109/DCC.1999.755663.

David Minnen, Johannes Ballé, and George Toderici. Joint autoregressive and hierarchical priors for learned image compression. In *Neural Information Processing Systems*, pages 10794–10803, 2018.

J. Schalkwijk. A coding scheme for additive noise channels with feedback—II: Band-limited signals. *IEEE Transactions on Information Theory*, 12(2):183–189, 1966.

J. Schalkwijk and L. Bluestein. Transmission of analog waveforms through channels with feedback (Corresp.). *IEEE Transactions on Information Theory*, 13(4):617–619, 1967.

Gesualdo Scutari, Daniel P. Palomar, and Sergio Barbarossa. The MIMO iterative waterfilling algorithm. *IEEE Transactions on Signal Processing*, 57(5):1917–1935, 2009.

Claude Elwood Shannon. A mathematical theory of communication. *The Bell System Technical Journal*, 27(3):379–423, 1948.

Claude Shannon. The zero error capacity of a noisy channel. *IRE Transactions on Information Theory*, 2(3):8–19, 1956.

Tze-Yang Tung and Deniz Gndz. Deepwive: Deep-learning-aided wireless video transmission. *IEEE Journal on Selected Areas in Communications*, 40(9):2570–2583, 2022. doi: 10.1109/JSAC.2022.3191354.

Sridhar Vembu, Sergio Verdu, and Yossef Steinberg. The source-channel separation theorem revisited. *IEEE Transactions on Information Theory*, 41(1):44–54, 1995.

Haotian Wu, Yulin Shao, Krystian Mikolajczyk, and Deniz Gündüz. Channel-adaptive wireless image transmission with OFDM. *IEEE Wireless Communications Letters*, 11(11):2400–2404, 2022.

Haotian Wu, Yulin Shao, Chenghong Bian, Krystian Mikolajczyk, and Deniz Gündüz. Vision transformer for adaptive image transmission over MIMO channels. In *2023 IEEE International Conference on Communications (ICC)*, 2023a.

Haotian Wu, Yulin Shao, Chenghong Bian, Krystian Mikolajczyk, and Deniz Gündüz. Deep joint source-channel coding for adaptive image transmission over MIMO channels. arXiv preprint arXiv:2309.00470, 2023b.

Haotian Wu, Yulin Shao, Emre Ozfatura, Krystian Mikolajczyk, and Deniz Gündüz. Transformer-aided wireless image transmission with channel feedback. arXiv preprint arXiv:2306.09101, 2023c.

Mingyu Yang, Chenghong Bian, and Hun-Seok Kim. OFDM-guided deep joint source channel coding for wireless multipath fading channels. *IEEE Transactions on Cognitive Communications and Networking*, 8(2):584–599, 2022. doi: 10.1109/TCCN.2022.3151935.

Selim F. Yilmaz, Can Karamanlı, and Deniz Gündüz. Distributed deep joint source-channel coding over a multiple access channel. In *ICC 2023-IEEE International Conference on Communications*, pages 1400–1405. IEEE, 2023.

Taesang Yoo and Andrea Goldsmith. Capacity and power allocation for fading MIMO channels with channel estimation error. *IEEE Transactions on Information Theory*, 52(5):2203–2214, 2006.

5

When Information Is a Function of Data – Some Information Theoretic Perspectives on Semantic Communications

Alexander Mariona[1], Homa Esfahanizadeh[2], Rafael Gregorio Lucas D'Oliveira[3], and Muriel Médard[1]

[1]*Research Laboratory of Electronics, Massachusetts Institute of Technology, Cambridge, MA, USA*
[2]*Radio Systems Research Group, Nokia Bell Labs, Murray Hill, NJ, USA*
[3]*School of Mathematical and Statistical Sciences, Clemson University, Clemson, SC, USA*

For many applications, the most pragmatic and well-developed mathematical framework for describing uncertainty and randomness is probability theory. There is a long history of formalism and philosophy, which attempts to justify the axioms of probability, grounding them in a consistent and logical framework, which coincides with our intuitive notions of uncertainty [Reichenbach, 1971; Hájek, 2023]. Although these arguments have their critics, the usefulness of probability theory is indisputable. Probability theory gives us a way to formalize the idea of a distribution of outcomes, some of which may be more likely than others.

These distributions can be very high-dimensional objects, can be defined over continuous or discrete domains, and can sometimes be difficult to describe in closed form. Despite this complexity and heterogeneity, the problem of quantifying the similarity between probability distributions arises in many practical and theoretical settings.

Due in part to the wide variety of problem settings, there is a correspondingly varied set of methods to compare distributions. One natural way is to define a metric on the space of distributions, such as the total variation (TV) distance. Informally, the TV distance between two distributions p and q is defined to be equal to the largest difference in probability that they assign to the same event. As another example, the Kullback–Liebler (KL) divergence, or relative entropy, is ubiquitous in information theory [Cover and Thomas, 2006]. The KL divergence captures the expected number of additional bits needed to encode the data distributed according to p using a code designed for q. In the Bayesian inference setting, the KL divergence measures the increase in information when updating from the prior belief

Foundations of Semantic Communication Networks, First Edition.
Edited by Walid Saad, Christina Chaccour, Christo Kurisummoottil Thomas, and Merouane Debbah.

distribution q to the posterior distribution p. Although technically not a metric, the KL divergence nonetheless distills some notion of the similarity between two distributions down to a single real number.

In some cases, however, we may not be able to simply compute the distance or divergence between two distributions. Perhaps we have an incomplete characterization, knowing only, e.g., some finite number of moments rather than the entire density function. Perhaps we only have a collection of samples drawn from two different distributions, or maybe we are not interested in the distribution of a random variable itself, but rather in the distribution of some function of that random variable. These more general settings call for more general tools to compare distributions.

5.1 The Central Limit Theorem

One of the most foundational results in probability theory, the central limit theorem (CLT), states that the sample mean of a collection of independent random variables has a distribution similar to that of a Gaussian random variable. This is perhaps the simplest example of a function of random variables. Suppose that we draw n independent observations from the same distribution p. Each observation $X_1, \dots, X_n$ is a random variable with the same mean μ and variance σ^2. The sample mean,

$$S_n = \frac{1}{n}\sum_{i=1}^{n} X_n,$$

is a function of the sequence $X_1, \dots, X_n$ and is itself a random variable. As we collect more samples, i.e., as n grows, the distribution of S_n will change, maybe dramatically at first, but more gradually as n becomes sufficiently large. The simplest version of the CLT, which we cite here, treats this setting.

Theorem 5.1 ***(Lindeberg-Lévy CLT)*** Let $X_1, \dots, X_n$ be a sequence of i.i.d. random variables with $\mathbb{E}\left[X_i\right] = \mu$ and $\text{Var}[X_i] = \sigma^2 < \infty$ for all $i \in [n]$. Then, as n goes to infinity, the random variable $\frac{\sqrt{n}(S_n - \mu)}{\sigma}$ converges in distribution to a standard normal random variable:

$$\lim_{n\to\infty} \mathbb{P}\left[\frac{\sqrt{n}(S_n - \mu)}{\sigma} \le z\right] = \Phi(z),$$

where $\Phi(z)$ is the cumulative distribution function (CDF) of $\mathcal{N}(0, 1)$ evaluated at z.

This version of the CLT has rather strong assumptions on the distribution of the summands, but there are more general versions, which relax these assumptions.

Indeed, there are a variety of other limit theorems, which handle dependent random variables [Dvoretzky, 1972]; local limit theorems, which are stronger than the central variants [Szewczak and Weber, 2022]; and non-CLTs for stochastic processes, which do not produce a Gaussian limit [Nourdin and Taqqu, 2014]. In general, limit theorems describe the asymptotic behavior of processes, but do not quantitatively describe how they approach the limit.

5.2 Quantitative Bounds

To obtain an upper bound on how close the distribution of S_n is to its Gaussian limit for a given value of n, we can turn to a result such as the Berry–Esseen theorem [Berry, 1941; Esseen, 1942].

Theorem 5.2 ***(Berry–Esseen)*** Let $X_1, \dots, X_n$ be a sequence of i.i.d. random variables with $\mathbb{E}\left[X_i\right] = 0$, $\text{Var}[X_i] = \sigma^2 > 0$, and $\mathbb{E}\left[|X_i|^3\right] = \rho < \infty$ for all $i \in [n]$. Then, there exists some constant $C > 0$ such that, for all z and n,

$$\left|\mathbb{P}\left[\frac{\sqrt{n}(S_n - \mu)}{\sigma} \leq z\right] - \Phi(z)\right| \leq \frac{C\rho}{\sigma^3\sqrt{n}}.$$

We have cited here a version of the Berry–Esseen theorem with essentially the same assumptions as the version of the CLT given in 5.1. In addition to giving a uniform upper bound on the CDF-closeness of S_n and its limit, the Berry–Esseen theorem also clarifies the factors, which affect the rate of convergence in the CLT. First and foremost, the convergence occurs at a rate of $1/\sqrt{n}$. Second, increasing the variance σ^2 also increases the rate. This is because more "spread out" distributions are already closer to Gaussians than very peaky distributions, which are heavily concentrated around their mean. Similarly, the rate decreases with the (absolute) third moment, since Gaussians are symmetric, i.e., they have no skew. As for the constant C, at the time of writing, the best value for this uniform version of the theorem is $C = 0.4748$ [Shevtsova, 2011], and Esseen himself showed that $C > 0.4097$ [Esseen, 1956].

As with the CLT, there are a number of ways to prove the Berry–Esseen theorem, including Stein's method [Stein, 1972, 1986], which uses characteristic functions, and Lindeberg's method [Lindeberg, 1922]. The latter is more readily understood and easily extended to prove more general theorems. The key idea is that, instead of analyzing the CDF of $\sqrt{n}(S_n - \mu)/\sigma$ directly, we can make use of the structural form of S_n and show that

$$X_1 + X_2 + \cdots + X_n \approx Z_1 + Z_2 + \cdots + Z_n, \tag{5.1}$$

where each Z_i is an independent Gaussian random variable. Lindeberg's method is sometimes called the "replacement method" because it is based on repeatedly replacing each term on the left-hand side of (5.1) with a term on the right-hand side. The overall bound on the quality of the approximation is then obtained by bounding the error at each step.

In fact, perhaps the most straightforward way to apply the replacement method results in a slightly different statement than the Berry–Esseen theorem. The CDF-closeness in the Berry–Esseen theorem can be thought of as a special case of measuring the expected difference between $\psi(S_n)$ and $\psi(Z)$, where Z is the appropriate Gaussian random variable and ψ is an arbitrary "test" or "measurement" function. For this measurement in particular, we would like to choose $\psi(z) = \mathbb{1}_{(-\infty,z]}$. The replacement method makes this kind of argument easy when ψ is sufficiently smooth, but with some additional analytical techniques, the indicator function can be treated. This argument follows that given in O'Donnell [2014a] and Khoshnevisan [2007].

Theorem 5.3 Let $X_1, \dots, X_n$ be a sequence of i.i.d. random variables with $\mathbb{E}\left[X_i\right] = 0$, $\mathrm{Var}[X_i] = \sigma^2 > 0$, and $\mathbb{E}\left[|X_i|^3\right] = \rho < \infty$ for all $i \in [n]$. Let the function f be $\mathscr{C}^3$, i.e., the derivatives f', f'', and f''' exist, are continuous, and bounded. Then, there exists some constant $C > 0$ such that, for all n,

$$\left|\mathbb{E}\left[f(S_n)\right] - \mathbb{E}\left[f(Z)\right]\right| \leq CM_f \cdot \rho n,$$

where $Z \sim \mathcal{N}(0, n\sigma^2)$ and $M_f = \sup_z |f'''(z)|$.

Proof: Since f is sufficiently differentiable, we can take a third-order Taylor expansion around S_{n-1}, obtaining

$$f(S_n) = f(S_{n-1}) + X_n f'(S_{n-1}) + \frac{X_n^2}{2} f''(S_{n-1}) + R(X_n),$$

where $|R(X_n)| \leq \frac{M_f}{6} \cdot |X_n|^3$. We may then take expectations and use the independence of $X_1, \dots, X_n$ to bound the quality of that approximation as

$$\left|\mathbb{E}\left[f(S_n)\right] - \mathbb{E}\left[f(S_{n-1})\right] - \frac{\sigma^2}{2}\mathbb{E}\left[f''(S_{n-1})\right]\right| \leq \frac{M_f}{6} \cdot \rho. \tag{5.2}$$

Then, for a normal random variable Z_n, which has the same first and second moments as X_n but is independent of $X_1, \dots, X_n$, we can replace X_n by Z_n to obtain the sequence $X_1, \dots, X_{n-1}, Z_n$ and perform the same Taylor expansion. This yields

$$\left|\mathbb{E}\left[f(S_{n-1} + Z_n)\right] - \mathbb{E}\left[f(S_{n-1})\right] - \frac{\sigma^2}{2}\mathbb{E}\left[f''(S_{n-1})\right]\right| \leq \frac{M_f}{6}\mathbb{E}\left[|Z_n|^3\right]. \tag{5.3}$$

Combining (5.2) and (5.3) using the triangle inequality,

$$\left|\mathbb{E}\left[f(S_n)\right] - \mathbb{E}\left[f(S_{n-1} + Z_n)\right]\right| \leq \frac{M_f}{6}\left(\mathbb{E}\left[\left|Z_n\right|^3\right] + \rho\right). \tag{5.4}$$

Noting that $\sigma^3 \leq \mathbb{E}\left[\left|X_n\right|^3\right] = \rho$ and

$$\mathbb{E}\left[\left|Z_n\right|^3\right] = \sigma^3 \cdot \mathbb{E}\left[\left|\mathcal{N}(0,1)\right|^3\right] = \frac{2\sigma^3}{\sqrt{\pi/2}} > 1,$$

we can upper bound the right-hand side of (5.4) as

$$\left|\mathbb{E}\left[f(S_n)\right] - \mathbb{E}\left[f(S_{n-1} + Z_n)\right]\right| \leq \frac{2M_f}{3\sqrt{\pi/2}}\rho.$$

Repeating this replacement process inductively, we obtain

$$\left|\mathbb{E}\left[f(S_n)\right] - \mathbb{E}\left[f\left(\sum_{i=1}^{n} Z_i\right)\right]\right| \leq \frac{2M_f}{3\sqrt{\pi/2}}\rho n.$$

Since $\sum_{i=1}^{n} Z_i = \mathcal{N}(0, n\sigma^2)$, this gives the desired result with $C = \frac{2}{3\sqrt{\pi/2}}$.

Now, if we could take f to be the indicator function $\mathbf{1}_{(-\infty,z]}$, then we could recover the Berry–Esseen theorem from 5.3. Unfortunately, such a function is not differentiable, but we can "smooth out" the discontinuities and keep track of the approximation losses to get a similar bound, although the resulting constants or parameter dependencies meters may vary based on the exact techniques used. See O'Donnell [2014a] for an argument, which extends 5.3 to Lipschitz functions and then uses an anti-concentration inequality for Gaussians to handle the approximation at the discontinuity of the indicator function.

5.3 General Polynomials

The CLT and Berry–Esseen theorem, as we have presented both, describe how the distribution of the sum of independent random variables behaves. There are of course many functions other than the simple sum, which may be of interest, and the same methods, which lead to the Berry–Esseen theorem, can be generalized to handle more complex functions. In general, however, we do not have a limit theorem from which to begin, so neither the "target" distribution nor the correct metric for comparison are obvious. Nonetheless, the generality of the replacement method makes it easy to obtain results for a wide variety of different measurement functions, even if some obvious choices, like the CDF distance, require a little bit of extra work.

The goal of this section is to generalize the Berry–Esseen theorem, moving from the simplest linear function to general multilinear functions, i.e., functions of the form

$$F(\mathbf{x}) = \sum_{S\subseteq[n],|S|\leq k} \hat{F}(S)\prod_{i\in S} x_i,$$

where k is the degree of F. Such functions are called “multilinear” because each term may include multiple coordinates of the input $\mathbf{x} = (x_1, \dots, x_n)$, but no nonlinear function is applied to any of the coordinates. The notation $\hat{F}(S)$ denotes the coefficient of the term in F, which includes exactly the coordinates indexed by S. For example, if $n = 7$ and $S = \{1, 3, 5\}$, then $\hat{F}(S)$ is the coefficient for the term containing $x_1x_3x_5$. Note that $\hat{F}(S)$ is unique because F is multilinear.

In the context of the Berry–Esseen theorem, the variance and skewness of the random variables in question affect the quality of the approximation (or, from the CLT perspective, how quickly the distribution converges to its limit). This kind of condition can be interpreted as formalizing the notion of a distribution being “reasonable.” Naturally, in the CLT context, the most reasonable distributions are those that are similar to Gaussians. There must be some kind of assumption on reasonableness, and the appropriate generalization of this concept for multilinear polynomials is that no single variable in the sequence $X_1, \dots, X_n$ can be too “influential” on the value of F.

The concept of the influence of inputs on a function originated in the social sciences in the context of voting theory [Penrose, 1946] and has been extensively developed in the study of Boolean functions [O’Donnell, 2014b]. For our purposes, we will use the following real-valued generalization, which can be simply taken as a definition.

Definition 5.1 The *influence* of coordinate $i \in [n]$ on $f : \mathbb{R}^n \to \mathbb{R}$ is

$$\mathbf{Inf}_i[F] = \sum_{S\ni i} \hat{F}(S)^2.$$

Intuitively, $\mathbf{Inf}_i[F]$ quantifies the significance of the terms in F, which depend on the ith coordinate. To be precise, this definition coincides with the general concept of influence only when F is applied to a sequence of random variables $(X_1, \dots, X_n)$ for which $\mathbb{E}\left[X_i\right] = 0$ and $\mathbb{E}\left[X_i^2\right] = 1$, but this will be the case in all settings we discuss. In particular, we will generally assume that random variables satisfy the following set of assumptions, which amounts to saying that a random variable is sufficiently well-concentrated around its mean.

Assumption 5.1 The random variable X satisfies $\mathbb{E}\left[X\right] = 0$, $\mathbb{E}\left[X^2\right] = 1$, $\mathbb{E}\left[X^3\right] = 0$, and $\mathbb{E}\left[X^4\right] \leq 9$.

The generalization we give of the Berry–Esseen theorem is referred to as the basic invariance principle (BIP) [Mossel et al., 2005]. This kind of result is called an "invariance principle" because it describes the extent to which the distribution of a function of random variables is "invariant" to the details of the distribution of its inputs. The Berry–Esseen theorem could also be called an invariance principle – it says that the sum of a sequence of random variables is invariant to the distribution of its summands as long as they have sufficient variability and relatively low skewness.

The proof we give of the BIP follows that given in O'Donnell [2014a]. It utilizes the replacement method in much the same way as our proof of the (variant) Berry–Esseen theorem, except that we take a third-order Taylor expansion in order to make use of a "hypercontractivity" result, which essentially states that low-degree multilinear polynomials are also reasonable when applied to sequences of reasonable random variables.

Theorem 5.4 *(BIP)* Let F be an n-variate multilinear polynomial of degree at most k. Let $\mathbf{X} = (X_1, \dots, X_n)$ and $\mathbf{Y} = (Y_1, \dots, Y_n)$ be sequences of independent random variables each satisfying 5.1. Let $\psi : \mathbb{R} \to \mathbb{R}$ be $\mathscr{C}^4$, i.e., the derivatives $\psi', \dots, \psi''''$ exist and are continuous with $||\psi''''||_\infty \leq C$. Then,

$$|\mathbb{E}\,[\psi(F(\mathbf{X}))] - \mathbb{E}\,[\psi(F(\mathbf{Y}))]| \leq \frac{C}{12} \cdot 9^k \cdot \sum_{i=1}^{n} \mathbf{Inf}_i[F]^2.$$

Proof: Toward an application of the replacement method, define the intermediate polynomials $H_t = F(Y_1, \dots, Y_t, X_{t+1}, \dots, X_n)$ for all $t \in \{0, \dots, n\}$, such that $H_0 = F(X)$ and $H_n = F(Y)$. We show that, for all $t \in [n]$,

$$\left|\mathbb{E}\left[\psi(H_{t-1}) - \psi(H_t)\right]\right| \leq \frac{C}{12} \cdot 9^k \cdot \mathbf{Inf}_t[F]^2. \tag{5.5}$$

Summing over t and applying the triangle inequality will complete the proof.

Define the functions E_tF and D_tF such that $F(x) = E_tF(x) + x_tD_tF(x)$ for all $t \in [n]$, i.e.,

$$E_tF(x) = \sum_{S \not\ni t} \hat{F}(S) \prod_{i \in S} x_i, \qquad D_tF(x) = \sum_{S \ni t} \hat{F}(S) \prod_{i \in S\setminus\{t\}} x_i.$$

Since neither E_tF nor D_tF depends on coordinate t of the input, we can define

$$U_t = E_tF(Y_1, \dots, Y_{t-1}, 0, X_{t+1}, \dots, X_n),$$
$$\Delta_t = D_tF(Y_1, \dots, Y_{t-1}, 0, X_{t+1}, \dots, X_n).$$

We choose to assign the value 0 to coordinate t of the inputs simply for concreteness; the actual value does not matter. Then, $H_{t-1} = U_t + \Delta_tX_t$ and $H_t = U_t + \Delta_tY_t$.

We can now bound (5.5) by taking third-order Taylor expansions of $\psi(H_{t-1})$ and $\psi(H_t)$ around U_t. We have that

$$\psi(H_{t-1}) = \psi(U_t) + \psi'(U_t)\Delta_t X_t + \tfrac{1}{2}\psi''(U_t)\Delta_t^2 X_t^2 + \tfrac{1}{6}\psi'''(U_t)\Delta_t^3 X_t^3 + \tfrac{1}{24}\psi''''(U_t^*)\Delta_t^4 X_t^4, \tag{5.6}$$

$$\psi(H_t) = \psi(U_t) + \psi'(U_t)\Delta_t Y_t + \tfrac{1}{2}\psi''(U_t)\Delta_t^2 Y_t^2 + \tfrac{1}{6}\psi'''(U_t)\Delta_t^3 Y_t^3 + \tfrac{1}{24}\psi''''(U_t^{**})\Delta_t^4 Y_t^4. \tag{5.7}$$

for some random variables U_t^* and U_t^{**}. Taking expectations and subtracting (5.7) from (5.6), it is easy to see that the zeroth-order terms cancel directly. As for the first-order difference,

$$\begin{aligned}\mathbb{E}\left[\psi'(U_t)\Delta_t X_t - \psi'(U_t)\Delta_t Y_t\right] &= \mathbb{E}\left[\psi'(U_t)\Delta_t(X_t - Y_t)\right] \\ &= \mathbb{E}\left[\psi'(U_t)\Delta_t\right]\mathbb{E}\left[X_t - Y_t\right] &\quad (5.8)\\ &= 0, &\quad (5.9)\end{aligned}$$

where (5.8) follows from the fact that X_t and Y_t are independent of U_t and Δ_t by construction, and (5.9) follows from the fact that X_t and Y_t have matching first moments by 5.1. The same argument applies to the second- and third-order differences. For the fourth-order difference, we apply the triangle inequality and the assumption that $||\psi''''||_\infty \leq C$ to obtain

$$\left|\mathbb{E}\left[\psi(H_{t-1}) - \psi(H_t)\right]\right| \leq \frac{C}{24}\left(\mathbb{E}\left[(\Delta_t X_t)^4\right] + \mathbb{E}\left[(\Delta_t Y_t)^4\right]\right).$$

To finish, we must bound each of $\mathbb{E}\left[(\Delta_t X_t)^4\right]$ and $\mathbb{E}\left[(\Delta_t Y_t)^4\right]$ by $9^k\mathbf{Inf}_t[F]^2$. First, note that $\Delta_t x_t = x_t D_t F(y_1, \dots, y_{t-1}, x_t, x_{t+1}, \dots, x_n)$ has degree at most k. As a result, we can apply a hypercontractivity result due to [Bonami, 1970] to obtain

$$\mathbb{E}\left[(\Delta_t x_t)^4\right] \leq 9^k\mathbb{E}\left[(\Delta_t x_t)^2\right]^2. \tag{5.10}$$

For brevity, we omit the proof of this fact, but it holds whenever a polynomial of degree at most k is applied to a sequence of independent random variables satisfying 5.1. Furthermore, by the same assumptions,

$$\mathbb{E}\left[(\Delta_t x_t)^2\right] = \sum_{S\subseteq[n]} x_t\hat{D}_t F(S)^2 = \sum_{S\ni t}\hat{F}(S)^2 = \mathbf{Inf}_t[F]. \tag{5.11}$$

We will not elaborate on this result either, but it follows from the fact that, under 5.1, the monomials $r_S(x) = \prod_{i\in S} x_i$ are orthonormal. Combining (5.10) and (5.11), we have $\mathbb{E}\left[(\Delta_t x_t)^4\right] \leq 9^k\mathbf{Inf}_t[F]^2$ as desired. The argument for $\mathbb{E}\left[(\Delta_t Y_t)^4\right]$ is identical.

The most notable qualitative difference between the bound offered by the Berry–Esseen theorem and the bound offered by the BIP is that the former depends on the input distribution and the function, whereas the latter only depends on the function. In particular, the BIP, as we have given it, can be interpreted as stating that low-degree and low-influence functions are invariant to the choice of input distribution. Although useful, this should nonetheless be too surprising – if the function were dramatically dependent on a few of its inputs or if there were lots of interaction between the inputs, then the details of the input distribution would likely have a larger effect on the output distribution. Furthermore, this version of the BIP has strong assumptions input distribution. The versions with weaker assumptions do have dependencies on those distributions as well as on the function [Mossel et al., 2005].

5.4 Examples and Applications

The invariance principles discussed in this chapter have found applications in fields such as statistics, quality control and manufacturing, finance, investing, and medicine. The CLT has been extensively utilized for estimating unknown distribution parameters using sample statistics. The Berry–Esseen theorem has been applied to problems of statistical inference to assess convergence behavior and model evaluation. The BIP has found applications in theoretical computer science and social choice theory. Here, we present a toy example showing the kind of characterization offered by the BIP. Following that, we will explore how the BIP has recently been utilized in characterizing communication channels.

Example 5.1 Consider the following multilinear function with degree $k = 3$,

$$F(X_1, \dots, X_n) = \sum_{i=1}^{n-2} \frac{1}{n} X_i X_{i+1} X_{i+2}.$$

Suppose that we are interested in the measurement function $\psi(a) = a^4/24$, which does indeed have continuous derivatives. Since $\psi'''' = 1$, we will be able to choose $C = 1$ when we apply the BIP. Next, we can compute influences on F:

$$\mathbf{Inf}_i[F] = \sum_{S \ni i} \hat{F}(S)^2 = \begin{cases} 3/n^2, & i \in \{3, \dots, n-2\}, \\ 2/n^2, & i \in \{2, n-1\}, \\ 1/n^2, & i \in \{1, n\}. \end{cases}$$

We now describe two random sequences $\mathbf{X} = (X_1, \dots, X_n)$ and $\mathbf{Y} = (Y_1, \dots, Y_n)$, both of which are i.i.d. sequences. In particular, suppose that each random variable X_i is drawn from a Gaussian distribution with mean μ and variance $\sigma^2 = 0.75$, where μ itself is a binary random variable taking values in $\{-0.5, 0.5\}$ uniformly. Each random variable Y_i will be drawn uniformly from a binary distribution taking

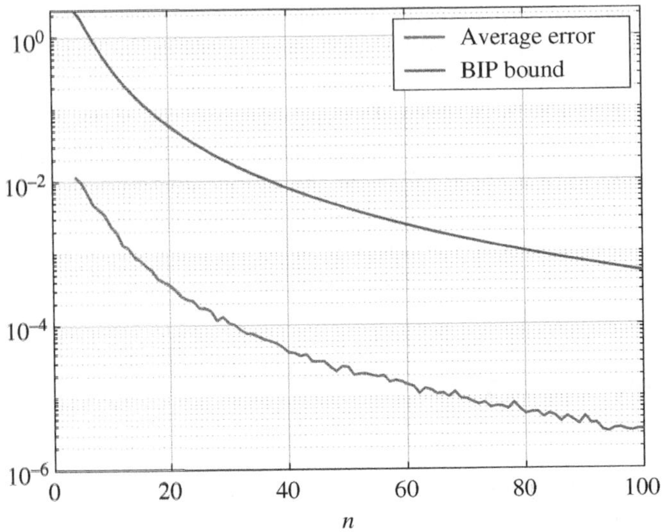

Figure 5.1 Evaluation of average replacement error, obtained empirically and using BIP bound.

values in $\{-1, 1\}$. It is straightforward to show that $E[X_i] = E[X_i^3] = 0$, $E[X_i^2] = 1$, and $E[X_i^4] = 2.875$. Furthermore, $E[Y_i] = E[Y_i^3] = 0$, and $E[Y_i^2] = E[Y_i^4] = 1$. The BIP then implies that (for simplicity, we assume $n > 3$),

$$|\mathbb{E}[\psi(F(\mathbf{X}))] - \mathbb{E}[\psi(F(\mathbf{Y}))]| \leq \frac{9^3}{12}\left((n-4)\frac{9}{n^4} + 2\frac{4}{n^4} + 2\frac{1}{n^4}\right).$$

Figure 5.1 compares this bound to the empirical difference observed from computing $\psi(F(\cdot))$ over 10^5 samples of $\mathbf{X}$ and $\mathbf{Y}$.

5.4.1 The Computational Wiretap Channel

An important part of semantic communication is information security. A situation which often arises is that certain features of a transmitted message need to be protected against unauthorized access, while other, non-sensitive information still needs to be delivered to authorized users.

An example of this setting is the wiretap channel [Wyner, 1975], shown in Figure 5.2a, in which the communication link between the sender, Alice, and an adversary, Eve, is noisier than the one between the sender and the intended receiver, Bob. A variant of this setup is the computational wiretap channel, shown in Figure 5.2b in which the adversary and the receiver both intercept the same transmission from the sender without any noise, but each party is interested in a different function of the information in the message. The goal is then for the

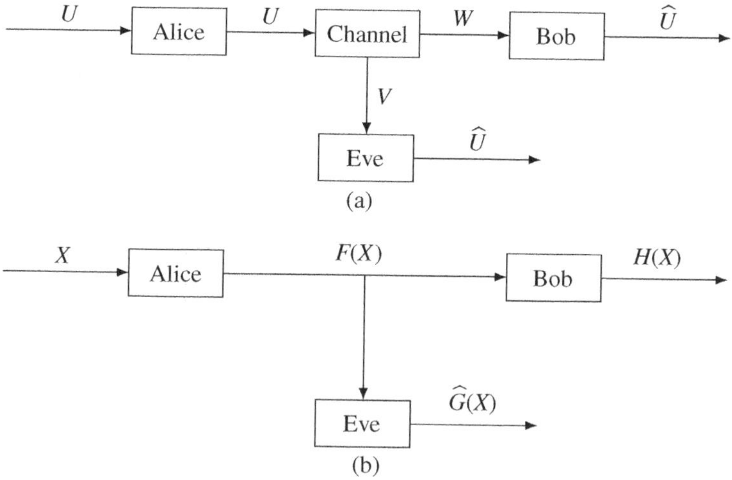

Figure 5.2 The computational versus the classic wiretap channel. Under certain conditions on F and G, the computational wiretap channel can be approximated as a classic wiretap channel from Eve's point of view. (a) The Classic Wiretap Channel: Alice sends a message U through a channel which outputs possibly distinct noisy versions W to Bob and V to Eve, Bob and Eve both compute an estimate $\widehat{U}$ of U. (b) The Computational Wiretap Channel: Alice sends a computation $F(X)$ of her data, where F is a sufficient statistic for H. Bob is able to perfectly recover $H(X)$ and Eve can only compute an estimate $\widehat{G}(X)$ of $G(X)$.

sender to transmit a function of its information, which is a sufficient statistic for receiver's function but insufficient for the adversary.

It was shown in D'Oliveira et al. [2018] that the classic wiretap channel and computational wiretap channel are equivalent using invariance principles. In particular, the computational wiretap can be modeled as a classic wiretap channel where the adversary observes a noisier version of the data observed by the receiver. When the functions used in the computational wiretap channel are low-influence functions and the information being sent over the channel satisfies 5.1, the BIP states that one can effectively treat this modeled noise as independent of the transmitted information, even though it is not, strictly speaking, independent. From this perspective, both channels are equivalent for Eve, as she receives a version of her quantity of interest, which has been corrupted by independent noise.

For example, let $U = G(\mathbf{X})$ be the input to the classic wiretap channel and let $N(\mathbf{X}) = F(\mathbf{X}) - G(\mathbf{X})$ define the noise applied by the channel. Then, for both the classic and computational wiretap channels, the adversary sees $U + N(\mathbf{X})$ and she estimates U given $U + N(\mathbf{X})$. The BIP is used to show that $N(\mathbf{X})$ can be modeled by a function, which is independent of $\mathbf{X}$. In particular, if $\mathbf{X}$ is a uniformly distributed binary sequence, it can be replaced by a sequence of independent standard normal

Gaussian random variables **Z**. Then, via a corollary of the BIP (which we have not explicitly stated here),

$$|\mathbb{E}[\psi N(\mathbf{X})] - \mathbb{E}[\psi N(\mathbf{Z})]| \le \frac{C}{3} k 9^k \epsilon,$$

where k depends on the degrees of the functions F and G in the computational wiretap channel, and ϵ is an upper bound on the influences of those functions. This can be interpreted as saying that, for many reasonable ways of measuring the similarity between the distribution of truly independent noise and this "model" noise that we have defined for the computational wiretap, the two are very similar when F and G are low-degree, low-influence functions.

5.5 Further Generalizations

The version of the BIP that we gave above is, as its name suggests, only the simplest of a family of similar results. When it originally appeared in Mossel et al. [2005], multiple versions were given, with varying assumptions on the distributions of the random variables in question. In O'Donnell [2014a], multiple corollaries are listed, such as results relaxing the $\mathscr{C}^4$ assumption to a Lipschitz condition, giving alternative dependencies in the upper bound, and even a direct generalization of the Berry–Esseen theorem, which bounds the CDF-closeness.

For the purposes of illustrating one way that the BIP can be generalized to a kind of "multivariate" case, in which two sequences of random variables are replaced with two different sequences of random variables, we highlight here a bivariate invariance principle for a setting in which it gives a better bound than the BIP.

Theorem 5.5 ***(Mariona et al., 2023)*** Let F be a multilinear polynomial of the form $F(x_1, x_2) = f(x_1) + g(x_2) + h(x_1 x_2)$, where f, g, and h are each multilinear polynomials of degree at most k. Let x_1, x_2, y_1, and y_2 be n-length sequences of independent random variables satisfying 5.1. Let $\psi : \mathbb{R} \to \mathbb{R}$ be $\mathscr{C}^4$ with $||\psi''''||_\infty \le C$. Then,

$$\left|\mathbb{E}\left[\psi(F(x_1, x_2))\right] - \mathbb{E}\left[\psi(F(y_1, y_2))\right]\right| \le \frac{2C}{3} \cdot 9^k \cdot \sum_{i=1}^{n} \left(\mathbf{Inf}_i[f]^2 + \mathbf{Inf}_i[g]^2 + 2\mathbf{Inf}_i[h]^2\right).$$

The proof of this bivariate invariance principle follows from separating out the dependency of F on x_1 and x_2 and treating the resulting functions as usual, multilinear polynomials, but with random coefficients. Then, using the same structure as in the proof of the BIP with the intermediate tools adapted to these random functions, we obtain analogous results to the BIP in terms of the *expected* influences of the random functions. When F is of the form given in 5.5, those expected influences are easily computed. For these functions, this result is asymptotically tighter than the BIP by a factor of 9^k.

As another example of the kind of results that the BIP and the techniques used to prove it can lead to, we offer below an invariance principle for the TV distance. Although the CLT states that the sample mean distribution converges to that of a normal distribution, this is not the same as stating that the two distributions converge with respect to all statistical metrics, such as the TV distance. It turns out that convergence of the sample mean to a normal random variable in TV distance can be shown, under sufficient conditions. Following the BIP, this was extended to a result on multilinear polynomials, which we state informally for illustrative purposes.

Theorem 5.6 ***(Nourdin and Poly, 2015, informal)*** Let F be a sufficiently low-influence multilinear polynomial and let $\mathbf{X} = (X_1, \dots, X_n)$ and $\mathbf{Y} = (Y_1, \dots, Y_n)$ be sequences of independent random variables with sufficiently reasonable distributions. Then, $\lim_{n\to\infty} d_{TV}(F(\mathbf{X}), F(\mathbf{Y})) = 0$.

Bibliography

Andrew C. Berry. The accuracy of the Gaussian approximation to the sum of independent variates. *Transactions of the American Mathematical Society*, 49(1):122—136, 1941.

Aline Bonami. Étude des coefficients de Fourier des fonctions de $L^p(G)$. *Annales de l'institut Fourier*, 20:335—402, 1970.

Thomas M. Cover and Joy A. Thomas. *Elements of Information Theory*. John Wiley & Sons, 2006.

Rafael G. L. D'Oliveira, Salim El Rouayheb, and Muriel Médard. The computational wiretap channel. In *56th Annual Allerton Conference on Communication, Control, and Computing*, pages 1136—1140, 2018.

Aryeh Dvoretzky. Central limit theorems for dependent random variables. In *Proceedings of the 6th Berkeley Symposium on Mathematical Statistics and Probability*, volume 2, pages 513—535, 1972.

Carl-Gustav Esseen. On the Liapounoff limit of error in the theory of probability. *Arkiv för Matematik, Astronomi och Fysik*, 28(9):1—19, 1942.

Carl-Gustav Esseen. A moment inequality with an application to the central limit theorem. *Scandinavian Actuarial Journal*, 1956(2):160—170, 1956.

Alan Hájek. Interpretations of probability. In Edward N. Zalta and Uri Nodelman, editors, *The Stanford Encyclopedia of Philosophy*. Metaphysics Research Lab, Stanford University, 2023.

Davar Khoshnevisan. *Probability*. American Mathematical Society, 2007.

Jarl Waldemar Lindeberg. Eine neue Herleitung des Exponentialgesetzes in der Wahrscheinlichkeitsrechnung. *Mathematische Zeitschrift*, 15(1):211—225, 1922.

Alexander Mariona, Homa Esfahanizadeh, Rafael G. L. D'Oliveira, and Muriel Médard. A non-asymptotic analysis of mismatched guesswork. In *2023 IEEE International Symposium on Information Theory*, pages 2505—2510, 2023.

Elchanan Mossel, Ryan O'Donnell, and Krzysztof Oleszkiewicz. Noise stability of functions with low influences: Invariance and optimality. In *46th Annual IEEE Symposium on Foundations of Computer Science*, pages 21—30, 2005.

Ivan Nourdin and Guillaume Poly. An invariance principle under the total variation distance. *Stochastic Processes and their Applications*, 125(6):2190—2205, 2015.

Ivan Nourdin and Murad S. Taqqu. Central and non-central limit theorems in a free probability setting. *Journal of Theoretical Probability*, 27:220—248, 2014.

Ryan O'Donnell. *Analysis of Boolean Functions*. Cambridge University Press, 2014a.

Ryan O'Donnell. Social choice, computational complexity, Gaussian geometry, and Boolean functions. arXiv preprint arXiv:1407.7763, 2014b.

Lionel S. Penrose. The elementary statistics of majority voting. *Journal of the Royal Statistical Society*, 109(1):53—57, 1946.

Hans Reichenbach. *The Theory of Probability*. University of California Press, 1971.

Irina Shevtsova. On the absolute constants in the Berry-Esseen type inequalities for identically distributed summands, 2011.

Charles Stein. A bound for the error in the normal approximation to the distribution of a sum of dependent random variables. In *Proceedings of the 6th Berkeley Symposium on Mathematical Statistics and Probability*, volume 2, pages 583—603, 1972.

Charles Stein. Approximate computation of expectations. In *Institute of Mathematical Statistics Lecture Notes*. Institute of Mathematical Statistics, 1986.

Zbigniew Szewczak and Michel Weber. Classical and almost sure local limit theorems, 2022.

Aaron D. Wyner. The wire-tap channel. *The Bell System Technical Journal*, 54(8):1355—1387, 1975.

6

Interoperability and Coexistence of 6G Semantic, Goal-Oriented, and Legacy Systems

Emilio Calvanese Strinati, Mohamed Sana, Mattia Merluzzi, and Tomás Huttebraucker

CEA-Leti, Université Grenoble Alpes, Grenoble, France

6.1 Introduction

At present, artificial intelligence (AI) and 5G systems are designed, developed, and deployed as distinct entities, yet they critically depend on each other for long-term sustainable success. Communication systems are instrumental to the establishment of AI as a cornerstone of our society. Massive data are collected, communicated, and processed in distributed locations, then fed to (edge) AI reasoning engines, thanks to the current and next generation communication systems. In addition, new concepts of *training while communicating* will enable continuous learning functionalities [Strinati and Barbarossa, 2021; Chaccour et al., 2022]. With 6G, new semantic aware communication architecture and protocols will enable efficient AI-native cooperation and distributed operations [Strinati et al., 2024]. On the other hand, emerging (generative) AI and machine learning (ML) methods are expected to enhance 5G/6G functionalities and operations [Bariah et al., 2024] while imposing parsimony to avoid unsustainable use of computing, memory, and energy resources [Strinati et al., 2024]. This heterogeneity of data, intelligence, and resources will ineluctably face a critical issue related to the interoperability and coexistence with other systems, including legacy services.

This issue of interoperability is at the core of 3rd generation partnership project (3GPP) and European Telecommunications Standards Institute (ETSI) standardization bodies, for instance in the discussions related to Open-RAN (O-RAN) [Strinati et al., 2024]. Interoperability poses serious challenges to effective AI

Foundations of Semantic Communication Networks, First Edition.
Edited by Walid Saad, Christina Chaccour, Christo Kurisummoottil Thomas, and Merouane Debbah.

cooperation, especially when cooperating agents are incompatible in their training, neural architectures, and embedded logics used to extract, manipulate, communicate, and interpret information. This creates operational bottlenecks, where theoretical performance diverge from real measured ones. Then, effective interoperability requires: (i) *Standardized formats, protocols, and interfaces* to be compatible to reliably exchange information between AI and 6G systems; (ii) *semantic alignment*, ensuring that the meaning of data, concepts, or terms is consistent and compatible across different AI systems, applications, and domains. This requires adoption of compatible ontology and logic that shape how semantics are associated with data, and knowledge is represented; and (iii) *coexistence mechanisms* to enable autonomous AI agents to concurrently achieve common or divergent objectives.

Besides the challenges of interoperability, the coexistence of legacy and 6G semantic, goal-oriented systems hinges on establishing standardized protocols and interfaces to facilitate seamless communication and information exchange. Additionally, for cohesive operation, effective sharing of communication and computation resources among the systems is essential. Consequently, a trade-off emerges between *goal-effectiveness* and *goal cost*, shaping the strategies for the coexistence of heterogeneous systems [Merluzzi et al., 2024].

This chapter focuses on challenges and solutions in the state of the art for the interoperability faced by agents communicating through semantic strategies, and the coexistence between legacy (*i.e.*, data-oriented) and 6G semantic, goal-oriented (*i.e.*, semantic-oriented) services.

6.2 Interoperability Issue in Goal-oriented and Semantic Systems

6G networks are set to become a world of connected and generalized intelligence, requiring efficient interaction with each other to support efficient operations or to complete a variety of tasks as illustrated in Figure 6.1. Such interactions will critically face the major challenge of interoperability. It is the ability of different systems, devices, or applications to seamlessly work together, communicate, and exchange reliable and meaningful information. In the context of AI, interoperable AI models or systems can understand the same data and concepts in the same way. They can cooperate with each other through the exchange of semantics to effectively solve complex tasks. However, achieving interoperability is a challenging task. In general, a common or shared language (either natural or artificial) is assumed between interacting systems. Such a language can be learned a priori, e.g. , via a (supervised) learning procedure. Common language can also emerge from an interactive consensus based on, e.g., reinforcement learning (RL) to

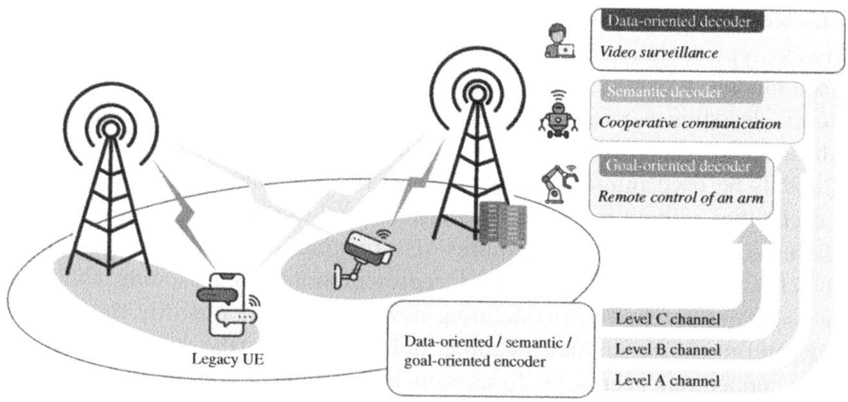

Figure 6.1 Coexistence and interoperability between heterogeneous agents, sharing the same spectrum resource in a 6G network.

establish a communication protocol (an effective language to attain a goal-driven cooperation) [Farshbafan et al., 2023]. This consensus requires communication attempts, which incurs errors and communication costs [Foerster et al., 2016; Ding et al., 2020]. Alternatively, a common language can be suggested by the environment, where agents ground their language in representations of the observed world [Lin et al., 2021].

The crucial role of languages in semantic communications is often overlooked. When this is not the case, agent languages are assumed compatible and unambiguously interoperable, ignoring practical limitations that may arise due to language mismatching [Sana and Strinati, 2023]. In practice, distributed systems maintain different models employing different languages as a result of a training procedure based on a private dataset. In addition, even if two AI agents are trained on the same data, they might not be able to "talk" to each other due to different training conditions (random weights initialization, training hyper-parameters, etc.), which entail different logic (internal representations) in learned languages [Moschella et al., 2022]. However, when agents employ distinct languages (a different set of rules), Sana and Strinati [2022] show that communication is prone to semantic noise that might stimulate interpretation errors, causing defective cooperation strategies [Wang et al., 2020], which need to be circumvented.

To understand how models can be made compatible, Lenc and Vedaldi [2015] introduce *model stitching*, which allows studying the internal representations of neural networks (NN). Stitching layers allow swapping parts of different models to make them compatible. Similarly, Gygli et al. [2021] propose a mechanism, which directly learns compatible and reusable model components. Such an approach assumes availability of the data and requires a new learning procedure. To tackle this problem, Moschella et al. [2022] introduce *relative representation* as an alternative representation of latent space, which enforces *zero-shot* compatibility between models, providing a way to explain and compare models representation. However, decoding a relative representation also requires a *relative decoder*, which needs an additional expensive training procedure. Henceforth, how can two models be used interchangeably? How can they communicate seamlessly with each other without requiring an additional end-to-end learning procedure?

Answering these questions requires (i) studying the impact of language mismatch (which governs knowledge representation and reasoning abilities) in messages interpretation, (ii) defining new measures to quantify the risk of misinterpreting the transmitted meaning, and (iii) modeling the semantic channel between cooperating agents. To do so, Sana and Strinati [2023] propose to translate language mismatch into knowledge representation mismatch, captured by measurable transformations introduced into the semantic representation space. Such transformations can be modeled as transport plans, which can be learned to build a codebook of transformations using methods from optimal transport theory [Perrot et al., 2016]. With the learned codebook of transformations, compensating for language mismatch becomes a problem of identifying and implementing the right transformation during communication.

6.2.1 Language in Multiuser Communication

A good language representation is instrumental in structuring the construction and consolidation of knowledge, influencing conceptual representation and semantic extraction and interpretation.

A *Language generator* is a mapping function $\lambda : \mathcal{M} \rightarrow \mathcal{X}$, possibly parameterized, which endows a message or observation $- \in \mathcal{M}$ of the world with semantic meaning $x \in \mathcal{X}$. In the context of AI, a language generator defines a way to partition a semantic space and associates to each partition a semantic meaning. For instance, let $P = \{P_1, P_2, \dots\}$ be a partition of the probability space $(\mathcal{X}, \mathfrak{B}, \mu)$, where μ is a probability measure over $\mathcal{X}$ and $\mathfrak{B}$ is a σ-algebra on $\mathcal{X}$ (*i.e.*, a non-empty collection of subsets of $\mathcal{X}$ closed under complement, countable unions and intersections). We will often think of a partition of $\mathcal{X}$ as a list of measurable sets that cover $\mathcal{X}$. We refer to P_i as the ith atom of the partition P. Thus, $\mu(P_i)$ is the probability measure of atom P_i. An atom may capture a certain level of semantics. For instance, for an image classification task, an atom may be related to the features (semantic symbols) associated to a certain class of an image. Different languages may partition the semantic space differently. Also, different partitioning of the semantic space captures different level of semantics. Depending on the meaning, the transmitter assigns to each transmitted data (the atom to which – belongs to), generated semantic symbols get interpreted by the language interpreter using a mapping function, $\gamma : \mathcal{X} \rightarrow \mathcal{A}$. Here, $\mathcal{A}$ defines the target message space, which may differ from $\mathcal{M}$. For example, $\mathcal{A}$ may define the set of classes or actions associated with each message. Hence, $(\mathcal{M}, \lambda, \gamma, \mathcal{A}, \mathcal{X}, \mathfrak{B}, \mu)$ fully defines a language, whose expressivity is crucial for effective communication. Indeed, language may be limited: Either (i) not all messages of the world have a meaning in this language or (ii) can be efficiently represented by the associated language generator, or (iii) the meaning accurately recovered by the language interpreter.

6.2.2 A Measure of Semantic Mismatch

Consider communication between TX k and RX l. Let $P = \{P_1, P_2, \dots\}$ and $Q = \{Q_1, Q_2, \dots\}$ be the two finite partitions of the probability space $(\mathcal{X}, \mathfrak{B}, \mu)$ defined by their language generators λ_k and λ_l, respectively. Given a message $- \in \mathcal{M}$, TX k encodes and transmits the associated meaning $x_k = \lambda_k(-) \in P_i$ to RX l, which interprets x_k in its own language as $\gamma_l(x_k)$. However, in RX l language, data "–" maps to $y_l = \lambda_l(-) \in Q_j$. Thus, the atom P_i in TX k language is equivalent to atom Q_j in RX l language. As semantics are associated with atoms of partitions, RX l may misinterpret received symbol x_k, especially when $x_k \notin Q_j$. More specifically, when TX k communicates with RX l, only the observations that get mapped into $P_i \cap Q_j$ can be correctly interpreted in RX l language. Therefore, a mismatch of language translates to a mismatch of atoms of language partitions. Hereafter, we propose to model theses mismatches with *measurable transformations.*

Definition 6.1 Let $(\mathcal{X}, \mathfrak{B})$ be a measurable space where $\mathfrak{B}$ denotes a σ-algebra on $\mathcal{X}$. A function $T : \mathcal{X} \rightarrow \mathcal{X}$ is said to be a measurable transformation if $\forall B \in \mathfrak{B}$, **the preimage of** B **under** T **denoted** $T^{-1}(B)$ is in $\mathfrak{B}$; *i.e.*,

$$T^{-1}(B) := \{x \in \mathcal{X} \mid T(x) \in B\} \in \mathfrak{B}. \tag{6.1}$$

Let $P = \{P_1, P_2, \dots\}$ and $Q = \{Q_1, Q_2, \dots\}$ define two finite partitions of a probability space $(\mathcal{X}, \mathfrak{B}, \mu)$ such that $\mu(P_i) > 0$, $\forall P_i \in P$. Let $T \in \mathcal{T}$ be a measurable transformation. We define, $\forall\ (P_i, Q_j) \in P \times Q$:

$$\rho_{P_i \rightarrow Q_j}(T) = \frac{\mu(T^{-1}(Q_j) \cap P_i)}{\mu(P_i)}. \tag{6.2}$$

Intuitively, $\rho_{P_i \rightarrow Q_j}(T)$ measures the amount of information being transported from P_i to Q_j under one iterate of T [Sinha et al., 2019]. Since μ is a probability measure, we have $0 \leq \mu(T^{-1}(Q_j) \cap P_i) \leq \mu(P_i)$, thus, $\rho_{P_i \rightarrow Q_j}(T) \in [0, 1]$.

To align atom P_i with atom Q_j, it is sufficient to find a transformation T that transports $x \in P_i$ into its equivalent semantic atom $T(x) \in Q_j$:

$$\underset{T \in \mathcal{T}}{\arg\max}\ \rho_{P_i \rightarrow Q_j}(T). \tag{6.3}$$

Transformation T can be viewed as a language "translator" bridging the semantic gap between the source and target atom.

Definition 6.2 ***(Language mismatch)*** Let J_P and J_Q define finite index set of two language partitions $P = \{P_1, P_2, \dots\}$ and $Q = \{Q_1, Q_2, \dots\}$, respectively. Let $\kappa : J_P \rightarrow J_Q$ define a function, which associates each atom $i \in J_P$ of partition P to the corresponding semantic atom $j \in \kappa(i) \subset J_Q$ in partition Q. Here, $\kappa(\cdot)$ can but is not required to be a one-to-one mapping. It can also be a many-to-one (*e.g.*, the digits $\{1, 3, 5, 7, 9\}$ are all odd numbers) or one-to-many mapping (*e.g.*, an even digit can be $\{0, 2, 4, 6, 8\}$). Let $T \in \mathcal{T}$ be a measurable transformation. If the source emits a message m with a probability $p_{\mathcal{M}}(m)$, the a priori (before syntactic channel) risk of misinterpretation at the RX by the target language when we employ the transformation T for compensating the semantic channel to transmit data m reads as

$$\mathcal{E}_T(-) = 1 - \sum_{i \in J_P} \sum_{j \in \kappa(i)} \mu(P_i|-) \rho_{P_i \rightarrow Q_j}(T). \tag{6.4}$$

Here, $\mu(P_i|m)$ is the probability that the semantic symbol $x = \lambda(-)$ associated to data m belongs to atom P_i of the source language partition P. Thus, $\bar{\mathcal{E}}_T = \mathbb{E}_{m \sim p_{\mathcal{M}}(m)}[\mathcal{E}_T(-)]$ measures the average mismatch of source and target language when we employ transformation T. In particular, when T is the identity function, $\bar{\mathcal{E}}_T$ translates to the average mismatch between source and target language. In Eq. (6.4), the correspondence mapping κ is not always known. Instead, a goal-driven measure $\omega : P \times Q \rightarrow [0, 1]$, either deterministic or probabilistic may be available or estimated, such that $\omega(P_i, Q_j)$ quantifies how good[1] it is to map P_i into Q_j to convey the desired meaning associated to data m to the RX. In this case, we can rewrite Eq. (6.4) as

$$\mathcal{E}_T(-) = 1 - \sum_{i \in J_P} \sum_{j \in J_Q} \mu(P_i|-) \omega(P_i, Q_j) \rho_{P_i \rightarrow Q_j}(T). \tag{6.5}$$

Thus, to minimize the risk of misinterpretation by the RX of the meaning intended by the TX, either the RX manages to compensate for transformations introduced by the semantic channel (postequalization), or the TX intentionally distorts intended meaning by selecting another symbol (preequalization).

6.2.3 Semantic Channel Equalization

To equalize the semantic channel between source and destination, one can either (i) learn a single, yet complex transformation $T \in \mathcal{T}$, which minimizes the average

1 For instance, in a RL problem, ω could be the normalized Q-values measure.

risk of misinterpretation $\bar{\mathcal{E}}_T$, or (ii) operate a low-complexity codebook $\mathcal{C} \in \mathcal{T}$ of biased linear transformations obtained by solving Eq. (6.3) for each pair of atoms (P_i, Q_j). Such transformations can be modeled as transport plans and learned using optimal transport theory as in Sana and Strinati [2023]. Then, compensating for language mismatch becomes a problem of identifying and implementing the right transformation $T \in \mathcal{C}$ that minimizes $\mathcal{E}_T(m)$ for transmitted data m.

Example 6.1 ***(The scout and the treasure)***
A treasure and a scout are randomly placed in a discrete grid. At each time step, a controller observes the state of the environment and transmits a semantic symbol $x \in \mathcal{X} = \mathbb{R}^2$ through a Gaussian noisy channel. A scout receives the noisy symbol and takes one of four actions $\mathcal{A} = \{\text{right, down, left, up}\}$. The episode ends when the scout reaches the treasure or the maximum number of steps is attained.

Figure 6.2 shows the partition of the semantic space for the controller and the scout languages learned under different training conditions (random weight initialization) using RL [Huttebraucker et al., 2024]. As we can see, the two languages partition the semantic space differently. Such misalignment stimulates misinterpretation error if the controller naively communicates with the scout without any preequalization, thus leading to goal failure as shown in Figure 6.3. In contrast, when the semantic channel is equalized using the approach proposed in Sana and Strinati [2023] to minimize Eq. (6.4), the performance of the system substantially improves, as the episode length – the time taken to

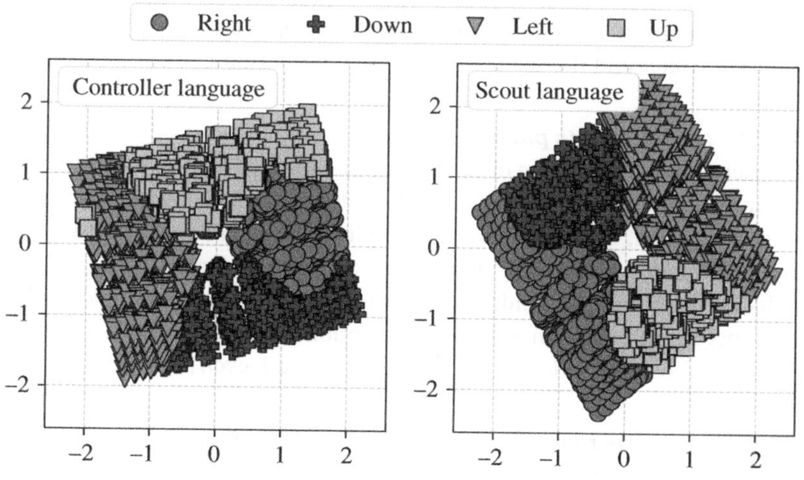

Figure 6.2 Latent space representations of the controller and the scout languages in Example 6.1.

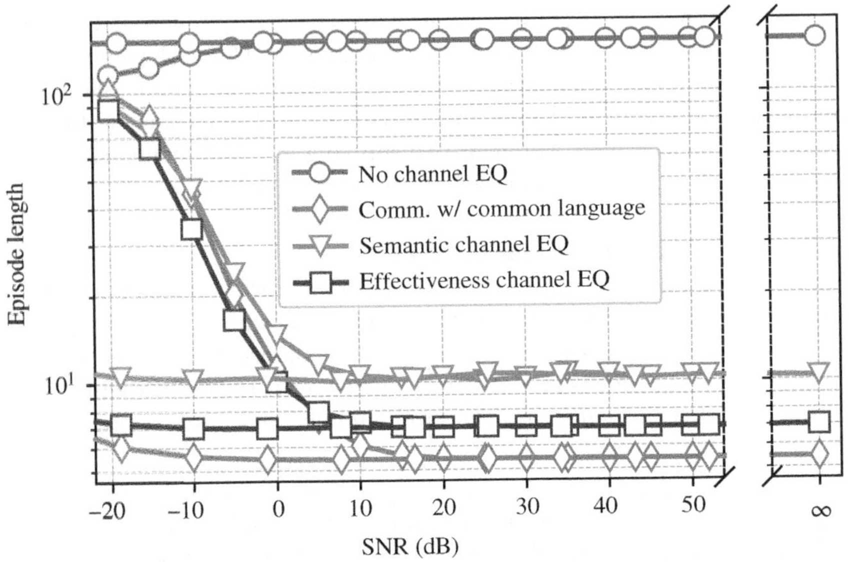

Figure 6.3 Performance comparison of different equalization techniques.

complete the goal – is divided by 10. However, as the objective is to optimize a goal-driven performance, perfect semantic alignment is not sufficient. Instead, equalizing the effectiveness channel between the controller and the scout by minimizing Eq. (6.5) is more efficient as it directly impacts the goal performance [Huttebraucker et al., 2024]. In this case, the resulting solution approaches the performance of a communication system, which adopts a common language between the controller and the scout with practical limitations.

Example 6.2 ***(From digit to parity)***
A sender observes the pixels value of random color-Modified National Institute of Standards and Technology (MNIST) image [Foerster et al., 2016], while the color label (either red, green, blue, or yellow) and digit value are hidden, and encodes underlying information $x \in \mathcal{X} = \mathbb{R}^2$ for a receiver, which infers the digit on the image or its parity. Here, the sender and the receiver have incompatible goals. The former encodes information in 10 possible options $\mathcal{A} = \{0, 1, \dots, 9\}$, whereas the latter has two possible options for decoding $\mathcal{A} = \{\text{odd}, \text{even}\}$.

Figure 6.5 shows the performance comparison between different transmission approaches. In particular, when the sender and the receiver use a common language, Figure 6.5 shows that the communication is very sensitive to syntactic channel noise exhibiting poor performance for low signal to noise ratio (SNR)

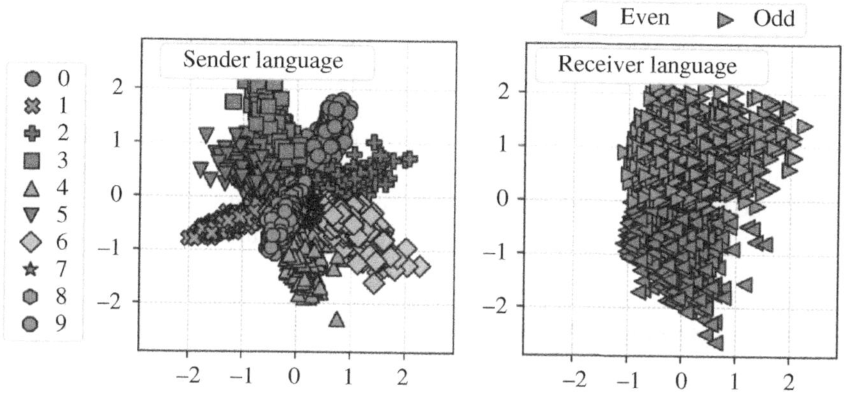

Figure 6.4 Latent space representations of the sender and the receiver languages in Example 6.2.

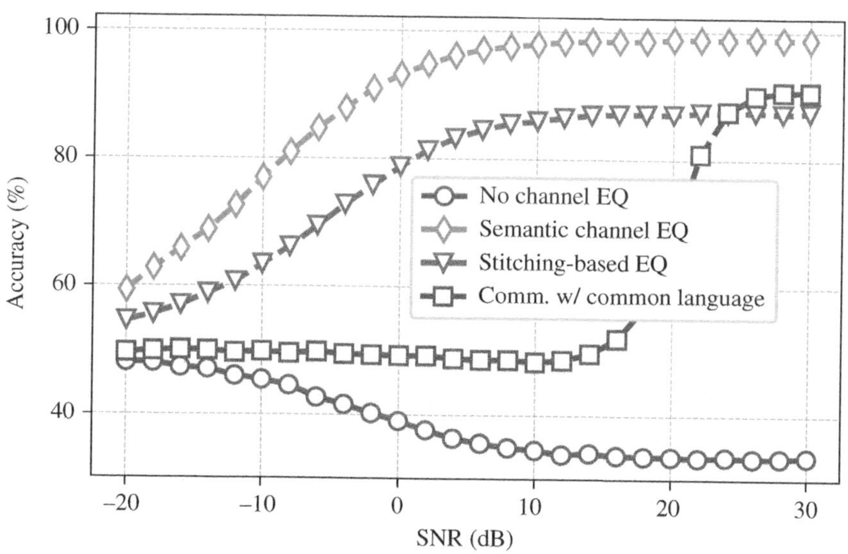

Figure 6.5 Performance comparison of different equalization techniques.

regime. However, even under error-free syntactic channel, a naive communication between the sender and the receiver results in more 65% misinterpretation rate due to language mismatch (see Figure 6.4). This performance is even below the 50% rate expected if the receiver randomly guesses the parity of the digit for a uniformly distributed symbol in received constellation. Thus, language mismatch needs compensation. If a new training procedure is possible, an NN-based stitching layer can be learned to align the sender's encoder with the receiver's decoder, which are kept fixed during the training. Thus, the performance of this method is limited to the expressivity of the sender and receiver languages [Sana and Strinati, 2023]. In contrast, the method proposed in Sana and Strinati [2023] does not require a complex additional training procedure. It uses a codebook of linear transformations, operating over the latent semantic space to compensate for language mismatch, exhibiting 30% improvement in interpretation accuracy compared to the solution employing a stitching layer (Figure 6.5).

6.3 Coexistence of Semantic, Goal-Oriented, and Legacy Services in 6G

6G will have to accommodate several services with extremely diverse requirements on the same network infrastructure, due to the ever-increasing demand for flexibility, frugality and resource re-use, needed for the fulfillment of the sustainable development goals [Le Blanc, 2015]. To do so, the integration of semantic and goal-oriented services into wireless communication networks cannot correspond to a complete rethinking of network infrastructures, but rather to the exploitation of the already existing network as much as possible.

This rises new challenges related to the coexistence of semantic, goal-oriented, and legacy (data-oriented) communication services, sharing the same physical resources (hardware, bandwidth, etc.). This coexistence comes with two mutually contradicting facets. On the one hand, new semantic and goal-oriented services will demand network resources, whose utilization could negatively affect the performance of existing legacy services. Also, the message format and communication protocols might potentially need to be revisited with respect to today's standardized solutions. On the other hand, a beneficial effect will certainly arise from the semantic approach, thanks to a more efficient use of resources, or, in other words, the possibility of reducing resource demand while not necessarily sacrificing performance in terms of goal achievement (effectiveness). In this section, we focus on this second aspect, and in particular on the potentials of *goal-oriented resource allocation*. A few works started investigating this topic. In particular work of, Mu and Liu [2023] delves into the innovative approach of exploiting semantic communication techniques to enhance the efficiency of non-orthogonal multiple access (NOMA) in wireless networks. By incorporating semantic information into the communication process, such as context or user intent, NOMA systems can achieve improved spectral efficiency and better resource allocation, leading to enhanced performance and user experience in wireless communication environments. A similar approach is proposed in Mu and Liu [2022], in which the authors propose to integrate semi-non-orthogonal multiple access (Semi-NOMA) with semantic and bit communications. This approach enables the simultaneous transmission of semantic information alongside traditional bit-level data. Finally, in Merluzzi et al. [2024], we investigated the exploitation of the goal-oriented approach in a scenario in which a goal-oriented and a legacy (data-oriented) communication system access different access points but share the same spectrum resources, *i.e.*, they interfere with each other, as schematically illustrated in Figure 6.1. However, unlike legacy systems, goal-oriented and semantic communication is generally designed to be more robust to noise (and interference), so that better spectrum efficiency is achieved via an effective interference management.

6.3.1 Goal-Oriented Resource Allocation

In the considered scenario, a legacy user aims to upload data to an access point with maximum data rate, with full buffer assumption. At the same time, a *goal-oriented user (GO user)* uploads data to an access point to absolve a task of targeted *goal-effectiveness*. The latter is defined as the probability of fulfilling a predefined set of (beyond communication) requirements related to the task. A task can generally involve control actions (e.g., Example 6.1 in Section 6.2.3), inference on data collected by the device (e.g., Example 6.2 in Section 6.2.3), etc. Inspired by cognitive radio settings, in which an interference temperature is set as a constraint while maximizing performance of an opportunistic user jumping into a licensed spectrum, we proposed a long-term formulation in which the main constraint is set on the goal-effectiveness of the GO user. The difference with respect to a legacy system is that *no communication-related key performance indicator (KPI) is set a priori*, but resource allocation and interference tolerance are rather *learned and adapted based on goal fulfillment*. The aim is then to optimize resource allocation to maximize the data rate of the legacy user and, at the same time, guarantee a target goal-effectiveness to the GO user. This paradigm of *goal-oriented interference management* can be applied to several use cases and scenarios, including the *scout and treasure* problem (Example 6.1 introduced in Section 6.2.3). In fact, in the latter, not only the semantic information can be extracted and transmitted to the scout, but wireless resources can be allocated for communication to affect its performance in terms of treasure hunting, as well as the performance of other users in the network, being them GO or legacy users.

6.3.2 Goal-Driven Measures for Edge Inference

In this section, as a goal/task, we consider the *timely and confident edge inference* on data uploaded by the GO user. In particular, denoting by $\boldsymbol{\Psi}_t$ the vector of allocated resources in the network (bandwidth, transmit power, etc.) at time t, and by

$\boldsymbol{\Omega}_t$ the vector containing the current network states (wireless channels, computing resource availability, etc.), the delay for uploading data is generally $D_t = D(\boldsymbol{\Psi}_t, \boldsymbol{\Omega}_t)$. The inference confidence is also dependent on network states and resource allocation (both affecting message distortion). Here, we focus on classification, and as a measure of confidence, we exploit to entropy H_t at the output of the classifier. The lower is the entropy, the higher is the confidence. Then, overall, the goal (timely and confident inference) is achieved if both the events $\{D_t \leq D_{max}\}$ and $\{H_t \leq H_{max}\}$ occur at the same time. Finally, the goal-effectiveness is defined as follows:

$$\mathcal{G} = \lim_{T\to\infty} \frac{1}{T} \sum_{t=0}^{T-1} \mathbb{E}\{\mathbf{1}_{\{D_t \leq D_{max}\}} \times \mathbf{1}_{\{H_t \leq H_{max}\}}\}, \tag{6.6}$$

where $\mathbf{1}_A$ denotes the indicator function of event A. At the same time, we can define the *goal cost* as the relative data rate loss experienced by the legacy user with respect to the case the GO service is not requested (*i.e.*, in the absence of interference). Denoting by $R_{l,t} = R_l(\boldsymbol{\Psi}_t, \boldsymbol{\Omega}_t)$ the data rate of the legacy user at time t, and by $R_{l,max}$ its maximum value, the goal cost reads as follows:

$$C = \lim_{T\to\infty} \frac{1}{T} \sum_{t=0}^{T-1} \mathbb{E}\left\{\frac{R_{l,max} - R_{l,t}}{R_{l,max}}\right\}. \tag{6.7}$$

The objective is then to minimize the goal cost while guaranteeing a defined goal-effectiveness constraint (*i.e.*, $\mathcal{G} \geq \mathcal{G}_{th}$ – with $\mathcal{G}_{th}$ denoting the threshold), by dynamically and adaptively orchestrating wireless resources (*i.e.*, managing interference), however, without setting a priori KPIs for communication. The interested reader is reminded to Merluzzi et al. [2024], which details the solution of the goal-oriented resource management problem, based on ad-hoc defined dynamic Lagrange multipliers that track the behavior of the system in terms of constraint violations (*i.e.*, goal-effectiveness). Consequently, a goal-oriented optimization engine drives the system toward efficient and effective operations, through the balance among goal-effectiveness and goal cost. The performance of the proposed goal-oriented resource allocation are shown in Figure 6.6, in which the goal-effectiveness is plotted as a function of the goal cost, with the horizontal line representing $\mathcal{G}_{th}$. The simulation setting is detailed in Merluzzi et al. [2024, Sec. 5]. Three different entropy thresholds are considered, corresponding to high, medium, and low inference confidence. Also, the results of our proposed goal-oriented resource allocation (solid lines) are compared with two benchmarks: (i) The *Optimal* solution provided by a genie, *i.e.*, with an exhaustive search and a posteriori knowledge on the effect of decisions (dotted lines), and (ii) *BW splitting* solution, which assumes an orthogonal use of resources, in which the bandwidth is split among the two systems so as to guarantee the goal-effectiveness of the gal-oriented user (standalone markers). Obviously the

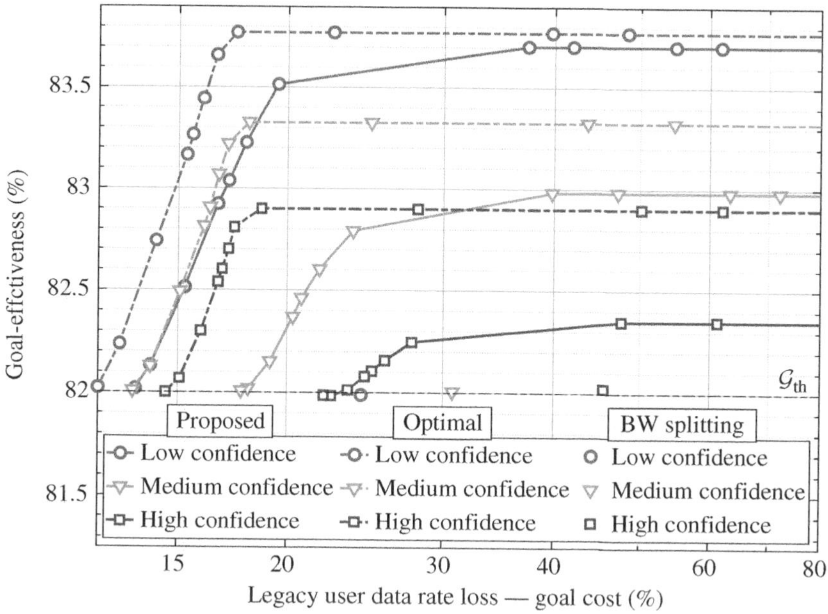

Figure 6.6 Trade-off between legacy user performance and goal-effectiveness.

first approach leads to the best performance as shown in Figure 6.6. However, the second approach, which does not assume any genie-aided information, achieve close goal costs. On the other hand, splitting the bandwidth (*i.e.*, fully avoiding interference) leads to the highest goal cost for the same goal-effectiveness.

6.4 Conclusion

Today, AI and the forthcoming 6G wireless communications are merging to form a unique system. Within this system, data is continually gathered, processed, and exchanged among diverse agents, presenting an opportunity to seamlessly connect various intelligent agents. This facilitates native collaboration on intricate cooperative tasks.

This chapter delves into the complexities of semantic and goal-oriented communication systems. It first discusses the central challenges related to the interoperability issues encountered by agents employing different semantic communication strategies whose languages incompatibility can cause defective cooperation. Then it introduces a recent, effective method for equalizing semantic channels in multiuser semantic communications. This method, known as semantic equalization, addresses semantic mismatches between cooperating agents, offering robust solutions against both syntactic and semantic channel noise. It significantly improves transmission accuracy while reducing complexity compared to traditional approaches. Eventually, state-of-the-art solutions for coexistence are succinctly presented, focusing on the effective sharing of spectrum resources between legacy data-oriented systems and emerging goal-oriented edge intelligence platforms.

To conclude, with an ultimate goal of making 6G a network of connected intelligence through the convergence of communication, computation, control, and learning, it is essential to take into account the issue related to the coexistence and the interoperability of such systems. A consideration of the semantic channel, goal-effectiveness, radio, and computing resources will lay the foundations for efficient 6G networks, supported by effective interaction and knowledge exchange between agents endowed with various forms of intelligence.

Acknowledgment

This work was supported by the EU Horizon programs 6G-GOALS (no. 101139232), the France 2030 ANR program "PEPR Networks of the Future" (ref. 22-PEFT-0010) and by the Marie Skłodowska-Curie ITN Greenedge (no. 953775).

Bibliography

Lina Bariah, Qiyang Zhao, Hang Zou, Yu Tian, Faouzi Bader, and Merouane Debbah. Large generative AI models for telecom: The next big thing? *IEEE Communications Magazine*, 1–7, 2024.

Christina Chaccour, Walid Saad, Merouane Debbah, Zhu Han, and H. Vincent Poor. Less data, more knowledge: Building next generation semantic communication networks. arXiv preprint arXiv:2211.14343v1, 2022.

Ziluo Ding, Tiejun Huang, and Zongqing Lu. Learning individually inferred communication for multi-agent cooperation. In *Advances in Neural Information Processing Systems 33*, pages 22069–22079, 2020.

Mohammad Karimzadeh Farshbafan, Walid Saad, and Merouane Debbah. Curriculum learning for goal-oriented semantic communications with a common language. *IEEE Transactions on Communications*, 71(3):1430–1446, 2023.

Jakob Foerster, Ioannis Alexandros Assael, Nando de Freitas, and Shimon Whiteson. Learning to communicate with deep multi-agent reinforcement learning. In *Advances in Neural Information Processing Systems 29*, 2016.

Michael Gygli, Jasper Uijlings, and Vittorio Ferrari. Towards reusable network components by learning compatible representations. In *Proceedings of the AAAI Conference on Artificial Intelligence*, volume 35, pages 7620–7629, 2021.

Tomás Huttebraucker, Mohamed Sana, and Emilio Calvanese Strinati. Pragmatic goal-oriented communications under semantic-effectiveness channel errors. *Accepted for publication in: IEEE Consumer Communications & Networking Conference*, 2024.

David Le Blanc. Towards integration at last? The sustainable development goals as a network of targets. *Sustainable Development*, 23(3):176–187, 2015.

Karel Lenc and Andrea Vedaldi. Understanding image representations by measuring their equivariance and equivalence. In *Proceedings of IEEE Conference on Computer Vision and Pattern Recognition*, pages 991–999, 2015.

Toru Lin, Jacob Huh, Christopher Stauffer, Ser Nam Lim, and Phillip Isola. Learning to ground multi-agent communication with autoencoders. In *Advances in Neural Information Processing Systems 34*, pages 15230–15242, 2021.

Mattia Merluzzi, Miltiadis C. Filippou, Leonardo Gomes Baltar, Markus Dominik Mueck, and Emilio Calvanese Strinati. 6G goal-oriented communications: How to coexist with legacy systems? *Telecom*, 5(1):65–97, Jan. 2024. ISSN 2673-4001.

Luca Moschella, Valentino Maiorca, Marco Fumero, Antonio Norelli, Francesco Locatello, and Emanuele Rodola. Relative representations enable zero-shot latent space communication. arXiv preprint arXiv:2209.15430, 2022.

Xidong Mu and Yuanwei Liu. Semi-NOMA enabled coexisting semantic and bit communications. In *Proceedings of International Symposium on Wireless Communication Systems (ISWCS)*, pages 1–6, 2022.

Xidong Mu and Yuanwei Liu. Exploiting semantic communication for non-orthogonal multiple access. *IEEE Journal on Selected Areas in Communications*, 41(8):2563–2576, 2023.

Michaël Perrot, Nicolas Courty, Rémi Flamary, and Amaury Habrard. Mapping estimation for discrete optimal transport. In *Advances in Neural Information Processing Systems 29*, 2016.

Mohamed Sana and E. Calvanese Strinati. Learning semantics: An opportunity for effective 6G communications. In *Proceedings of Consumer Communications & Networking Conference (CCNC)*, pages 631–636, 2022.

Mohamed Sana and Emilio Calvanese Strinati. Semantic channel equalizer: Modelling language mismatch in multi-user semantic communications. arXiv preprint arXiv:2308.03789, 2023.

Subhrajit Sinha, Umesh Vaidya, and E. Yeung. On information transfer in dynamical systems with applications in control of non-equilibrium dynamics. In *Proceedings of Indian Control Conference (ICC)*, pages 326–331, 2019.

E. Calvanese Strinati and S. Barbarossa. 6G networks: Beyond Shannon towards semantic and goal-oriented communications. *Computer Networks*, 190:107930, 2021.

Emilio Calvanese Strinati, Paolo Di Lorenzo, Vincenzo Sciancalepore, Adnan Aijaz, Marios Kountouris, Deniz Gündüz, Petar Popovski, Mohamed Sana, Photios A. Stavrou, Beatriz Soret, Nicola Cordeschi, Simone Scardapane, Mattia Merluzzi, Lanfranco Zanzi, Mauro Boldi Renato, Tony Quek, and Nicola di Pietro. Goal-oriented and semantic communication in 6G AI-native networks: The 6G-GOALS approach. arXiv preprint arXiv:2402.07573, Feb. 2024.

Pei Wang, Junqi Wang, Pushpi Paranamana, and Patrick Shafto. A mathematical theory of cooperative communication. In *Advances in Neural Information Processing Systems 33*, pages 17582–17593, 2020.

Part II

Semantic Communications Networking

In this section, the communication- and information-theoretic designs discussed in Part I transition into the distinctive multiple access and networking features of semantic communications. The first chapter outlines a multiuser semantic communication framework wherein a set of servers collaboratively transmit images to their associated users. Methods discussed here focuses on optimizing the transmission latency while meeting communication resource constraints. The following chapter focuses on the challenges of multiple access in semantic communications. Emphasis is placed on the coexistence of conventional bit transmission and semantic transmission, and various multiple access schemes facilitating this coexistence are examined. The chapter explores the performance limits of orthogonal and non-orthogonal multiple access schemes. The part concludes by highlighting the superior transmission efficiency achievable in semantic communications compared to classical wireless systems. One specific example is the utilization of contextual reasoning techniques. The subsequent section delves deeper into the significance of leveraging machine reasoning techniques in the context of wireless systems. This exploitation enables the boundary-pushing and unlocking of the full potential of wireless systems, serving as a bridge to the next part of the book. The subsequent chapter delves into the critical aspect of interoperability in multiuser semantic communications. When multiple users with AI models trained on distinct source and channel distributions are present in the system, semantic misalignment arises, impairing semantic reliability. To tackle this issue, a viable solution is proposed, leveraging a fusion of distributed learning and large language models.

Foundations of Semantic Communication Networks, First Edition.
Edited by Walid Saad, Christina Chaccour, Christo Kurisummoottil Thomas, and Merouane Debbah.

7

Optimization of Image Transmission in a Cooperative Semantic Communication Networks

Ye Hu[1] *and Mingzhe Chen*[2]

[1]*Department of Industrial and Systems Engineering, University of Miami, Coral Gables, FL, USA*
[2]*Department of Electrical and Computer Engineering, University of Miami, Coral Gables, FL, USA*

7.1 Introduction

Current communication technologies are trying to approach the Shannon physical capacity limit [Zhu et al., 2020; Chen et al., 2021; Mu et al., 2022]. The integration of communication and artificial intelligence (AI) technology promotes the development of communication to a higher level, i.e., from the technical level to the semantic level [Chaccour et al., 2022; Kalfa et al., 2022; Kang et al., 2022]. A paradigm called *semantic communication*, which shifts from rate-centric toward content-aware communication technologies has been proposed [Lin et al., 2022; Luo et al., 2022; Qin et al., 2022; Zou et al., 2022; Yang et al., 2023], to effectively transmit a fast-growing amount of data (i.e., image, video, and immersive data) over wireless networks [Saad et al., 2020; Xiao et al., 2021; Meng et al., 2022; Tandon et al., 2022]. Semantic communications enable devices to communicate with each other using the desired meaning of the original data so as to improve communication efficiency [Shi et al., 2021; Han et al., 2022; Jiang et al., 2022]. However, current semantic communication techniques are mostly studied for text and image data transmission. Compared to textual data where semantic information is explicitly represented by words, semantic information in an image is implicit and knowledge-based. Hence, AI technology has recently attracted significant interest due to its advantages in terms of extracting and exploiting the meaning of an image. However, the AI-based semantic extraction model can bring difficulty to the concerned optimization problem solving, e.g., resource allocation in semantic communication. Therefore, developing a semantic communication framework for image transmission faces several challenges including: (i) human-oriented semantic information representation, (ii) metric design for

Foundations of Semantic Communication Networks, First Edition.
Edited by Walid Saad, Christina Chaccour, Christo Kurisummoottil Thomas, and Merouane Debbah.

image semantic information, (iii) dynamic semantic information extraction based on users' service requirements, and (iv) semantic-aware resource allocation.

7.1.1 Related Works

Recently, several works in Bao et al. [2011], Basu et al. [2012], Zhang et al. [2022], Yang et al. [2021], Wang et al. [2022], and Lu et al. [2021] studied a number of problems related to semantic communications. Firstly, the works in Bao et al. [2011] and Basu et al. [2012] studied semantic information theory-related problems. In particular, the authors in Bao et al. [2011] investigated a logistic probability-based metric for semantic information measurement. The authors in Basu et al. [2012] defined the metric of semantic channel capacity of a semantic communication system as mutual information between semantic information and original data. However, both metrics designed in Bao et al. [2011] and Basu et al. [2012] measure only the received semantic information with logistic true without considering the completeness of the meaning that is expressed by the semantic information. Then, the authors in Zhang et al. [2022] and Yang et al. [2021] investigated the deep learning-based semantic communication system. Both works in Zhang et al. [2022] and Yang et al. [2021] compress original data into vectors and consider the compressed vectors as semantic information. However, these vectors do not have any practical meanings and are incomprehensible for human receivers. Finally, the works in evaluating the performance of the semantic communication system have been studied in Wang et al. [2022] and Lu et al. [2021]. In Wang et al. [2022], the authors introduced a text-semantic communication framework that seeks to maximize the semantic similarity between original data and semantic information. In Lu et al. [2021], the authors used the accuracy of the receive semantic information to measure the performance of the proposed semantic communication system. However, the metrics defined in Wang et al. [2022] and Lu et al. [2021] are based on the consistency of textual data in a word level, which cannot be used for image data.

The works in Xie et al. [2022], Wang et al. [2021], Huang et al. [2022], and Li et al. [2021] studied the use of semantic communication techniques for image transmission. At first, the works in Wang et al. [2021] and Huang et al. [2022] investigated the semantic coding for image data. In particular, the authors in Wang et al. [2021] the designed an image semantic coding model with Laplacian pyramid aiming to improve image compression ratio. The authors in Huang et al. [2022] introduced an reinforcement learning (RL)-based adaptive image semantic coding method to improve their rate-perception-distortion metric that is defined to measure the image reconstruction. Then, several works in Xie et al. [2022] and Li et al. [2021] studied the task-oriented image semantic communication system. In Xie et al. [2022], the authors proposed a deep neural network (DNN)-enabled semantic communication system to execute the visual question answering task. In Li et al. [2021], the authors investigated an RL-enabled image semantic communication framework for different tasks. However, these works in Xie et al. [2022], Wang et al. [2021], Huang et al. [2022], and Li et al. [2021] modeled the semantic information of an image as uninterpretable feature vectors that cannot be directly utilized and understood by human receivers. Hence, the receivers in these works Xie et al. [2022], Wang et al. [2021], Huang et al. [2022], and Li et al. [2021] need to reconstruct original images, which is inefficient and complicated since the receivers need to use neural networks to interpret received data into explainable and meaningful information. Meanwhile, most of these existing works studied the use of DNNs for semantic information extraction. However, when they introduced DNNs for semantic communications, the relationship between the extracted semantic information and the original data may not be represented by an exact equation such that we cannot use traditional optimization algorithms to optimize semantic communication performance.

Currently, a number of existing works studied the use of RL methods for solving semantic communication problems. In particular, the authors in Wang et al. [2022] utilized an based-RL algorithm to analyze the relationship between the original data and its semantic information. The authors in Lu et al. [2021] investigated a self-critic policy-gradient-enabled semantic communication system. The works in Huang et al. [2022] designed an RL-based adaptive semantic coding model. The works in Li et al. [2021] utilized RL to determine the quantization parameters of semantic coding in different tasks. However, these works Wang et al. [2022], Lu et al. [2021], Huang et al. [2022], and Li et al. [2021] do not consider the cooperation among different agents and hence each agent's performance will be affected by the actions of other agents thus reducing semantic communication performance achieved by RL. The authors in Hu et al. [2021] used a value decomposition based RL to design trajectory of a group of drones. However, the used RL method requires a large amount of users' historical experience due to its weak exploration ability to find a globally optimal solution.

7.2 Representative Work

7.2.1 System Model

Consider a cellular network in which a set $\mathcal{V}$ of V servers cooperatively transmit image data to a set $\mathcal{U}$ of U users using semantic communication techniques, as shown in Figure 7.2. Let $\mathcal{L}_v$ represent a set of the users that are located in the service area of server v. Here, the service areas of different servers may overlap. The procedure of the considered semantic communication of each server consists

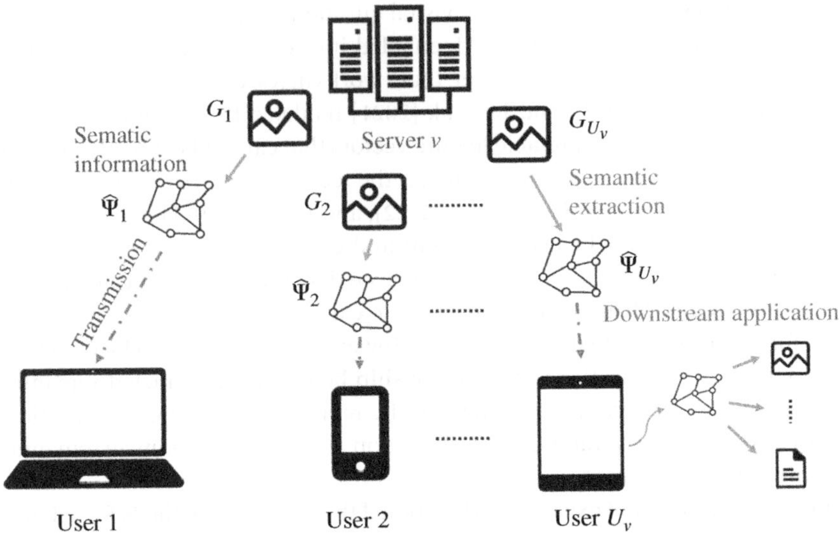

Figure 7.1 The image semantic transmission framework of each server.

of two phases (as shown in Figure 7.1): (i) semantic information extraction and (ii) semantic information transmission. Next, we introduce the process of the semantic information extraction. Then, we present a multimodal metric for the proposed image semantic communication framework which can evaluate the semantic similarity between the original image and its extracted semantic information (Figure 7.2).

7.2.2 Semantic Information Extraction

In our model, we assume that the semantic information of an image consists of the objects and their relationships in the image. Hence, the semantic information of each image is modeled by a scene graph defined by a set of nodes and edges, where a node represents an object (e.g., *a man*) and an edge represents the relationship between two objects, as shown in Figure 7.3. The semantic triple is a basic component of semantic information, which consists of two objects and the relationship between them. For example, a semantic triple in Figure 7.3 is ([“*man*”], [“*riding on*”], [“*bicycle*”]), where [“*man*”] and [“*bicycle*”] are objects and [“*riding on*”] is their relationship. An image that a server needs to transmit can be described by multiple semantic triples.

We use a scene graph extraction model in Tang et al. [2020] to extract the semantic information from the original image. The semantic information extraction process has two steps which are object identification and relationship capture.

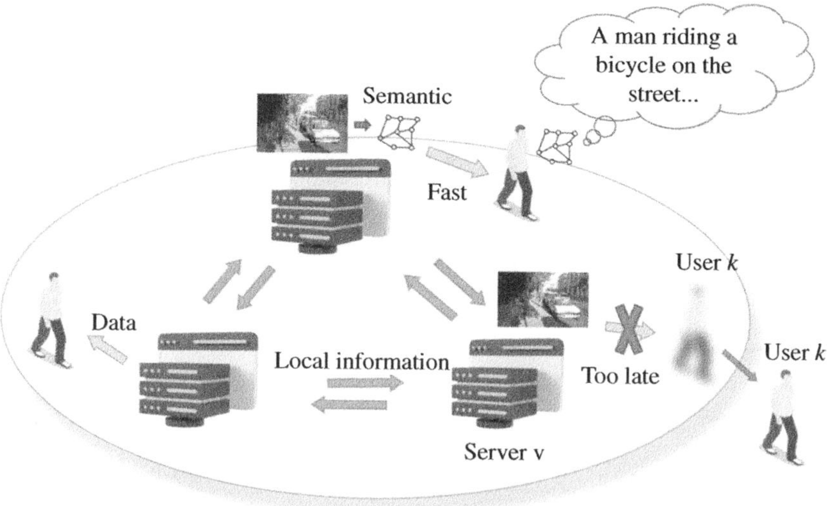

Figure 7.2 The cooperative multi-server image semantic communication wireless network. Source: Dickson et al. (2002)/IEEE.

Specifically, a server first uses the model to detects, locates, and categorizes all the objects in the image. Then, the model deduces the relationship between all the objects according to their geometry and logical correlation and output them in form of triples. The semantic information of an image G_k that is extracted by server v and transmitted to user k can be expressed as

$$\boldsymbol{\Psi}_{vk} = \left\{ \boldsymbol{\psi}_{vk}^{1}, \boldsymbol{\psi}_{vk}^{2}, \ldots, \boldsymbol{\psi}_{vk}^{n}, \ldots, \boldsymbol{\psi}_{vk}^{N_{vk}} \right\}, \tag{7.1}$$

where $\boldsymbol{\psi}_{vk}^{n} = \left(e_{vk,i}^{n}, l_{vk,ij}^{n}, e_{vk,j}^{n} \right)$ is a semantic triple and N_{vk} is the number of semantic triples in image G_k, $e_{vk,i}^{n}$ is the category of object i in image G_k, $l_{vk,ij}^{n}$ is the relationship between objects $e_{vk,i}^{n}$ and $e_{vk,j}^{n}$. Here, $l_{vk,ij}^{n}$ is directional and hence, $l_{vk,ij}^{n} \neq l_{vk,ji}^{n}$. To measure the size of the semantic information, we define $Z(\boldsymbol{x})$ as the number of letters in word sequence $\boldsymbol{x}$. Therefore, the total number of letters in each image semantic information $\boldsymbol{\Psi}_{vk}$ is

$$Z(\boldsymbol{\Psi}_{vk}) = \sum_{n=1}^{N_{vk}} \left(Z\left(e_{vk,i}^{n}\right) + Z\left(l_{vk,ij}^{n}\right) + Z\left(e_{vk,j}^{n}\right) \right). \tag{7.2}$$

For example, in Figure 7.3, the number of letters in semantic triple $\boldsymbol{\psi}_{vk}^{n} =$ ([“*man*”], [“*riding on*”], [“*bicycle*”]) is $Z\left(\boldsymbol{\psi}_{vk}^{n}\right) = Z\left(e_{vk,i}^{n}\right) + Z\left(l_{vk,ij}^{n}\right) + Z\left(e_{vk,j}^{n}\right) = 3 + 8 + 7 = 18$.

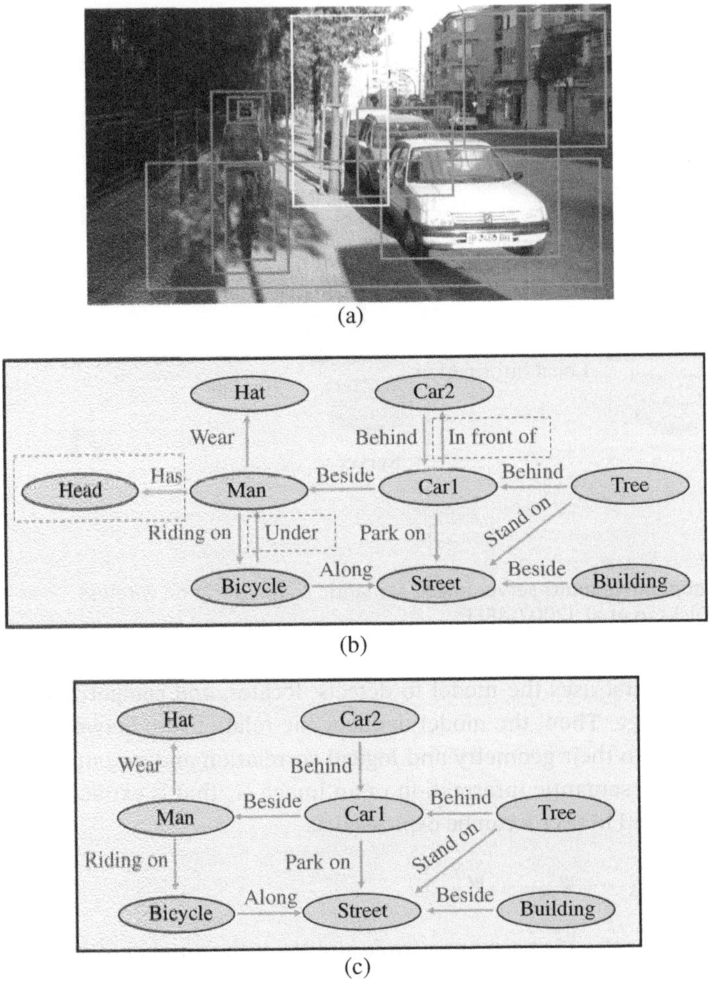

Figure 7.3 An example of semantic information extraction. (a) The extracted semantic information, (b) the extracted semantic information, and (c) the transmitted semantic information. Source: Dickson et al. (2002)/IEEE.

Note that some semantic triples in $\boldsymbol{\Psi}_{vk}$ may not contain useful information. For example, in Figure 7.3, we do not want to transmit the meaningless semantic triples such as ([“*man*”], [“*has*”], [“*head*”]) and redundant semantic triples such as ([“*bicycle*”], [“*under*”], [“*man*”]). In order to improve the efficiency of the considered semantic communication model, as shown in Figure 7.3c, each server v must transmit the semantic triples that contain the most significant meaning of an image. The partial semantic information that server v transmits to user k can be given as

$$\hat{\boldsymbol{\Psi}}_{vk} = \left\{ \hat{\boldsymbol{\psi}}_{vk}^{1}, \hat{\boldsymbol{\psi}}_{vk}^{2}, \ldots, \hat{\boldsymbol{\psi}}_{vk}^{n}, \ldots, \hat{\boldsymbol{\psi}}_{vk}^{\hat{N}_{vk}} \right\} \subset \boldsymbol{\Psi}_{vk}, \tag{7.3}$$

where $\hat{N}_{vk}$ is the number of selected semantic triples in $\hat{\boldsymbol{\Psi}}_{vk}$.

7.2.3 Transmission Model

We assume that an orthogonal frequency division multiple access (OFDMA) technique is adopted. A set $\mathcal{Q}$ of Q downlink orthogonal resource blocks (RBs) are allocated to serve users. The servers can reuse all these RBs and thus each server can allocate Q RBs to its associated users. The downlink rate of a server transmitting partial semantic information $\hat{\boldsymbol{\Psi}}_{vk}$ to user k is given as

$$c_k\left(\boldsymbol{a}_k\right) = \sum_{v=1}^{V}\sum_{q=1}^{Q} a_{vk}^{q} W \log_2\left(1 + \frac{Ph_{vk}^{q}}{I_{vk}^{q} + WN_0}\right), \tag{7.4}$$

where P is the transmit power of server v, W is the bandwidth of RB q which is assumed to be equal for all RBs, $h_{vk}^{q} = \gamma_{vk}^{q} d_{vk}^{-2}$ is the channel gain between server v and user k with γ_{vk}^{q} being the Rayleigh fading parameter and d_{vk} being the distance between server v and user k, $I_{vk}^{q} = \sum_{s\in\mathcal{V}_q, s\neq v} Ph_{s_k}^{q}$ represents the interference caused by other servers with $\mathcal{V}_q$ being the set of servers that use RB q, N_0 is the noise power spectral density, and $\boldsymbol{a}_k = \left[\boldsymbol{a}_{1k}, \ldots, \boldsymbol{a}_{vk}, \ldots, \boldsymbol{a}_{Vk}\right]$ with $\boldsymbol{a}_{vk} = \left[a_{vk}^{1}, \ldots, a_{vk}^{Q}\right]$ is an RB allocation vector for user k of server v with $a_{vk}^{q} \in \{0, 1\}$ being the user-server connection index. In particular, $a_{vk}^{q} = 1$ implies that server v transmits semantic information to user k using RB q, and $a_{vk}^{q} = 0$, otherwise. Here, we consider each user can only be served by one server with one RB, and each RB of a server can only be allocated to one user for simplicity. Then, we have $\sum_{v=1}^{V}\sum_{q=1}^{Q} a_{vk}^{q} \leqslant 1, \forall k \in \mathcal{U}$ and $\sum_{k\in\mathcal{U}} a_{vk}^{q} \leqslant 1, \forall v \in \mathcal{V}, \forall q \in \mathcal{Q}$. According to (7.2), (7.3), and (7.4), the transmission latency of server v transmitting selected partial semantic information $\hat{\boldsymbol{\Psi}}_{vk}$ to user k can be given as

$$T\left(\hat{\boldsymbol{\Psi}}_{vk}, \boldsymbol{a}_{vk}\right) = \frac{Z\left(\hat{\boldsymbol{\Psi}}_{vk}\right)}{c_{vk}\left(\boldsymbol{a}_{vk}\right)}, \tag{7.5}$$

where $c_{vk}\left(\boldsymbol{a}_{vk}\right) = \sum_{q=1}^{Q} a_{vk}^{q} W \log_2\left(1 + \frac{Ph_{vk}^{q}}{I_{vk}^{q} + WN_0}\right)$ is the transmitting rate. Here, we note that only the transmission latency of associated user are considered and calculated. From (7.5), we see that the transmission latency of semantic information depends on user association, RB allocation, and the data size of the transmitted partial semantic information. Hence, for a certain user, if its associated server changes, its received semantic information extracted from the same image will be different. Moreover, changes of the wireless communication environment such as dynamic channel will affect its received semantic information.

7.2.4 Image Semantic Similarity Model

To evaluate the performance of image semantic communication, we introduce a metric called image-to-graph semantic similarity (ISS). Different from conventional metrics, such as structural similarity (SSIM) [Zhou et al., 2004], that measure the differences in a set of pixels, the introduced metric can capture the correlation of the meaning between the extracted semantic information and its original image. We first use a pre-trained DNN as vectorization function to construct the relationship between the extracted semantic information and its original image. The DNN is trained by Webimagetext Radford et al. [2021] which is a dataset of 400 million image-text pairs collected from the internet. After the training, we can directly use the DNN to obtain the image vector and its semantic information vectors. The vectorized image data is $\boldsymbol{C}(G_k)$ and the vectorized semantic information is $\boldsymbol{O}_{vk} = \left\{\boldsymbol{C}\left(\hat{\psi}_{vk}^{1}\right), \dots, \boldsymbol{C}\left(\hat{\psi}_{vk}^{n}\right), \dots, \boldsymbol{C}\left(\hat{\psi}_{vk}^{\hat{N}_{vk}}\right)\right\}$, where $\boldsymbol{C}(\cdot)$ is the DNN based vectorization function.

The introduced ISS metric is defined as the included angle cosine between an image vector and its normalized semantic triple vectors, which is calculated by the projection of image vector on semantic information vector set. To build the basis of the semantic information vector set, the Gram–Schmidt algorithm is used to orthogonalize the semantic information vectors, which is given by $\overline{\boldsymbol{O}_{vk}} = \{\overline{\boldsymbol{C}(\hat{\psi}_{vk}^{1})}, \dots, \overline{\boldsymbol{C}(\hat{\psi}_{vk}^{n})}, \dots, \overline{\boldsymbol{C}(\hat{\psi}_{vk}^{\hat{N}_{vk}})}\}$. Then, the ISS of semantic information $\hat{\boldsymbol{\Psi}}_{vk}$ that transmitted from server v to user k is defined as:

$$E\left(\hat{\boldsymbol{\Psi}}_{vk}, \boldsymbol{a}_{vk}\right) = \sum_{q=1}^{Q} a_{vk}^{q} \frac{\| \sum_{n=1}^{\hat{N}_{vk}} \overline{\boldsymbol{C}\left(\hat{\psi}_{vk}^{n}\right)} \boldsymbol{C}(G_k)^T \overline{\boldsymbol{C}\left(\hat{\psi}_{vk}^{n}\right)} \|}{\|\boldsymbol{C}\left(G_k\right)\|}, \tag{7.6}$$

where $\sum_{q=1}^{Q} a_{vk}^{q}$ represents RB allocation of user k and $\overline{\boldsymbol{C}\left(\hat{\psi}_{vk}^{n}\right)} \boldsymbol{C}(G_k)^T \overline{\boldsymbol{C}\left(\hat{\psi}_{vk}^{n}\right)}$ is the projection vector of $\boldsymbol{C}\left(G_k\right)$ on $\overline{\boldsymbol{C}\left(\hat{\psi}_{vk}^{n}\right)}$. From (7.6), we see that the value of the ISS increases as the number of transmitted semantic triples $\hat{N}_{vk}$ increases, which is consistent with the objective human cognition [Corbetta and Shulman, 2002]. Hence, if the server transmits more triples, the image semantic transmission performance can be improved.

In the introduced framework, each server v transmits only partial semantic information and, hence, the received semantic information includes a part of meaning of the image. We define the minimum acceptable ISS of each user as

ξ. Then, the probability of the received partial semantic information satisfying $E\left(\hat{\boldsymbol{\Psi}}_{vk}, \boldsymbol{a}_{vk}\right) \geq \xi$ is defined as the semantic reliability, which is given by

$$P\left(E\left(\hat{\boldsymbol{\Psi}}_{vk}, \boldsymbol{a}_{vk}\right) \geq \xi\right) \geq \epsilon, \tag{7.7}$$

where ϵ is the semantic reliability threshold that is used to adjust the probability of reliable semantic transmission. The threshold ϵ and ξ are determined by the system operator in practice. In particular, if we want semantic transmission to be reliable, we can increase the semantic reliability threshold ϵ. If we want to improve the quality of each semantic transmission, we can increase the minimum acceptable semantic similarity ξ. For example, $\xi = 0.6$ and $\epsilon = 0.9$ represent that at least 90% semantic information transmission must satisfy $E\left(\hat{\boldsymbol{\Psi}}_{vk}, \boldsymbol{a}_{vk}\right) \geq 0.6$.

7.2.5 Problem Formulation

Given the defined system model, our objective is to minimize the average transmission latency of all users while satisfying the semantic reliability requirement. This minimization problem includes optimizing the user association, RB allocation, and determining the part of semantic information to transmit. The average transmission latency minimization problem is formulated as follows:

$$\min_{\hat{\boldsymbol{\Psi}}_{vk}, \boldsymbol{a}_{vk}} \frac{\sum\limits_{v \in \mathcal{V}} \sum\limits_{k \in \mathcal{U}_v} T\left(\hat{\boldsymbol{\Psi}}_{vk}, \boldsymbol{a}_{vk}\right)}{\sum\limits_{v \in \mathcal{V}} |\mathcal{U}_v|} \tag{7.8}$$

$$\text{s.t. } a_{vk}^q \in \{0, 1\}, \quad \forall k \in \mathcal{U}_v, \forall v \in \mathcal{V}, \forall q \in \mathcal{Q}, \tag{7.8a}$$

$$\sum_{v \in \mathcal{V}} \sum_{q \in \mathcal{Q}} a_{vk}^q \leqslant 1, \quad \forall k \in \mathcal{U}_v, \tag{7.8b}$$

$$\sum_{k \in \mathcal{U}_v} a_{vk}^q \leqslant 1, \forall v \in \mathcal{V}, \quad \forall q \in \mathcal{Q}, \tag{7.8c}$$

$$\mathcal{U}_v \subset \mathcal{L}_v, \quad \forall v \in \mathcal{V}, \tag{7.8d}$$

$$P\left(E\left(\hat{\boldsymbol{\Psi}}_{vk}, \boldsymbol{a}_{vk}\right) \geqslant \xi\right) \geqslant \epsilon, \quad \forall k \in \mathcal{U}_v, \forall \hat{\boldsymbol{\Psi}}_{vk} \subset \boldsymbol{\Psi}_{\boldsymbol{vk}}, \tag{7.8e}$$

where $\mathcal{U}_v$ is the set of users associated with server v and $\mathcal{L}_v$ is the set of users located in the service area of server v. Constraints (7.8a), (7.8b), and (7.8c) ensure that each server can allocate one RB to each associated user and an RB can only be occupied by one user for image semantic information transmission. Constraint (7.8e) is the semantic reliability requirement of each user. From (7.8), we can see that constraint (7.8e) is non-convex and the semantic information extraction depends on DNN models. Hence, the relationship between the objective function and optimization variables cannot be represented by an extract equation. In consequence, the problem (7.8) cannot be solved by traditional optimization algorithms. Furthermore, a single server cannot observe the global wireless communication environment and the information of users associated with other servers. Hence, the centralized RL algorithms (e.g., DQN) can only minimize the transmission latency of the implemented server based on the partial observation. To solve problem (7.8) which aims to minimize the sum of the average transmission latency of all users, we introduce a multi-agent RL algorithm that enables all servers cooperatively optimize the resource allocation of the considered semantic communication network.

7.3 Value-Decomposition-based Entropy-Maximized Multi-Agent RL Method

To effectively solve problem (7.8), we introduce a value-decomposition-based [Sunehag et al., 2017] entropy-maximized multi-agent RL (VD-ERL) algorithm to minimize the average transmission latency of all servers instead of individual servers. We first introduce the components of the introduced VD-ERL method. Then, we introduce the training procedure of the introduced VD-ERL method.

7.3.1 Components of VD-ERL Method

In this section, we introduce the fundamental components of the introduced VD-ERL method as follows:

- **Agent**: The agents are the servers that determine the RB allocation and the set of semantic triples that need to be transmitted to its associated users.
- **States**: The state is defined as $\boldsymbol{s} = \left[\boldsymbol{s}_1, \ldots, \boldsymbol{s}_v, \ldots, \boldsymbol{s}_V\right]$ where $\boldsymbol{s}_v = \left[\boldsymbol{\gamma}_v, \boldsymbol{\beta}_v\right]$ represents the partial state of server v. $\boldsymbol{\gamma}_v = \left[\gamma_v^1, \ldots, \gamma_v^Q\right]$ is the vector of available RBs where $\gamma_v^q = 0$ represents that RB q has been allocated, and $\gamma_v^q = 1$, otherwise. $\boldsymbol{\beta}_v = \left[\boldsymbol{\beta}_{v_1}, \ldots, \boldsymbol{\beta}_{vk}, \ldots, \boldsymbol{\beta}_{v_{|\mathcal{L}_v|}}\right]$ is the semantic triple score matrix of the users located in the coverage of server v and is used to evaluate the semantic reliability, where $|\mathcal{L}_v|$ is the number of users in the service area of server v and $\boldsymbol{\beta}_{vk} = \left[\beta\left(\boldsymbol{\psi}_{vk}^1\right), \ldots, \beta\left(\boldsymbol{\psi}_{vk}^n\right), \ldots, \beta\left(\boldsymbol{\psi}_{vk}^{N_{vk}}\right)\right]$ is the vector of scores of all semantic triples in semantic information $\boldsymbol{\Psi}_{vk}$. The score of each semantic triple $\boldsymbol{\psi}_{vk}^n$ can be given as

$$\beta\left(\boldsymbol{\psi}_{vk}^{n}\right)=\frac{\exp\left(\mu\left(e_{vk,i}^{n}\right)\mu\left(l_{vk,ij}^{n}\right)\mu\left(e_{vk,j}^{n}\right)\right)}{\sum_{n=1}^{N_{vk}}\exp\left(\mu\left(e_{vk,i}^{n}\right)\mu\left(l_{vk,ij}^{n}\right)\mu\left(e_{vk,j}^{n}\right)\right)},\tag{7.9}$$

where $\mu\left(e_{vk,i}^{n}\right)$ is the probability of object $e_{vk,i}^{n}$ being detected from image G_{vk} and $\mu\left(l_{vk,ij}^{n}\right)$ is the conditional probability of relationship $l_{vk,ij}^{n}$ being deduced given objects $e_{vk,i}^{n}$ and $e_{vk,j}^{n}$. In (7.9), $\mu\left(e_{vk,i}^{n}\right)$ and $\mu\left(l_{vk,ij}^{n}\right)$ can be obtained by a scene graph generation model [Tang et al., 2020]. The score of each semantic triple $\boldsymbol{\psi}_{vk}^{n}$ represents the probability of extracting triple $\boldsymbol{\psi}_{vk}^{n}$ from the original image and will be used to the selection of the partial semantic information to be transmitted. In particular, the semantic triple that has a high score can contribute more to the semantic information. Here, we note that each server v can only observe its partial state $\boldsymbol{s}_{v}$.

- **Actions**: Each action $\boldsymbol{\alpha}_{v}$ of server v is the RB allocation, which is given by:

$$\boldsymbol{\alpha}_{v}=\left[\boldsymbol{a}_{v1},\ldots,\boldsymbol{a}_{vk},\ldots,\boldsymbol{a}_{v|\mathcal{L}_{v}|}\right],\tag{7.10}$$

where $\boldsymbol{a}_{vk}$ representing RB allocation vector is the variable of problem (7.8). Then, the vector of all distributed servers' actions is $\boldsymbol{\alpha}=\left[\boldsymbol{\alpha}_{1},\ldots,\boldsymbol{\alpha}_{v},\ldots,\boldsymbol{\alpha}_{V}\right]$. Here, we note that the introduced VD-ERL algorithm can be extended to the scenarios where a server can allocate more than one RB to a user. Here, we only need to change the constraint of the action from $\sum_{q=1}^{Q}\boldsymbol{a}_{vk}^{q}\leqslant 1$ to $\sum_{q=1}^{Q}\boldsymbol{a}_{vk}^{q}\leqslant Q$.

- **Policy**: The policy is the conditional probability of each agent choosing an action $\boldsymbol{\alpha}_{v}$ in a given partial state $\boldsymbol{s}_{v}$. The policy is implemented by the DNN with parameter $\boldsymbol{\phi}_{v}$, which establishes the relation between the semantic triple scores, the ISS, and the transmission latency of each user. Then, the conditional probability of each agent taking action $\boldsymbol{\alpha}_{v}$ in a given partial state $\boldsymbol{s}_{v}$ can be expressed as $\boldsymbol{\pi}_{\boldsymbol{\phi}_{v}}\left(\boldsymbol{\alpha}_{v}\mid\boldsymbol{s}_{v}\right)$. To improve the action exploration, the policy networks are trained to maximize not only the expected reward, but also the entropy of actions $\mathcal{H}\left(\boldsymbol{\pi}_{\boldsymbol{\phi}_{v}}\left(\boldsymbol{\alpha}_{v}\mid\boldsymbol{s}_{v}\right)\right)$ which drives the agent to choose actions more randomly.
- **Reward**: The reward of each server is used to capture the benefits of a selected action in terms of semantic reliability and transmission latency. To calculate the reward of each server v, we first need to determine the partial semantic information $\hat{\boldsymbol{\Psi}}_{vk}$ that will be transmitted to user k. In particular, based on the state $\boldsymbol{s}_{v}$ and action $\boldsymbol{\alpha}_{v}$, we can sort the semantic triples $\boldsymbol{\Psi}_{vk}$ according to the score vector $\boldsymbol{\beta}_{vk}$. In particular, the sorted semantic triple vector is $\left[\hat{\boldsymbol{\psi}}_{vk}^{1},\ldots,\hat{\boldsymbol{\psi}}_{vk}^{N_{vk}}\right]$ where $\hat{\boldsymbol{\psi}}_{vk}^{1}$ is the triple with the highest score while $\hat{\boldsymbol{\psi}}_{vk}^{N_{vk}}$ is the triple with the lowest

score. Given this sorted triple vector, we use an iterative algorithm to select several triples to satisfy constraint (7.8e) while minimizing the transmission time. The iterative algorithm used to determine the selected triples $[\hat{\psi}^1_{vk}, \dots, \hat{\psi}^n_{vk}]$ to generate semantic information is called **Algorithm 1**.

Then, the reward of each user k associated with server v is given as

$$r_{vk}\left(\boldsymbol{s}_v, \boldsymbol{a}_{vk}\right) = \sum_{q=1}^{Q} a^q_{vk}\left[\mathbb{1}_{\{E(\hat{\boldsymbol{\Psi}}_{vk}, \boldsymbol{a}_{vk}) \geq \xi\}}\eta - T\left(\hat{\boldsymbol{\Psi}}_{vk}, \boldsymbol{a}_{vk}\right)\right], \tag{7.11}$$

where η is a constant bias, $T\left(\hat{\boldsymbol{\Psi}}_{vk}, \boldsymbol{a}_{vk}\right)$ is the transmission latency, and $\mathbb{1}_{\{E(\hat{\boldsymbol{\Psi}}_{vk}, \boldsymbol{a}_{vk}) \geq \xi\}}$ is a function that indicates if the received semantic information $\hat{\boldsymbol{\Psi}}_{vk}$ satisfies the semantic similarity requirement defined in constraint (7.8e). From (7.11), we can see that if the transmitted semantic information cannot satisfy semantic similarity requirement, i.e., $\mathbb{1}_{\{E(\hat{\boldsymbol{\Psi}}_{vk}, \boldsymbol{a}_{vk}) \geq \xi\}} = 0$, the reward of this server will be negative.

Since servers allocate RB resources to users in a distributed manner and each server does not know the RB allocation schemes of other servers, several servers may allocate their RB to one user and this user can use the RB of only one server thus wasting the RB of other servers. To improve RB usage, we add a negative penalty $\rho < 0$ for multiple servers allocating RBs to one user to the reward function. In particular, the total reward of all servers is given as

$$\begin{aligned} r(\boldsymbol{s}, \boldsymbol{\alpha}) &= \sum_{v=1}^{V} r_v\left(\boldsymbol{s}_v, \boldsymbol{\alpha}_v\right) \\ &= \sum_{v=1}^{V}\sum_{k \in \mathcal{U}_v}\left[\left(1 - \mathbb{1}_{\left\{\sum\limits_{\zeta \neq v, \zeta \in \mathcal{V}}\sum\limits_{q=1}^{Q}\boldsymbol{a}^q_{vk}=0\right\}}\right)\rho \right. \\ &\left. + \mathbb{1}_{\left\{\sum\limits_{\zeta \neq v, \zeta \in \mathcal{V}}\sum\limits_{q=1}^{Q}\boldsymbol{a}^q_{vk}=0\right\}} r_{vk}\left(\boldsymbol{s}_v, \boldsymbol{a}_{vk}\right)\right], \end{aligned} \tag{7.12}$$

where $r_v\left(\boldsymbol{s}_v, \boldsymbol{\alpha}_v\right)$ is the reward of server v and $\mathbb{1}_{\left\{\sum_{\zeta \neq v, \zeta \in \mathcal{V}}\sum_{q=1}^{Q}\boldsymbol{a}^q_{vk}=0\right\}}$ is a function that indicates whether user k is served by other servers. In particular, $\mathbb{1}_{\left\{\sum_{\zeta \neq v, \zeta \in \mathcal{V}}\sum_{q=1}^{Q}\boldsymbol{a}^q_{vk}=0\right\}} = 0$ implies that user k is served by other servers, and $\mathbb{1}_{\left\{\sum_{\zeta \neq v, \zeta \in \mathcal{V}}\sum_{q=1}^{Q}\boldsymbol{a}^q_{vk}=0\right\}} = 1$, otherwise. From (7.12), we can see that when multiple servers allocate RBs to one user, the reward will be ρ.

- **Individual Q value function**: The individual Q value function of each server v is defined as $Q_{\boldsymbol{\theta}_v}\left(\boldsymbol{s}_v, \boldsymbol{\alpha}_v\right)$, which is used to estimate the expected reward under a given partial state $\boldsymbol{s}_v$ of server v and a selected action $\boldsymbol{\alpha}_v$. Each server v uses a DNN with parameter $\boldsymbol{\theta}_v$ to approximate the individual Q value function. Since each server can observe only the state of the users located in its service area, each server will transmit its individual Q value to other servers for the estimation of global Q value function, which will be explained in the next bullet.

- **Global Q value function**: The global Q value function is defined as $Q_{tot}(\boldsymbol{s}, \boldsymbol{\alpha})$, which is used to estimate the total expected reward of all distributed servers. For the introduced VD-ERL algorithm, we assume that the global Q value of all servers is equal to the sum of the individual Q value of each server, which is given by Sunehag et al. [2017]

$$Q_{tot}(\boldsymbol{s}, \boldsymbol{\alpha}) = \sum_{v=1}^{V} Q_{\theta_v}\left(\boldsymbol{s}_v, \boldsymbol{\alpha}_v\right). \tag{7.13}$$

The goal of each server v is to cooperatively maximize the total expected reward, i.e., maximize the global Q value by training its policy network. After training, each server can find the optimal policy based on the global Q value function so as to minimize the sum of the transmission latency of all users while satisfying their semantic reliability requirements.

7.3.2 VD-ERL Algorithm for Semantic-Oriented Resource Allocation

Next, we introduce how the servers use the introduced VD-ERL algorithm to cooperatively minimize the sum of the average semantic information transmission latency. At first, each agent first collects local information that includes partial states $\boldsymbol{s}_v$ and actions $\boldsymbol{\alpha}_v$. Then, each agent transmits its local information to other agents to calculate its server reward $r_v\left(\boldsymbol{s}_v, \boldsymbol{\alpha}_v\right)$ and total reward $r(\boldsymbol{s}, \boldsymbol{\alpha}) = \sum_{v=1}^{V} r_v\left(\boldsymbol{s}_v, \boldsymbol{\alpha}_v\right)$ of all servers. Finally, as shown in Figure 7.4, based on the total reward and global Q value function, each agent updates the its policy

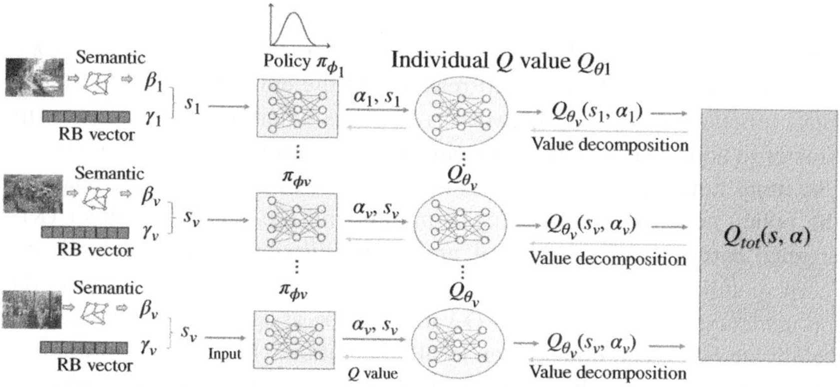

Figure 7.4 The training process of the introduced VD-ERL algorithm. Source: Dickson et al. (2002)/IEEE.

network and individual Q value function. In particular, each agent first collects a set of trajectories $\mathcal{D}_v = \{\boldsymbol{\tau}_v^1, \ldots, \boldsymbol{\tau}_v^d, \ldots, \boldsymbol{\tau}_v^D\}$ with $\boldsymbol{\tau}_v^d = [\boldsymbol{\alpha}_v^d, \boldsymbol{s}_v^d, r_v^d]$ based on the current policy $\boldsymbol{\pi}_{\phi_v}(\boldsymbol{s}_v^d, \boldsymbol{\alpha}_v^d)$. Then, each agent samples a batch size of trajectories from $\mathcal{D}_v$ and calculate total reward and global Q value to train individual Q value function Q_{θ_v} and policy network $\boldsymbol{\pi}_{\phi_v}$. Finally, each server samples action $\boldsymbol{\alpha}_v$ based on updated policy network $\boldsymbol{\pi}_{\phi_v}$ under given state $\boldsymbol{s}_v$ to collect new trajectories for next iteration. The loss function of global Q value function $Q_{tot}(\boldsymbol{s}, \boldsymbol{\alpha})$ is defined as follows

$$J(\theta_1, \ldots, \theta_V) = \mathbb{E}_{\tau^d \in D}\left[Q_{tot}(\boldsymbol{s}, \boldsymbol{\alpha}) - r(\boldsymbol{s}, \boldsymbol{\alpha}) - \max_{\boldsymbol{\alpha}'} Q_{tot}(\boldsymbol{s}', \boldsymbol{\alpha}')\right]^2, \tag{7.14}$$

where $\max_{\boldsymbol{\alpha}'} Q_{tot}(\boldsymbol{s}', \boldsymbol{\alpha}')$ is the maximal global Q value of next state $\boldsymbol{s}'$. The global Q value monotonically increases as each individual Q value increases, i.e., an action of an agent with a high individual Q value is also valuable for entire wireless networks. Hence, the goal that each server trains its individual Q value function is to maximize the global Q value. The individual Q value function $Q_{\theta_v}(\boldsymbol{s}_v, \boldsymbol{\alpha}_v)$ of each server can be updated using a gradient descent method as follows:

$$\theta_v \leftarrow \theta_v - \lambda_{\theta_v} \nabla_{\theta_v} J(\theta_1, \ldots, \theta_V), \tag{7.15}$$

where λ_{θ_v} is the updating rate and $\nabla_{\theta_v} J(\theta_1, \ldots, \theta_V)$ is the gradient of the global Q value function, which is given by

$$\begin{aligned} &\nabla_{\theta_v} J(\theta_1, \ldots, \theta_V) \\ &= \nabla_{\theta_v}\left[Q_{tot}(\boldsymbol{s}^d, \boldsymbol{\alpha}^d) - r(\boldsymbol{s}^d, \boldsymbol{\alpha}^d) - \max_{\boldsymbol{\alpha}'} Q_{tot}(\boldsymbol{s}', \boldsymbol{\alpha}')\right]^2 \\ &= 2\Delta Q_{tot} \nabla_{\theta_v} Q_{tot}(\boldsymbol{s}^d, \boldsymbol{\alpha}^d) \cdot \nabla_{\theta_v} Q_{\theta_v}(\boldsymbol{s}_v^d, \boldsymbol{\alpha}_v^d), \end{aligned} \tag{7.16}$$

where $\Delta Q_{tot} = Q_{tot}(\boldsymbol{s}^d, \boldsymbol{\alpha}^d) - r(\boldsymbol{s}^d, \boldsymbol{\alpha}^d) - \max_{\boldsymbol{\alpha}'} Q_{tot}(\boldsymbol{s}', \boldsymbol{\alpha}')$. Combined with entropy-maximization RL Haarnoja et al. [2018], the objective of the policy $\boldsymbol{\pi}_{\phi_v}$ is the weighted sum of the expected reward and the entropy of actions. Hence, the loss function of a policy network is given as where δ is the temperature parameter to adjust the weight of the entropy term and the policy of each server will be more random as δ increases. From (7.17), we can see that the objective of the policy network is equivalent to minimizing the Kullback–Leibler (KL) divergence between the conditional probability distribution $\boldsymbol{\pi}_{\phi_v}(\boldsymbol{\alpha}_v \mid \boldsymbol{s}_v^d)$ and the corresponding trained individual Q value function $Q_{\theta_v}(\boldsymbol{\alpha}_v \mid \boldsymbol{s}_v^d)$. Hence, the action achieving higher individual Q value will be assigned a higher selection probability to be chosen under given local observation. However, the introduced entropy-maximization enables other potential valuable actions with low selection probability to be taken properly. Therefore, the valuable action exploration ability of each server and the probability of finding an optimal RB allocation scheme are improved. Finally, the policy π_{ϕ_v} can be updated using gradient descent method as follows:

$$\phi_v \leftarrow \phi_v - \lambda_{\phi_v} \nabla J_\pi \left(\phi_v\right), \tag{7.17}$$

where λ_{ϕ_v} is the learning rate. The specific training procedure of the introduced VD-ERL algorithm is referred as **Algorithm 2**.

7.3.3 Complexity and Convergence of the Introduced Algorithm

In this section, we analyze the complexity and convergence of the introduced VD-ERL algorithm for semantic-oriented RB allocation. The complexity of the VD-ERL algorithm lies in semantic triple selection and determining the resource allocation of each server. First, from **Algorithm 1**, the complexity of semantic triple selection of user k is $\mathcal{O}\left(N_{vk}\right)$. Hence, the complexity of semantic triple selection of all users is $\mathcal{O}\left(\sum_{v\in\mathcal{V}}\sum_{k=1}^{|\mathcal{U}_v|} N_{vk}\right) = \mathcal{O}(N)$. Then, we explain the complexity of training policy and individual Q value networks of each server, which are two fully connected networks that consist of an input layer, hidden layers, and an output layer. Hence, the time-complexity of training networks of each server depends on the number of neurons in each layer [Wang et al., 2020]. The time-complexity of training policy network and individual Q value network of each server is given by

$$\mathcal{O}\left(\sum_{i=1}^{I_x-1} w_i^x w_{i+1}^x + \left(|\mathcal{L}_v| N_{vk} + Q\right) w_1^x + |\mathcal{A}_v| w_{I_x} \right.$$
$$\left. + \sum_{i=1}^{I_y-1} w_i^y w_{i+1}^y + \left(|\mathcal{L}_v| \left(N_{vk} + Q\right) + Q\right) w_1^y + w_{I_y}\right), \tag{7.18}$$

where w_i^x and w_i^y are the number of neurons in the i-th hidden layer, I_x and I_y are the number of hidden layers in policy and individual Q value network, respectively, $|\mathcal{L}_v| N_{vk} + Q$ and $|\mathcal{A}_v|$ represent the dimension of input and output layer of policy network, respectively, and $\left(|\mathcal{L}_v| \left(N_{vk} + Q\right) + Q\right)$ represents the dimension of input layer of individual Q value network with an one-dimension output layer. The introduced algorithm can be trained offline. Therefore, **Algorithms 1** and **2** are executed in the training stage. After training, we only need to implement **Algorithm 1** for RB allocation with complexity $\mathcal{O}(N)$.

Next, using the result of Haarnoja et al. [2018, Theorem 1], we can prove that the introduced VD-ERL algorithm is guaranteed to converge to a locally optimal solution of problem (7.8), as shown in the following lemma.

Lemma 7.1 The introduced VD-ERL algorithm is guaranteed to converge if the following conditions are satisfied: (1) Individual Q value function $Q_{\theta_v}\left(\boldsymbol{s}_v, \boldsymbol{\alpha}_v\right)$ is bounded. (2) $Q_{\theta_v}^{\pi_{\phi_v}^N}(s_v, \alpha_v) \geqslant Q_{\theta_v}^{\pi_{\phi_v}^O}(s_v, \alpha_v)$ holds for any state s_v and action α_v, where $\pi_{\phi_v}^N$ is the optimized policy-based on (7.17) with current individual Q value function in each iteration.

Proof: Next, we prove that the introduced VD-ERL algorithm satisfies these two conditions. Since the number of actions α in the introduced VD-ERL algorithm is finite, the global Q value function $Q_{tot}(s, \alpha)$ can be proved to be bounded using [Haarnoja et al., 2018, Lemma 1]. Hence, condition (1) is satisfied. For condition (2), from (7.17), the new policy $\pi_{\phi_v}^N$ satisfies the following inequality equation for any old policy $\pi_{\phi_v}^O$:

$$\begin{aligned} D_{KL} & \left[\pi_{\phi_v}^N(\alpha_v \mid s_v) \,\|\, \exp\left(Q_{\theta_v}(s_v, \alpha_v)\right)\right] \\ & \leqslant D_{KL}\left[\pi_{\phi_v}^O(\alpha_v \mid s_v) \,\|\, \exp\left(Q_{\theta_v}(s_v, \alpha_v)\right)\right]. \end{aligned} \tag{7.19}$$

Given (7.19), we can prove that the introduced method satisfies condition (2) using the result of Haarnoja et al. [2018, Lemma 2].

7.4 Simulation Results and Analysis

For our simulations, we consider a circular wireless network area. In the considered network, five servers are deployed around the center to transmit image data to $U = 50$ uniformly distributed users. Other system parameters are listed in Table 7.1. We use the scene graph generation model in Tang et al. [2020] for semantic information extraction and the multimodal data embedded model in Radford et al. [2021] for vectorization of semantic information and images. The visual genome (VG) [Krishna et al., 2016] dataset is used to train the introduced algorithm. For comparison purposes, we consider three baselines of RB allocation methods, which are given as follows:

Table 7.1 System parameters.

Parameter	Value	Parameter	Value
Q	05	W	2 MHz
V	5	U	50
P	1 W	N_0	−174 dBm/Hz
η	3	ρ	−3
ϵ	0.9	ϵ	0.5

- **(a) Random method**: In this method, each server allocates RB randomly to each user.
- **(b) Independent deep Q learning (IQL)**: In this method, each server does not consider the actions of other servers and trains its deep Q learning network [Tampuu et al., 2015] independently to maximize its own reward.
- **(c) Value decomposition-based deep Q learning network (VD-DQN)**: In this method, each server trains its deep Q learning network by cooperating with other servers based on the partial observation of environment. Different from IQL algorithm, VD-DQN enables each server to learn its deep Q learning network by backpropagating gradients from the Q learning rule using the total reward of all the servers rather than its own reward.

All experimental results are averaged over a large number of independent runs.

Figure 7.5 shows the relationships between the semantic score distribution of semantic triples and the original image. In particular, as the semantic score increases, the color of that semantic triple changes from white to green. For example, the semantic score of the greenest semantic triple "*woman holding*

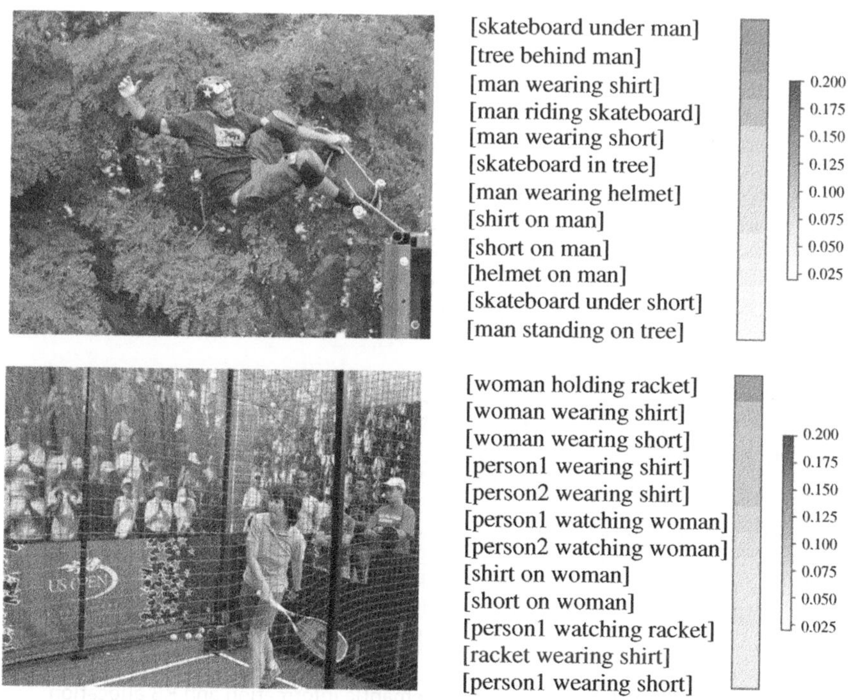

Figure 7.5 The semantic scores distribution of semantic triples. Source: Dickson et al. (2002)/IEEE.

racket" in Figure 7.5 is 0.1351. From Figure 7.5, we can see that the semantic triple with high semantic scores is more critical, e.g., the semantic triple "*man riding skateboard*" in Figure 7.5 and the semantic triple "*woman holding racket*" in Figure 7.5. In Figure 7.5, we can also see that, ranked by the semantic scores, the transmission priority of the redundant semantic triples, e.g., "*shirt on man*" and unreasonable semantic triples, e.g., "*man standing on tree*" is lower than other triples based on **Algorithm 1**.

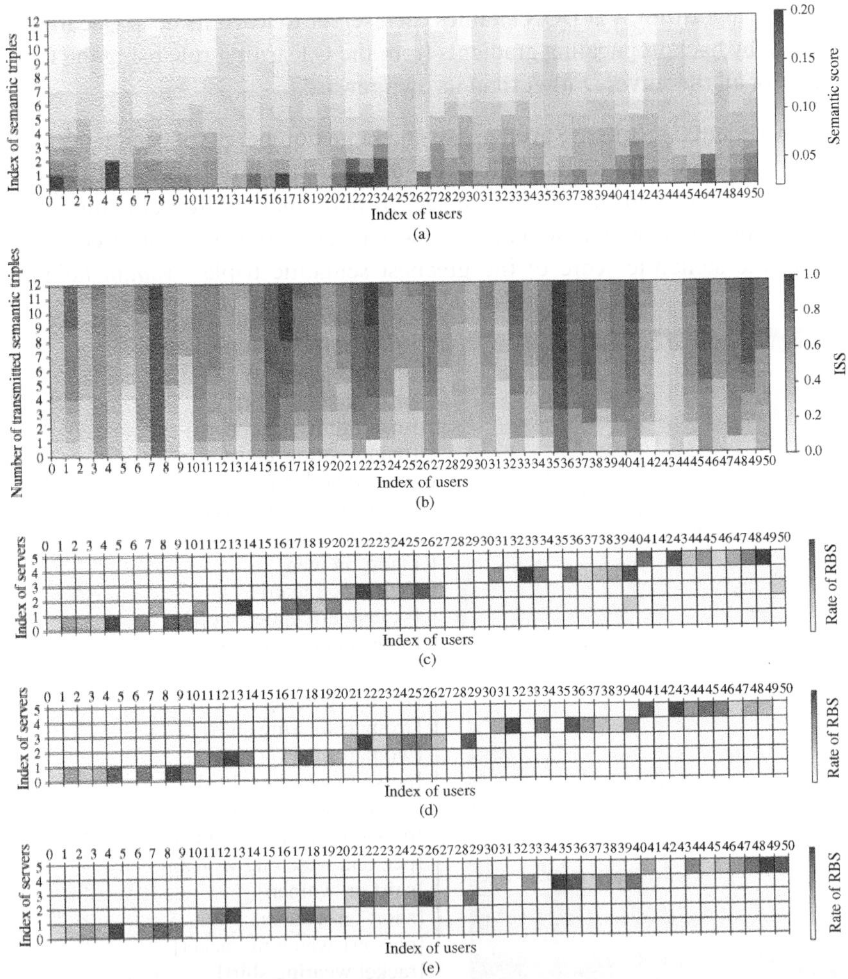

Figure 7.6 Correlation between transmitted semantic information and RB allocation policy: (a) Semantic scores distribution of users, (b) ISS varies as the number of transmitted semantic triples varies, (c) RB allocation of baseline b), (d) RB allocation of baseline c), and (e) RB allocation of the VD-ERL algorithm.

Figure 7.6 shows the correlation between the transmitted semantic information and RB allocation policies of the introduced VD-ERL algorithm and baselines. In particular, in Figure 7.6a, as the semantic score increases, the color of the semantic triple changes from white to green. Similarly, in Figure 7.6b, the color of the transmitted partial semantic triples changes from white to green as the ISS of transmitted semantic information increases. From Figure 7.6b, we can see that the ISS monotonically increases as the number of transmitted semantic triples increases. From Figure 7.6a,b, we can see that the semantic score can describe the importance of each semantic triple with small error. For example, the semantic scores of the semantic information transmitted to user 5 are smaller than other semantic information and its ISS is also smaller than that of other semantic information. Figure 7.6c–e show the RB allocation results of baseline (b), baseline (c), and the introduced VD-ERL algorithm, respectively. In particular, the user index is determined by the distance between the user and the nearby server. A user with minimum distance will have a smallest index. For example, users 0 to 9 are close to server 1, users 10 to 19 are close to server 2, users 20 to 29 are close to server 3, and so on. In these figures, as the rate of RB increases, the color of that RB becomes greener. Then, in Figure 7.6c–e, we can see that compared to independent deep Q learning algorithm, the introduced VD-ERL algorithm enables all servers to cooperatively determine RB allocation for each user. For example, as shown in Figure 7.6c, both server 2 and server 4 intend to allocate RBs to user 39, which causes a multi-RB allocation problem.

7.5 Conclusion

In this chapter, we have developed a novel image semantic communication framework that enables a set of servers to collaboratively transmit images to their associated users using semantic communication techniques. We have modeled the semantic information of each image as a scene graph that consists of a set of objects and relationships between them. We have introduced an ISS metric to evaluate the semantic similarity between the original image and its textual semantic information. Under the limited wireless resource constraints, each server must jointly determine the semantic information to be transmitted and the RB allocation scheme. This problem is formulated as an optimization problem whose goal is to minimize the average transmission latency while meeting the ISS requirement. To solve this problem, we have developed a value decomposition-based entropy-maximized multi-agent RL algorithm that enables servers to find an optimal cooperative RBs allocation scheme based on local observation of each server. Simulation results have shown that, compared with traditional multi-agent RL algorithms, the introduced algorithm significantly reduces the transmission latency and improves the convergence speed.

Bibliography

J. Bao, P. Basu, M. Dean, C. Partridge, A. Swami, W. Leland, and J. A. Hendler. Towards a theory of semantic communication. In *Proceedings of IEEE Network Science Workshop*, West Point, NY, USA, Jun. 2011. doi: 10.1109/NSW.2011.6004632.

P. Basu, J. Bao, M. Dean, and J. A. Hendler. Preserving quality of information by using semantic relationships. In *Proceedings of IEEE International Conference on Pervasive Computing and Communications Workshops*, Lugano, Switzerland, Mar. 2012. doi: 10.1109/PerComW.2012.6197583.

C. Chaccour, W. Saad, M. Debbah, Z. Han, and H. V. Poor. Less data, more knowledge: Building next generation semantic communication networks. URL https://arxiv.org/abs/2211.14343, 2022.

M. Chen, D. Gündüz, K. Huang, W. Saad, M. Bennis, A. V. Feljan, and H. V. Poor. Distributed learning in wireless networks: Recent progress and future challenges. *IEEE Journal on Selected Areas in Communications*, 39(12):3579–3605, Dec. 2021.

M. Corbetta and G. L. Shulman. Control of goal-directed and stimulus-driven attention in the brain. *Nature Reviews Neuroscience*, 3(3):201–215, Mar. 2002. doi: 10.1038/nrn755.

T. Haarnoja, A. Zhou, P. Abbeel, and S. Levine. Soft actor-critic: Off-policy maximum entropy deep reinforcement learning with a stochastic actor. URL http://arxiv.org/abs/1801.01290, Jan. 2018.

T. Han, Q. Yang, Z. Shi, S. He, and Z. Zhang. Semantic-preserved communication system for highly efficient speech transmission. URL https://arxiv.org/abs/2205.12727, 2022.

Y. Hu, M. Chen, W. Saad, H. V. Poor, and S. Cui. Distributed multi-agent meta learning for trajectory design in wireless drone networks. *IEEE Journal on Selected Areas in Communications*, 39(10):3177–3192, Oct. 2021. doi: 10.1109/JSAC.2021.3088689.

D. Huang, F. Gao, X. Tao, Q. Du, and J. Lu. Towards semantic communications: Deep learning-based image semantic coding. URL https://arxiv.org/abs/2208.04094, Aug. 2022.

P. Jiang, C. K. Wen, S. Jin, and G. Y. Li. Wireless semantic communications for video conferencing. URL https://arxiv.org/pdf/2204.07790v1, 2022.

M. Kalfa, S. Y. Yetim, A. Atalik, M. Gok, Y. Ge, R. Li, W. Tong, T. M. Duman, and O. Arikan. Reliable extraction of semantic information and rate of innovation estimation for graph signals. URL https://arxiv.org/abs/2211.05440, 2022.

J. Kang, H. Du, X. Li, Z. Xiong, S. Ma, D. Niyato, and Y. Li. Personalized saliency in task-oriented semantic communications: Image transmission and performance analysis. URL https://arxiv.org/abs/2209.12274, 2022.

R. Krishna, Y. Zhu, O. Groth, J. Johnson, K. Hata, J. Kravitz, S. Chen, Y. Kalantidis, L. Li, D. A. Shamma, M. S. Bernstein, and F. Li. Visual genome: Connecting language and vision using crowdsourced dense image annotations. URL https://arxiv.org/abs/1602.07332, Feb. 2016.

X. Li, J. Shi, and Z. Chen. Task-driven semantic coding via reinforcement learning. *IEEE Transactions on Image Processing*, 30:6307–6320, Jul. 2021. doi: 10.1109/TIP.2021.3091909.

Z. Lin, Y. Gong, and K. Huang. Distributed over-the-air computing for fast distributed optimization: Beamforming design and convergence analysis. URL https://arxiv.org/abs/2204.06876v1, 2022.

K. Lu, R. Li, X. Chen, Z. Zhao, and H. Zhang. Reinforcement learning-powered semantic communication via semantic similarity. URL https://arxiv.org/abs/2108.12121, Aug. 2021.

X. Luo, H. Chen, and Q. Guo. Semantic communications: Overview, open issues, and future research directions. *IEEE Wireless Communications*, 29(1):210–219, Jan. 2022. doi: 10.1109/MWC.101.2100269.

Z. Meng, C. She, G. Zhao, and D. D. Martini. Sampling, communication, and prediction co-design for synchronizing the real-world device and digital model in metaverse. URL https://arxiv.org/abs/2208.04233, 2022.

X. Mu, Y. Liu, L. Guo, and N. Al-Dhahir. Heterogeneous semantic and bit communications: A semi-NOMA scheme. URL https://arxiv.org/abs/2205.02620, 2022.

Z. Qin, X. Tao, J. Lu, and G. Y. Li. Semantic communications: Principles and challenges. URL https://arxiv.org/abs/2201.01389, 2022.

A. Radford, J. W. Kim, C. Hallacy, A. Ramesh, G. Goh, S. Agarwal, G. Sastry, A. Askell, P. Mishkin, J. Clark, G. Krueger, and I. Sutskever. Learning transferable visual models from natural language supervision. doi: 10.48550/ARXIV.2103.00020. URL https://arxiv.org/abs/2103.00020, Feb. 2021.

W. Saad, M. Bennis, and M. Chen. A vision of 6G wireless systems: Applications, trends, technologies, and open research problems. *IEEE Network*, 34(3):134–142, May. 2020. doi: 10.1109/MNET.001.1900287.

G. Shi, Y. Xiao, Y. Li, and X. Xie. From semantic communication to semantic-aware networking: Model, architecture, and open problems. *IEEE Communications Magazine*, 59(8):44–50, Sep. 2021. doi: 10.1109/MCOM.001.2001239.

P. Sunehag, G. Lever, A. Gruslys, W. M. Czarnecki, V. Zambaldi, M. Jaderberg, M. Lanctot, N. Sonnerat, J. Z. Leibo, K. Tuyls, and T. Graepel. Value-decomposition networks for cooperative multi-agent learning. URL http://arxiv.org/abs/1706.05296, Jun. 2017.

A. Tampuu, T. Matiisen, D. Kodelja, I. Kuzovkin, K. Korjus, J. Aru, J. Aru, and R. Vicente. Multiagent cooperation and competition with deep reinforcement learning. URL http://arxiv.org/abs/1511.08779, Nov. 2015.

P. Tandon, S. Chandak, P. Pataranutaporn, Y. Liu, A. M. Mapurange, P. Maes, T. Weissman, and M. Sra. Txt2Vid: Ultra-low bitrate compression of talking-head videos via text. URL https://arxiv.org/abs/2106.14014v3, 2022.

K. Tang, Y. Niu, J. Huang, J. Shi, and H. Zhang. Unbiased scene graph generation from biased training. In *Proceedings of IEEE/CVF Conference on Computer Vision and Pattern Recognition*, Seattle, WA, USA, Jun. 2020. doi: 10.1109/CVPR42600.2020.00377.

Y. Wang, M. Chen, Z. Yang, T. Luo, and W. Saad. Deep learning for optimal deployment of UAVs with visible light communications. *IEEE Transactions on Wireless Communications*, 19(11):7049–7063, Nov. 2020. doi: 10.1109/TWC.2020.3007804.

J. Wang, Y. Duan, X. Tao, M. Xu, and J. Lu. Semantic perceptual image compression with a Laplacian pyramid of convolutional networks. *IEEE Transactions on Image Processing*, 30:4225–4237, Mar. 2021. doi: 10.1109/TIP.2021.3065244.

Y. Wang, M. Chen, T. Luo, W. Saad, D. Niyato, H. V. Poor, and S. Cui. Performance optimization for semantic communications: An attention-based reinforcement learning approach. *IEEE Journal on Selected Areas in Communications*, 40(9):2598–2613, Jul. 2022. doi: 10.1109/JSAC.2022.3191112.

L. Xiao, Y. Ding, J. Huang, S. Liu, Y. Tang, and H. Dai. UAV anti-jamming video transmissions with QoE guarantee: A reinforcement learning-based approach. *IEEE Transactions on Communications*, 69(9):5933–5947, Jun. 2021. doi: 10.1109/TCOMM.2021.3087787.

H. Xie, Z. Qin, and G. Y. Li. Task-oriented multi-user semantic communications for VQA. *IEEE Wireless Communications Letters*, 11(3):553–557, Dec. 2022. doi: 10.1109/LWC.2021.3136045.

Y. Yang, C. Guo, F. Liu, C. Liu, L. Sun, Q. Sun, and J. Chen. Semantic communications with AI tasks. URL https://arxiv.org/abs/2109.14170, 2021.

Z. Yang, M. Chen, G. Li, Y. Yang, and Z. Zhang. Secure semantic communications: Fundamentals and challenges. URL https://arxiv.org/abs/2301.01421, 2023.

H. Zhang, S. Shao, M. Tao, X. Bi, and K. B. Letaief. Deep learning-enabled semantic communication systems with task-unaware transmitter and dynamic data. URL https://arxiv.org/abs/2205.00271, 2022.

W. Zhou, A. C. Bovik, H. R. Sheikh, and E. P. Simoncelli. Image quality assessment: From error visibility to structural similarity. *IEEE Transactions on Image Processing*, 13(4):600–612, Apr. 2004. doi: 10.1109/TIP.2003.819861.

G. Zhu, D. Liu, Y. Du, C. You, J. Zhang, and K. Huang. Toward an intelligent edge: Wireless communication meets machine learning. *IEEE Communications Magazine*, 58(1):19–25, 2020. doi: 10.1109/MCOM.001.1900103.

H. Zou, C. Zhang, S. Lasaulce, L. Saludjian, and H. V. Poor. Goal-oriented quantization: Analysis, design, and application to resource allocation. URL https://arxiv.org/abs/2209.15347, 2022.

8

Multiple Access Design for Joint Semantic and Classical Communications

Xidong Mu[1] *and Yuanwei Liu*[2]

[1] *School of Electronic Engineering and Computer Science, Queen Mary University of London, London, UK*
[2] *Department of Electrical and Electronic Engineering, The University of Hong Kong, Hong Kong, Island*

8.1 Introduction

Following the pioneering contributions of the Shannon classical information theory [Shannon and Weaver, 1949], wireless communications have significantly developed from the first generation (1G) to the fifth generation (5G) only in a few decades. For approaching the fundamental Shannon rate limit, diverse efficient technologies have been developed, such as massive multiple-input multiple-output (MIMO), mmWave/THz communication, and turbo codes [Boccardi et al., 2014]. Nevertheless, the development of wireless communication is far from the end and extensive research efforts have been devoted to developing the next generation (6G) wireless communication system [Liu et al., 2020; Tong and Li, 2022]. This is because, on the one hand, the requirements of communication performance are still explosively growing (e.g., Tbps-level data rates, 0.1 ms-level latency, 99.99999% reliability, etc.). On the other hand, the communication services have to be provided for both conventional human-type users and new machine-type users as well as Internet-of-Things (IoT) users. To achieve these goals, the effectiveness of relentlessly increasing of the size of transceivers and the frequencies of operating carriers is questionable, given the extremely high hardware cost and energy consumption. Moreover, one important observation is that current wireless communication designs still focus on the technical-level problem [Shannon and Weaver, 1949] using the Shannon classical information theory, i.e., *how accurately can the symbols of communication be transmitted?* There is a paucity of investigations on the other two semantic- and effectiveness-level problems [Shannon and Weaver, 1949], i.e., *how precisely do the transmitted symbols convey the desired meaning?* And *how effectively does the*

Foundations of Semantic Communication Networks, First Edition.
Edited by Walid Saad, Christina Chaccour, Christo Kurisummoottil Thomas, and Merouane Debbah.

received meaning affect conduct in the desired way?, which are more related to the ultimate goal of communications.

In response to the semantic- and effectiveness-level communication problems, semantic communications have recently attracted significant research attentions from both industry and academia [Qin et al., 2021; Gündüz et al., 2023]. In semantic communications, only the key information that contains the specific meaning/actions/goals relevant to the destinations needs to be transmitted. This is different from conventional bit-based communications, the entire sources have to be transmitted [Qin et al., 2021]. Figure 8.1 illustrates the difference between conventional bit-based communications and semantic communications in an end-to-end scenario. Here, we take the image recognition as an example. The transmitter in conventional bit-based communications sends the bit sequences converted from the entire image, which allows the receiver to fully recover the image for recognizing the cat. By contrast, the transmitter in semantic communications only needs to transmit the key features extracted by the semantic encoder. Upon receiving the semantic symbols, the semantic decoder enables the receiver to directly recognize that the received information is a cat. Compared to the conventional bit-based communications, the abundant information (e.g., image background and colors) for the image recognition is safely dropped in semantic communications, thus focusing on the semantic- and effectiveness-level design [Qin et al., 2021]. Given the significant breakthrough of artificial intelligence in semantic information extraction, many efficient approaches have been developed for facilitating semantic communications, such as text/speech/image/video transmission [Weng and Qin, 2021; Xie et al., 2021; Kang et al., 2023], image classification, machine translation, and visual question answering [Xie et al., 2022]. The salient advantages of semantic communications are listed as follows:

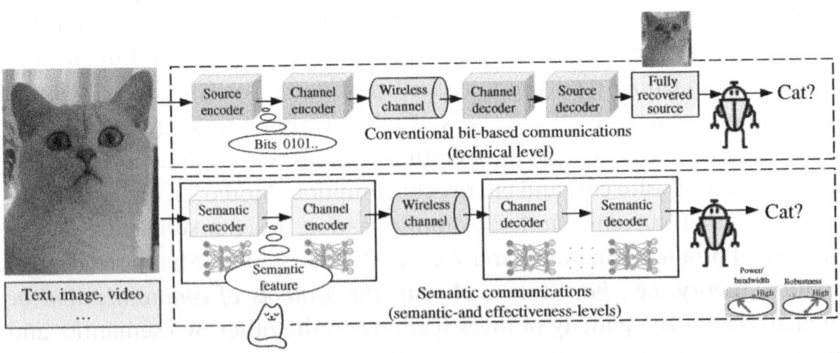

Figure 8.1 Conventional bit-based communications and new semantic communications.

- **Sustainability**: As only the necessary information needs to be transmitted, the power consumption and bandwidth required in semantic communications can be greatly reduced, thus leading to a sustainable and green communication paradigm.
- **Reliability**: Semantic communications exhibit robustness in undesired wireless environment, e.g., in the low signal-to-noise ratio (SNR) regime, thus ensuring the reliability of communication systems.
- **Intelligence**: The information transmitted in semantic communications depends on the specific tasks, thus achieving intelligent information exchange.

Given the above benefits, it is expected that semantic communications provide a new paradigm of information transmission for future wireless networks. However, the integration of semantic communications into multiuser networks leads to highly heterogeneous features. On the one hand, from the network perspective, both new semantic-based users (S-users) and existing bit-based users (B-users) have to be accommodated. On the other hand, from the user perspective, both semantic and bit communication approaches can be employed for information transmission. This calls for efficient multiple access and resource management for supporting joint semantic and classic bit-oriented communications, which provide the main motivation of developing this chapter.

The rest of this chapter is organized as follows. A heterogeneous semantic and bit multiuser network is discussed with the potential employment of orthogonal multiple access (OMA) or non-orthogonal multiple access (NOMA), followed by discussing the promising interplay between semantic communications and NOMA. (i) In the context of "NOMA enabled semantic communications," a semi-NOMA enabled heterogeneous semantic and bit multiuser communication design is proposed. (ii) In the context of "semantic communications enhanced NOMA," a novel opportunistic semantic and bit communication approach is proposed for alleviating the early-late rate disparity issue in NOMA. For each proposed design, numerical examples are provided for verifying the effectiveness. Finally, conclusions are drawn.

8.2 Heterogeneous Semantic and Bit Multiuser Network

Figure 8.2 illustrates the considered heterogeneous semantic and bit multiuser network, which consists of a semantics and bits supported base station (BS) to serve both the conventional B-users and the new emerging S-users for satisfying different communication requirements. As illustrated, S-users can be robotic users in smart factories for task executions, while B-users are conventional IoT

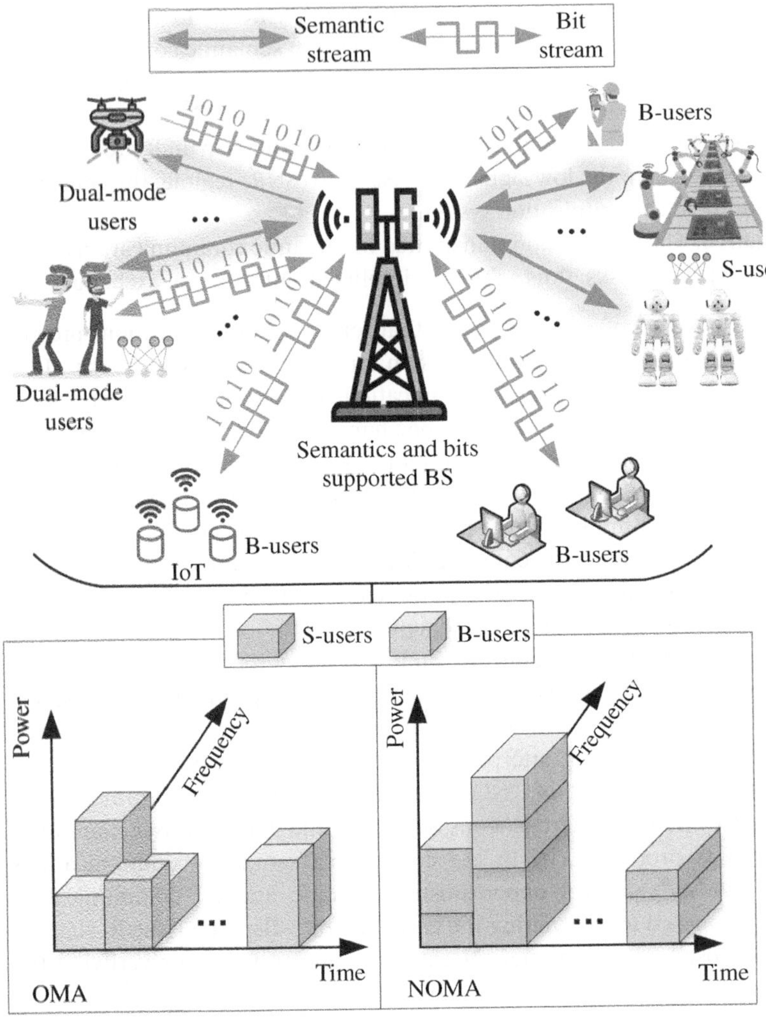

Figure 8.2 The considered heterogeneous semantic and bit multiuser network employing OMA and NOMA.

devices for data collections or laptops for file transmission. Moreover, users can also work in a dual semantic and bit communication mode. For example, the BS control an unmanned aerial vehicle (UAV) for video monitoring via semantic communications and the UAV sends the video back to the BS via bit communications considering the high-quality air-to-ground link. Moreover, for virtual reality (VR)/augmented reality (AR) users in immersive games, semantic and bit communications would coexist for supporting different information flows and realizing multimodal experience (e.g., video, control, haptic perception, etc.). Therefore, compared to current bit-only networks, the considered heterogeneous semantic and bit multiuser network enables the information transmission to be realized in the most efficient manner possible in terms of different applications.

8.2.1 Multiple Access for the Heterogeneous Semantic and Bit Multiuser Network

For facilitating the considered heterogeneous semantic and bit multiuser network, multiple access is one of the most fundamental enablers given the limited wireless resources. Here, we broadly classify the multiple access schemes into two categories, namely OMA and NOMA.

- **OMA**: On the one hand, OMA can be employed in the considered heterogeneous multiuser network by allocating S- and B-users with different orthogonal time/frequency resource blocks, as illustrated in Figure 8.2. Like conventional bit-only multiuser communications, the advantages of OMA are that there is no interference between each user and it is easy for implementation. Therefore, no redesigns are required in the individual semantic and bit communications. However, OMA usually has a low resource efficiency and cannot support massive connectivity. This shortcoming of OMA will become more significant when involving new S-users in the considered heterogeneous multiuser network.
- **NOMA**: On the other hand, as illustrated in Figure 8.2, NOMA can accommodate S- and B-users over the same time/frequency resource blocks and increase the connectivity performance for the considered heterogeneous multiuser network. Therefore, NOMA is more promising to be employed than OMA. However, note that the multiuser interference mitigation is the core task in NOMA [Ding et al., 2020]. How to efficiently use NOMA and design the corresponding multiuser interference mitigation in the heterogeneous network are challenging tasks.

8.2.2 Interplay Between Semantic Communications and NOMA

Against the above discussion, in this chapter, we focus our attentions on the employment of NOMA to efficiently support the above heterogeneous semantic and bit multiuser network. In particular, the attractive interplay between semantic communications and NOMA can be summarized as follows.

- **NOMA-enabled semantic communications**: On the one hand, NOMA provides efficient and flexible resource allocation for incorporating semantic communications and improving the resource efficiency and connectivity of the considered heterogeneous multiuser network. This constitutes the most foundation for facilitating diversified applications in future wireless networks, such as VR, AR, and metaverse. The details of NOMA-enabled heterogeneous semantic and bit multiuser communications will be discussed in Section 8.3.

- **Semantic communications-enhanced NOMA**: On the other hand, the new information transmission paradigm provided by semantic communications can also improve the performance of NOMA. As discussed above, semantic communications are more robust and sustainable than conventional bit communications. Applying semantic communications in NOMA can enhance the performance of users having a poor channel conditions and/or reduce the required transmit power at users, i.e., reducing the multiuser interference for NOMA. This enables us to employ semantic communications for improving NOMA performance, the details of which will be discussed in Section 8.4.

8.3 NOMA-Enabled Heterogeneous Semantic and Bit Multiuser Communications

In this section, we focus on "NOMA-enabled semantic communications." We will first introduce the new performance metric, namely semantic rate, for characterizing the performance of semantic communications. Then, we propose a novel semi-NOMA scheme for realizing the heterogeneous semantic and bit multiuser communications, which is capable of achieving the best semantic-versus-bit rate region.

8.3.1 Semantic Rate: A New Performance Metric

Conventional bit communications focus on the generalized performance metrics, such as symbol-error rate (SER) and bit-error rate (BER), regardless of the types of original sources. The classic Shannon capacity characterizes the performance upper bound of bit communications by ignoring the specific meaning contained in the bit sequences. However, as semantic communications focus on the correctness of information transmission in the semantic domain, new performance metrics have to be developed for characterizing its performance. Depending on the different types of sources (e.g., text, speech, image) and different tasks/goals to be accomplished, the corresponding performance metrics for semantic communications are in general different. This also imposes challenges in multiple access design and resource management when incorporating semantic communications into multiuser networks. As a first step, the authors of Guler and Yener [2014] formulated an index assignment problem and proposed to employ lexical taxonomy-based semantic distances to measure the semantic similarity between two words. More recently, based on one practical deep learning (DL)-enabled semantic communication (DeepSC) tool proposed in Xie et al. [2021] for text transmission, the authors Yan et al. [2022] proposed a new performance metric, namely semantic rate, to quantify the efficiency of semantic communications. In particular, the achieved semantic rate depends on the employed bandwidth and the achieved sentence semantic similarity, which is a function of the received SNR and the specific neural network structure used in the DeepSC. However, in Yan et al. [2022], there is no closed-form expressions provided for the sentence semantic similarity function, the value of which can only be obtained via experiments on DeepSC. To address this difficulty, we further proposed to employ the data regression method, where the "S-shape" sentence semantic similarity function was approximated by a generalized logistic function with a high accuracy [Mu et al., 2023]. This facilitates further theoretical studies and communication designs on semantic communications. It is also worth noting that the development of performance metrics for semantic communications is still in an initial stage, which requires further research efforts for different types of semantic communications.

8.3.2 Semi-NOMA: A Unified Multiple Access Scheme

As shown in Figure 8.3, we focus on the downlink case, where a BS aims to transmit the semantic stream and the bit stream to one S- and B-user over the given wireless resource. Based on the introduced semantic rate, we investigate how to efficiently support this heterogeneous multiuser communication with the employment of NOMA. Previously, the superiority of NOMA over OMA has been extensively demonstrated in bit-only multiuser communications. However, this conclusion may not hold in the new heterogeneous scenario and proper redesigns are required for exploiting the NOMA gain. Recalling the fact that for downlink NOMA in conventional bit communications, the successive interference cancellation (SIC) decoding order among users for mitigating the interference caused by spectrum sharing is determined by the order of users' channel conditions [Ding et al., 2020]. Users having higher channel conditions will first decode the signal of users having lower channel conditions and remove them from the superimposed signal. However, this "channel condition-based SIC ordering" cannot be applied to the new heterogeneous multiuser communication. This is because for semantic communications, the neural networks employed at the BS and the S-user should be jointly trained in advance [Xie et al., 2021], thus ensuring the successful implementation. As a result, the B-user in the heterogeneous transmission is almost impossible to decode the semantic signal from the superimposed semantic and bit signal and thus the SIC cannot be carried out. Nevertheless, as the bit communications do not require the prior training, S-user is still capable of decoding the bit signal from the superimposed semantic and bit

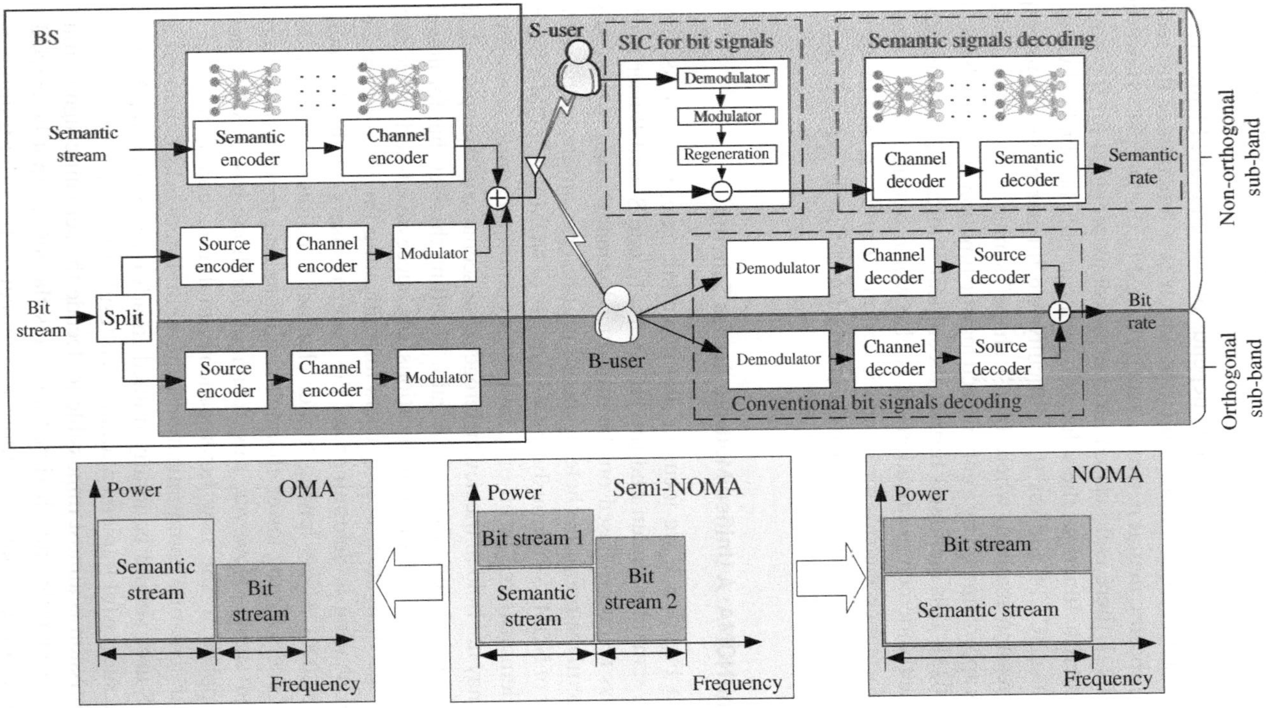

Figure 8.3 Semi-NOMA for heterogeneous semantic and bit communications.

signal. This leads to a "bits-to-semantics SIC ordering" when applying NOMA in the heterogeneous multiuser communication. However, despite the S-user can decode its intended semantic signal in an interference-free manner with the aid of SIC, the drawback is that the B-user will always suffer from the interference from the S-user. Therefore, if NOMA is directly applied in the heterogeneous multiuser communication (termed as pure NOMA), the performance of the B-user might cannot be well guaranteed.

To guarantee the performance of both S- and B-users, we propose a novel semi-NOMA scheme for realizing the heterogeneous multiuser communication [Mu et al., 2023]. As shown in Figure 8.3, the total bandwidth is divided into one non-orthogonal sub-band and one orthogonal sub-band. Based on this bandwidth allocation, the operating mechanism of each node in the proposed semi-NOMA scheme is summarized as follows.

- **BS**: As illustrated, the bit stream for the B-user should be split into two sub-streams at the BS, where one bit sub-stream is superimposed with the semantic stream and transmitted over the non-orthogonal sub-band while the other bit sub-stream is transmitted alone over the orthogonal sub-band.
- **S-user**: Upon receiving the superimposed from the non-orthogonal sub-band, the S-user decodes the superimposed signal following the aforementioned bits-to-semantics SIC ordering. As a result, the semantic signal is still decoded in an interference-free manner in semi-NOMA.
- **B-user**: Accordingly, the B-user decodes the two bit sub-streams received from the non-orthogonal and orthogonal sub-bands and combines the decoded results. It can be observed that compared to pure NOMA, partial information transmission of B-user in semi-NOMA can also be achieved in an interference-free manner.

The resource allocation among S- and B-users in the proposed semi-NOMA is illustrated at the bottom of Figure 8.3. It can be observed that by adjusting the bandwidth and power allocation, the proposed semi-NOMA can be simplified into OMA and NOMA. Notably, semi-NOMA also provides more transmission options, which cannot be achieved by OMA and NOMA. Therefore, the salient advantage of semi-NOMA is that it provides a unified multiple access scheme for facilitating the heterogeneous semantic and bit multiuser communication.

8.3.3 Fundamental Limit: Semantic-Versus-Bit Rate Region

We continue to investigate the fundamental performance limit of the considered heterogeneous semantic and bit multiuser communication, namely the semantic-versus-bit rate region, which characterizes all the semantic and bit rate-pairs achieved by S- and B-users for any given transmit power and bandwidth

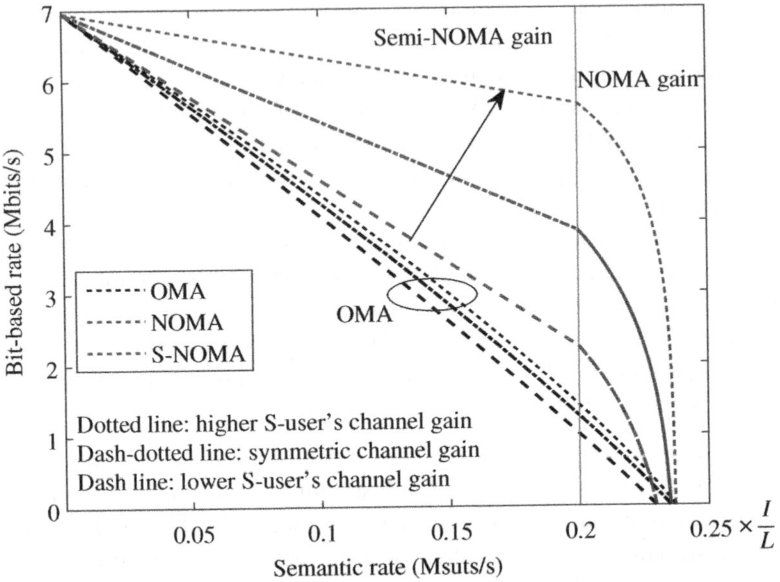

Figure 8.4 The achieved semantic-versus-bit rate region achieved by different multiple access schemes. The system parameter settings can be found in Mu et al. [2023] Section VI]. Source: Mu et al. [2023]/With permission of IEEE.

[Mu et al., 2023]. Figure 8.4 compares the semantic-versus-bit rate region achieved by the proposed semi-NOMA and the conventional OMA and NOMA. In the provided numerical results, we fix the channel condition of the B-user and vary the channel condition of the S-user [Mu et al., 2023]. Therefore, compared to the channel condition of the B-user, we have three cases: (i) a higher channel condition at the S-user, (ii) a symmetric channel condition, and (iii) a lower channel condition at the S-user. As can be observed from Figure 8.4, in all three cases, semi-NOMA achieves the largest semantic-versus-bit rate region, which strictly contains those achieved by OMA and NOMA. This is indeed expected since semi-NOMA unifies OMA and NOMA and provides enhanced degrees-of-freedom for resource allocation in the heterogeneous multiuser communication. It can be also observed that due to the full resource sharing, NOMA is only capable of achieving a reduced semantic-versus-bit rate region, where the maximum bit rate is degraded by the interference from the S-user. Moreover, an interesting result is that even for the case of symmetric channel condition, OMA is still strictly suboptimal. This is fundamentally different from the result in the conventional bit-only multiuser communication, where OMA achieves the same performance as NOMA when users have the same channel conditions. This reveals the importance of using NOMA-based schemes when including semantic communications in multiuser network. By comparing the achieved semantic-versus-bit rate region in the three cases, we can find that it is preferable to pair an S-user having a higher channel condition with a B-user having a lower channel condition in semi-NOMA, which leads to a larger semantic-versus-bit rate region. This provides a useful guidance for implementing semi-NOMA. The above results confirm the effectiveness of the proposed semi-NOMA scheme in the considered heterogeneous semantic and bit multiuser communication.

8.4 Semantic Communications-Enhanced NOMA

In this section, we continue to discuss the employment of semantic communications for improving the performance of NOMA, i.e., "semantic communications-enhanced NOMA." We first introduce the "early-late" rate disparity issue among NOMA users. Then, we explore semantic communications to alleviate this issue and propose a novel opportunistic semantic and bit communication approach for NOMA users.

8.4.1 Early-Late Rate Disparity Issue in NOMA

Due to the resource sharing nature of NOMA, the mitigation of multiuser interference is the core task for further exploiting the limited wireless resources and supporting massive connectivity. To this end, NOMA relies on the employment of SIC to deal with the multiuser interference [Ding et al., 2020]. For implementing the SIC, users should be priorly ordered in terms of specific criteria, such as the order of users' channel conditions and the order of users' quality-of-service (QoS) requirements [Ding et al., 2020]. As a result, for any given SIC decoding order, users who are assigned at earlier decoding orders will always suffer from more multiuser interference than those who are assigned at later decoding orders, which yields the early-late rate disparity issue among NOMA users. In other words, the communication performance achieved by each NOMA user significantly varies in terms of their SIC decoding orders.

To further demonstrate the limitation caused by the early-late rate disparity issue, as shown in Figure 8.5, we consider a fundamental two-user uplink NOMA system, which includes one primary user and one secondary user. At the BS, the signal of the primary user is firstly decoded and then removed using SIC before the signal of the secondary user is decoded. Therefore, on the one hand, it is discouraged for the primary user to participate in NOMA since its performance will be degraded by the interference from the secondary user. On the other hand, although the secondary user's signal can be decoded in an interference-free manner, its maximum achieved performance is also conditioned on the communication requirement of the primary user. For example, if the primary user has a

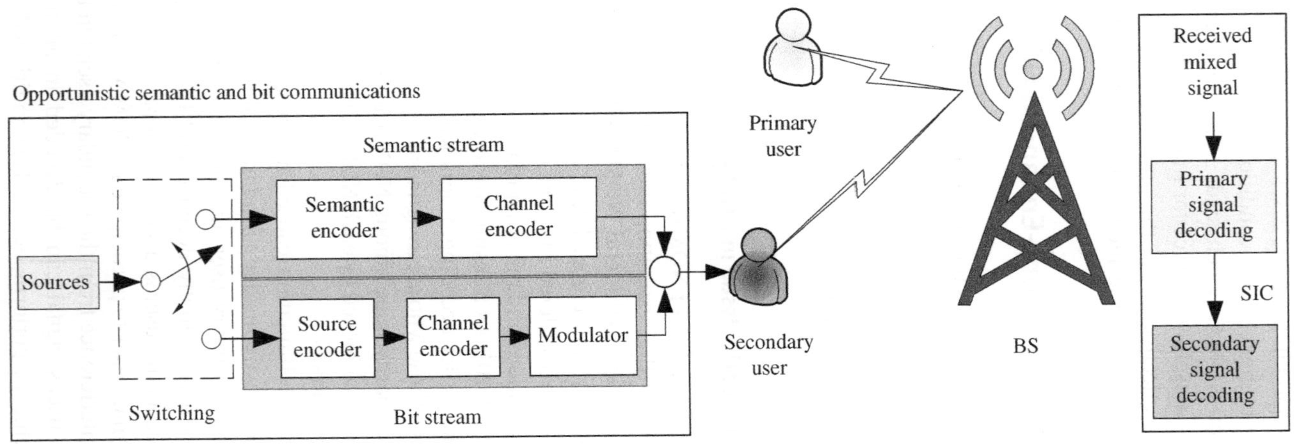

Figure 8.5 The proposed opportunistic semantic and bit communication approach for NOMA.

stringent communication requirement, the signal strength of the secondary user should be strictly capped for controlling the interference. Owing to the early-late rate disparity issue, it is impossible for the two NOMA users to simultaneously achieve high communication performance. As a result, previous works usually paired users with different QoS requirements together when using NOMA [Liu et al., 2017; Ding et al., 2020]. Nevertheless, the early-late rate disparity issue in NOMA has not been solved.

8.4.2 An Opportunistic Semantic and Bit Communication Approach for NOMA

Current research contributions have revealed that for achieving the same communication performance, semantic communications consume less transmit power than conventional bit communications [Qin et al., 2021]. It implies that semantic communications would cause less multiuser interference without degrading the communication performance. Motivated by this observation, as shown in Figure 8.5, we propose an opportunistic semantic and bit communication approach for NOMA users. By switching between the two modes, the original sources at the user's end are converted into either the semantic stream or the bit stream. The proposed opportunistic approach enables the user to employ the most suitable communication strategy for participating in NOMA with other users. Let us still take the aforementioned two-user as an example. Intuitively, if the primary user has a stringent communication requirement and/or the secondary user experiences a poor channel condition, the employment of semantic communications is desired. Otherwise, the employment of bit communications would be desired.

8.4.3 Numerical Case Studies

We continue to provide numerical examples to show the performance gain of employing semantic communications for NOMA and the effectiveness of the proposed opportunistic communication approach. The simulation results are obtained by considering a primary user using bit communications and a secondary user using the proposed opportunistic semantic and bit communication approach [Mu and Liu, 2023]. We study the NOMA performance over fading channels, where the secondary user has to choose the desired strategy in each fading state. For comparison, two baselines are also considered, where the secondary user merely uses semantic communications or bit communications. Moreover, if the secondary user employs bit communications, the corresponding achieved bit rate is converted into the equivalent semantic rate for a fair comparison. Figure 8.6 shows the ergodic (equivalent) semantic rate achieved

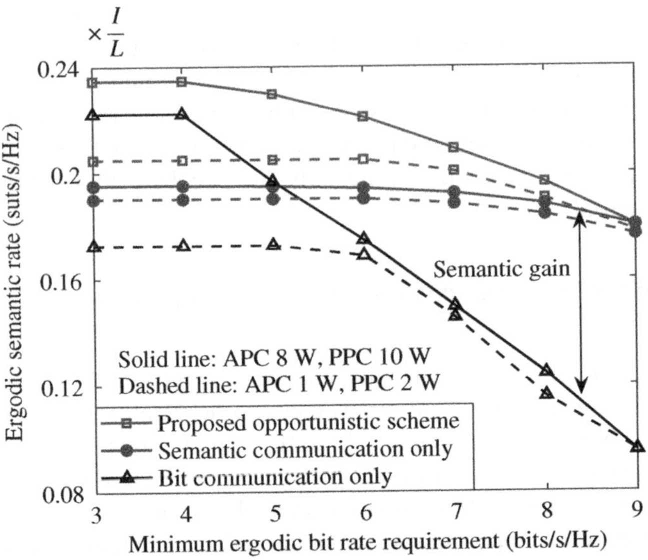

Figure 8.6 The ergodic (equivalent) semantic rate achieved by the secondary user versus the ergodic bit rate required by the primary user in different schemes. The system parameter settings can be found in Mu and Liu [2023, Section V]. Source: Mu and Liu [2023]/With permission of IEEE.

by the secondary user over fading channels versus the ergodic bit rate required by the primary user, subject to the average power constraint (APC) and the peak power constraint (PPC) of the secondary user. It can be observed that by reaping the benefits of both semantic and bit communications, the proposed opportunistic scheme achieves the best performance for NOMA. It allows the secondary user not only to efficiently control the interference imposed to the primary user but also to enhance it own communication performance. We can also observe that bit communications are preferable to be employed when the primary user has a low communication requirement and the secondary user has a sufficiently large power budget. Moreover, when the primary user has a stringent communication requirement and/or the secondary user has limited transmit power, the employment of semantic communications can achieve a significant performance gain for NOMA than bit communications. The provided results confirm the effectiveness of employing semantic communications for enhancing the performance of NOMA.

8.5 Concluding Remarks and Future Research

In this chapter, the interplay between NOMA and semantic communication was discussed to support the heterogeneous semantic and bit multiuser network, namely NOMA-enabled semantic communications and semantic communications-enhanced NOMA. For NOMA-enabled semantic communications, a semi-NOMA-enabled heterogeneous semantic and bit communication scheme was proposed, which unifies conventional OMA and NOMA as special cases and provides more flexible transmission options. It was also numerically shown that the proposed semi-NOMA scheme can achieve the largest semantic-versus-bit rate region. For semantic communications-enhanced NOMA, an opportunistic semantic and bit communication approach was conceived for alleviating the early-late rate disparity issue in NOMA. Numerical studies showed that the employment of semantic communications enables the NOMA users to better balance between their own achieved performance and the imposed interference, i.e., further improving the NOMA performance.

Bibliography

Federico Boccardi, Robert W. Heath, Angel Lozano, Thomas L. Marzetta, and Petar Popovski. Five disruptive technology directions for 5G. *IEEE Communications Magazine*, 52(2):74–80, 2014. doi: 10.1109/MCOM.2014.6736746.

Zhiguo Ding, Robert Schober, and H. Vincent Poor. Unveiling the importance of SIC in NOMA systems—Part 1: State of the art and recent findings. *IEEE Communications Letters*, 24(11):2373–2377, 2020. doi: 10.1109/LCOMM.2020.3012604.

Basak Guler and Aylin Yener. Semantic index assignment. In *2014 IEEE International Conference on Pervasive Computing and Communication Workshops (PERCOM WORKSHOPS)*, pages 431–436, 2014. doi: 10.1109/PerComW.2014.6815245.

Deniz Gündüz, Zhijin Qin, Inaki Estella Aguerri, Harpreet S. Dhillon, Zhaohui Yang, Aylin Yener, Kai Kit Wong, and Chan-Byoung Chae. Beyond transmitting bits: Context, semantics, and task-oriented communications. *IEEE Journal on Selected Areas in Communications*, 41(1):5–41, 2023. doi: 10.1109/JSAC.2022.3223408.

Jiawen Kang, Hongyang Du, Zonghang Li, Zehui Xiong, Shiyao Ma, Dusit Niyato, and Yuan Li. Personalized saliency in task-oriented semantic communications:

Image transmission and performance analysis. *IEEE Journal on Selected Areas in Communications*, 41(1):186–201, 2023. doi: 10.1109/JSAC.2022.3221990.

Yuanwei Liu, Zhijin Qin, Maged Elkashlan, Zhiguo Ding, Arumugam Nallanathan, and Lajos Hanzo. Nonorthogonal multiple access for 5G and beyond. *Proceedings of the IEEE*, 105(12):2347–2381, 2017. doi: 10.1109/JPROC.2017.2768666.

Guangyi Liu, Yuhong Huang, Na Li, Jing Dong, Jing Jin, Qixing Wang, and Nan Li. Vision, requirements and network architecture of 6G mobile network beyond 2030. *China Communications*, 17(9):92–104, 2020. doi: 10.23919/JCC.2020.09.008.

Xidong Mu and Yuanwei Liu. Exploiting semantic communication for non-orthogonal multiple access. *IEEE Journal on Selected Areas in Communications*, 41(8):2563–2576, 2023. doi: 10.1109/JSAC.2023.3288242.

Xidong Mu, Yuanwei Liu, Li Guo, and Naofal Al-Dhahir. Heterogeneous semantic and bit communications: A semi-NOMA scheme. *IEEE Journal on Selected Areas in Communications*, 41(1):155–169, 2023. doi: 10.1109/JSAC.2022.3222000.

Zhijin Qin, Xiaoming Tao, Jianhua Lu, Wen Tong, and Geoffrey Ye Li. Semantic communications: Principles and challenges. arXiv:2201.01389, 2021.

Claude E. Shannon and Warren Weaver. *The Mathematical Theory of Communication*. University of Illinois Press, 1949.

Wen Tong and Geoffrey Ye Li. Nine challenges in artificial intelligence and wireless communications for 6G. *IEEE Wireless Communications*, 29(4):140–145, 2022. doi: 10.1109/MWC.006.2100543.

Zhenzi Weng and Zhijin Qin. Semantic communication systems for speech transmission. *IEEE Journal on Selected Areas in Communications*, 39(8):2434–2444, 2021. doi: 10.1109/JSAC.2021.3087240.

Huiqiang Xie, Zhijin Qin, Geoffrey Ye Li, and Biing-Hwang Juang. Deep learning enabled semantic communication systems. *IEEE Transactions on Signal Processing*, 69:2663–2675, 2021. doi: 10.1109/TSP.2021.3071210.

Huiqiang Xie, Zhijin Qin, Xiaoming Tao, and Khaled B. Letaief. Task-oriented multi-user semantic communications. *IEEE Journal on Selected Areas in Communications*, 40(9):2584–2597, 2022. doi: 10.1109/JSAC.2022.3191326.

Lei Yan, Zhijin Qin, Rui Zhang, Yongzhao Li, and Geoffrey Ye Li. Resource allocation for text semantic communications. *IEEE Wireless Communications Letters*, 11(7):1394–1398, 2022. doi: 10.1109/LWC.2022.3170849.

9

Contextual Reasoning-based Semantics-Native Communication

Hyowoon Seo[1], Yoon Huh[2], Heekang Song[3], Wan Choi[2], and Mehdi Bennis[4]

[1]*Department of Electronics and Communications Engineering, Kwangwoon University, Seoul, Republic of Korea*
[2]*Department of Electrical and Computer Engineering, Seoul National University, Seoul, Republic of Korea*
[3]*School of Electrical Engineering, Korea Advanced Institute of Science and Technology, Daejeon, Republic of Korea*
[4]*Centre for Wireless Communications, University of Oulu, Oulu, Finland*

9.1 Semantics-Native Communication

In traditional communication systems, including those that employ semantic communication (SC) frameworks, it is customary to meticulously design encoders and decoders before their deployment to senders and receivers. This methodical approach enhances communication performance, ensuring efficiency, and effectiveness in executing tasks. However, this conventional methodology encounters limitations in situations lacking a central computing entity, necessitating that senders and receivers independently develop and implement their encoders and decoders. This aspect of independent encoder and decoder development, largely unexplored in current literature, presents a unique challenge. Furthermore, while machine learning (ML) solutions have sought to address these challenges, the opaque nature of ML processes obscures our understanding and optimization of SC operations.

This chapter seeks to clarify the complexities of SC among intelligent agents by introducing a mathematical model inspired by linguistics and information theory—the semantics-native communication (SNC) model. This model emphasizes the mapping of entities of interest into semantic concepts and symbols, forming a semantic representation (SR). This is a crucial step that precedes traditional encoding and decoding processes, aiming to recreate the intended action from the received SR. By integrating semantic coding with classical encoding and

Foundations of Semantic Communication Networks, First Edition.
Edited by Walid Saad, Christina Chaccour, Christo Kurisummoottil Thomas, and Merouane Debbah.

decoding, we gain valuable insights into their functionalities, thereby advancing the principles of SC.

A key innovation of the SNC model is the incorporation of contextual reasoning (CR), a process of internal simulation and iterative reasoning about the communication context of the counterpart. This process, inspired by the concept of "I think of you thinking of me thinking of you," and informed by recent literature, optimizes semantic coding for specific tasks and contexts, significantly reducing communication overhead by transmitting only the most pertinent semantics.

The primary objective of this chapter is to introduce the SNC model and its fundamental component, CR, highlighting their roles in enhancing communication efficiency and task execution. Sections 9.2–9.4 delve into the semantic coding system model of SNC, the intricacies of CR, and its practical applications in communication models. We explore how agents deduce each other's contexts in asynchronous communication and implement SNC based on these inferences. Moreover, the chapter examines the implementation of the information bottleneck (IB) principle through CR, concluding with a summary of the chapter's findings.

9.1.1 System Model

The basic SNC framework envisions a collaborative system involving two agents, a sender and a receiver, as depicted in Figure 9.1, engaged in a communication-dependent task. In order to thoroughly examine the characteristics of SNC, we focus intensively on the model related to the semantic coding aspect of SNC. The sender encodes the source data (or concepts) X into SRs (or symbols) Z before transmission, and the receiver subsequently receives and deduces relevant data (or target entity) Y. For instance, in an image classification task, the relevant data associated with the source data may denote labels corresponding to the images. Our primary focus is on developing an encoder that effectively transforms the given source data into a compressed representation

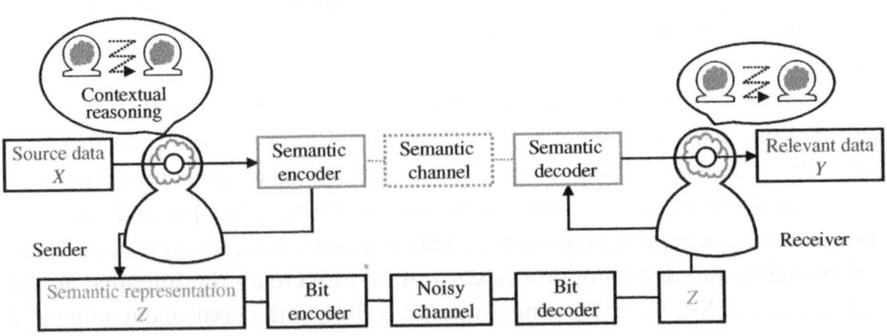

Figure 9.1 A schematic illustration of semantics-native communication system.

and a decoder that derives relevant data from this compressed representation. Specifically, our goal is to optimize both the encoder and decoder concurrently, aiming to maximize inference accuracy while simultaneously maximizing the compression rate of the compressed representation.

9.1.1.1 Information-Theoretic Model Description

Fix finite alphabets $\mathcal{X}$ and $\mathcal{Y}$, and consider a pair of random variables (X, Y), such that $X \in \mathcal{X}$ and $Y \in \mathcal{Y}$, following a fixed joint probability mass function (p.m.f.) $p(x, y)$ for all $(x, y) \in \mathcal{X} \times \mathcal{Y}$. As mentioned earlier, Y represents data relevant to X, indicating that the mutual information between them is positive, i.e., $I(X; Y) > 0$.

Suppose a source data x is given to the sender with a fixed p.m.f. $p(x)$ for some $x \in \mathcal{X}$. The sender transforms x into an SR $z \in \mathcal{Z}$ with a conditional p.m.f. $p(z|x)$, and transmits it to the receiver, where $\mathcal{Z}$ is a finite alphabet of SRs. Here, the p.m.f. $p(z|x)$ (or $p_s(z|x)$ for the sake of clarity) for all $z \in \mathcal{Z}$ can be seen as a semantic encoder that stochastically maps given data x to its SR at the sender. In addition, unless specifically stated otherwise, it is assumed that the prior distribution of Z, denoting the random variable of a SR, is unbiased and uniform, i.e., $p(z) = \frac{1}{|\mathcal{Z}|}$ for all $z \in \mathcal{Z}$. Upon receiving the SR z, the receiver infers the relevant data $Y = y$ based on a conditional p.m.f. $p(y|z)$ (or $p_r(y|z)$ for clarity), which can be viewed as a semantic decoder.

In the given scenario, the triple of random variables (X, Y, Z) exhibits joint distribution characteristics described by $p(x, y, z) = p(x)p(z|x)p(y|z)$ for all $(x, y, z) \in \mathcal{X} \times \mathcal{Y} \times \mathcal{Z}$. Consequently, the Markov chain $X \to Z \to Y$ is formed. This probability distribution, denoted as $p(x, y, z)$, encapsulates insights into the correlation among the source, the transmitted representation in the communication system under consideration, and the information required by the receiver for inference. Henceforth, we will specifically designate this as the *context*.

9.1.1.2 Motivation from Triangle of Meaning Model

The system modeling of SNC draws inspiration from Ogden and Richards' triangle of meanings [Ogden and Richards, 1923] in human cognition and communication, as depicted in Figure 9.2. Within this model, the vertices of the semantic triangle establish connections among the three domains of entities, concepts, and symbols. In this context, entities represent the subjects of communication, concepts denote the unit concepts disentangled by humans from the complex ideas of entities, and symbols are representations of these concepts.

While the traditional human communication model assumes a symmetrical situation where senders and receivers have reciprocal encoder and decoder functions, we extend this model as shown in Figure 9.3. Specifically, at the sender, we contemplate an encoder model that extracts concepts from entities (or source data X) and transmits their symbolic representation, referred to as the semantic

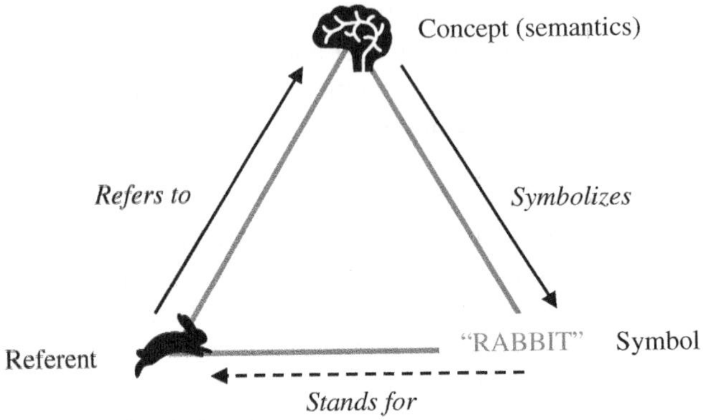

Figure 9.2 A schematic illustration of Ogden and Richards' triangle of meaning.

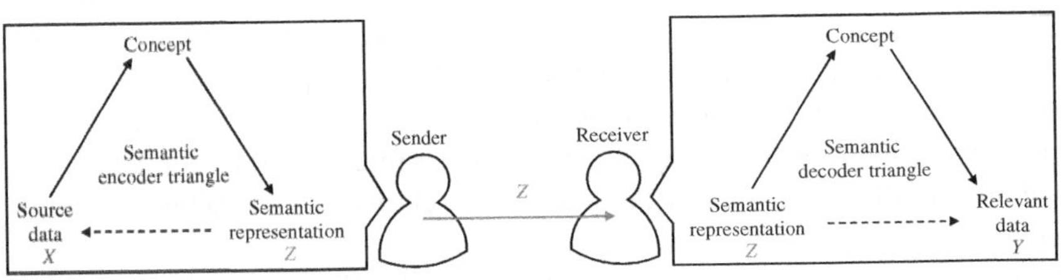

Figure 9.3 A schematic illustration of semantic encoder/decoder triangles.

representation Z. Simultaneously, at the receiver, we consider a decoder model that infers the relevant entity (or relevant data Y) from the symbolized concept. This expanded model offers insights into the asymmetric encoding and decoding processes involved in SNC.

9.2 Contextual Reasoning for Semantics-Native Communication

According to Bengio [2019], System 1 ML, often characterized by speed, intuition, and unconscious processing, revolves around recognizing patterns and statistical correlations in raw data space. In contrast, System 2 ML, characterized by slower, logical, and conscious processing, focuses on uncovering the underlying causation behind perceived correlations, engaging in reasoning on the data's SRs [Schölkopf et al., 2021]. The imminent shift from System 1 ML to System 2 ML-based communication systems, leveraging logical reasoning for communication, is upon us.

Within the realm of various types of reasoning [Goyal and Bengio, 2020; Schölkopf et al., 2021], our primary emphasis lies in incorporating the principles of CR into SNC. CR entails the human ability to decipher hidden meanings beyond communication, such as linguistic ambiguity and the intentions of the interlocutor, based on the local context of communication and social interactions [Grice, 1975; Bell, 1995]. A notable computational approach to CR is the rational speech act (RSA) framework [Frank and Goodman, 2012; Goodman and Stuhlmüller, 2013], which conceptualizes the speaker and listener as stochastic models and simulates CR-based human communication.

9.2.1 Motivation from Referential Game

Efficient communication through CR is evident in a referential game scenario. Imagine a setting with a sender and receiver, as depicted in Figure 9.4, and suppose $X = Y$ so that we only focus on the random variable pair (X, Z). The receiver has three potential entities to communicate: a rabbit sitting (x_1), a rabbit jumping (x_2), and a rabbit jumping into a ring (x_3). Each agent possesses a set of atomic symbolized concepts $\mathcal{Z} = \{z_{rabbit}, z_{jumping}, z_{ring}\}$. The relation between the entities and symbolized concepts, which is known as a context, is represented as a graph in Figure 9.4.

In a communication-limited context, the sender, given an entity to communicate, selects a symbolized concept from $\mathcal{Z}$ and aims to convey the entity to the receiver. Suppose the sender aims to communicate x_2. Then, a naïve sender, obligated to communicate all symbolized concepts, struggles to efficiently describe x_2 with a single symbolized concept. Conversely, a rational sender, assuming a rational receiver, selects $z_{jumping}$ to induce x_2. The rationale is clear: if z_{rabbit} is chosen, a rational receiver infers x_1, and if z_{ring} is chosen, x_3 is inferred. These counter-examples justify the rational choice of $z_{jumping}$. In contrast, a naive receiver, receiving $z_{jumping}$, can't identify the action without z_{rabbit}, while a rational receiver directly infers x_2. In summary, rational agents efficiently communicate x_2 using a single symbol, $z_{jumping}$, balancing efficiency and accuracy.

9.2.2 Single-Sided Contextual Reasoning

As mentioned earlier, the RSA model stands out as a highly intuitive computational model for CR [Frank and Goodman, 2012; Goodman and Stuhlmüller, 2013; Frank et al., 2016; Goodman and Frank, 2016]. Rooted in the Gricean view of language use [Grice, 1975], the RSA model assumes that individuals are rational agents capable of effective and efficient communication through reasoning. In addition, it assumes that there is a one-to-one mapping relationship between X and Y, which implies that, given X, Y is deterministic. In other words, the task of a decoder is to reconstruct the source data at the receiver. Consequently, our focus shifts to the random variable pair (X, Z), where the context is captured by $p(x, z)$

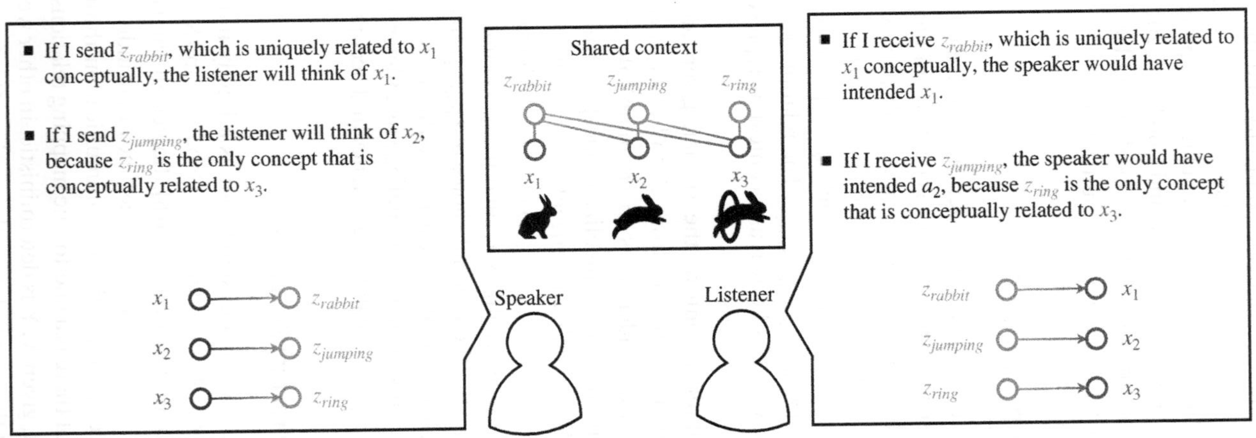

Figure 9.4 A schematic illustration of a contextual reasoning example in a referential game.

for all $(x, z) \in \mathcal{X} \times \mathcal{Z}$. If both the sender and receiver share a common grounding in the initial context $q^{[0]}(x, z)$ for all $(x, z) \in \mathcal{X} \times \mathcal{Z}$, the sender can align its encoder with the receiver's inference strategy, and reciprocally, the receiver can do the same for its decoder.

In pursuit of this objective, both the sender and receiver independently engage in a recursive process that leads to the emergence of semantic encoders and decoders [Zaslavsky et al., 2021]. The recursion is characterized by the following expressions:

$$\begin{cases} p_s^{[t]}(z|x) = \dfrac{\exp\left\{\alpha u^{[t-1]}(x, z)\right\}}{\sum_z \exp\left\{\alpha u^{[t-1]}(x, z)\right\}}, \\ p_r^{[t]}(x|z) = \dfrac{p_s^{[t]}(z|x)p(x)}{\sum_x p_s^{[t]}(z|x)p(x)}, \end{cases} \tag{9.1}$$

where, for all $(x, s) \in \mathcal{X} \times \mathcal{Z}$, the parameter $\alpha \geq 0$ dictates the rationality degree of the sender. Here, $t \geq 1$ represents the recursion depth, and $u^{[t]}(\cdot)$ denotes the sender's utility function at recursion depth t. The initialization of the encoder and decoder is given by: $p(x, z) = p_s^{[0]}(z|x)p(x) = p_r^{[0]}(x|z)p(z)$. Assuming that all agents possess the capability to compute (9.1), and as long as they share a common grounding within the same context, they can each derive identical semantic encoders and decoders at the same recursion depth. For simplicity, we henceforth represent $q_s^{[t]}(x, z) = \frac{p_s^{[t]}(z|x)p(x)}{\sum_{(x,z)} p_s^{[t]}(z|x)p(x)}$ and $q_r^{[t]}(x, z) = \frac{p_r^{[t]}(x|z)p(z)}{\sum_{(x,z)} p_r^{[t]}(x|z)p(z)}$, as the context at the sender and receiver, respectively, at recursion depth t. The same notation is used, whether or not the superscript denoting recursion depth is present.

The recursive nature of (9.1) yields diverse outcomes contingent upon the design of the utility function. One straightforward and intuitive instantiation of the utility function is through the logarithmic function. Let the utility function be defined as $u^{[t]}(x, z) = \log q_r^{[t]}(x, z)$. In this scenario, the recursion (9.1) manifests as follows:

$$\begin{cases} p_s^{[t]}(z|x) = \dfrac{\left\{q_r^{[t-1]}(x, z)\right\}^{\alpha}}{\sum_z \left\{q_r^{[t-1]}(x, z)\right\}^{\alpha}}, \\ p_r^{[t]}(x|z) = \dfrac{q_s^{[t]}(x, z)}{\sum_x q_s^{[t]}(x, z)}, \end{cases} \tag{9.2}$$

for all $(x, z) \in \mathcal{X} \times \mathcal{Z}$. It is noteworthy that when $\alpha = 1$, each step of the recursion in (9.2) closely resembles the Bayesian formula. By representing the encoder and decoder as matrices, and given that the prior distributions $p(z)$ and $p(x)$ are uniform, the recursion can be reformulated as a matrix scaling algorithm. Moreover, if the cardinalities of $\mathcal{X}$ and $\mathcal{Z}$ are the same, the recursion implements the Sinkhorn–Knopp's matrix scaling algorithm [Sinkhorn, 1964; Wang et al., 2020].

On the flip side, an inquiry into the optimization objectives of the RSA model becomes an unavoidable pursuit. This quest finds resolution through the lens of rate-distortion theory and the Blahut–Arimoto algorithm. Specifically, when interpreting the recursion of the RSA model as the outcome of alternating optimization, one may deduce that the original problem seeks to maximize

$$\mathcal{L}_{\text{SS}} = H(p_s) + \alpha H(q_s, q_r), \tag{9.3}$$

where $H(p_s) = -\sum_{(x,z)} q_s(x,z) \log p_s(z|x)$ denotes the conditional entropy of the encoder and $H(q_s, q_r) = -\sum_{(x,z)} q_s(x,z) \log q_r(x,z)$ is the cross-entropy of the sender and receiver's contexts. Moreover, grounded in this inference, it can be asserted that the recursion (9.2) converges. To be precise, one can obtain an encoder and decoder that converge toward a local maximum of $\mathcal{L}_{\text{SS}}$ as $t \to \infty$ [Zaslavsky et al., 2021].

9.2.3 Double-Sided Contextual Reasoning

As mentioned earlier, on the one hand, the RSA model is essentially a CR model proposed to make communication more rational from the perspective of the sender. On the other hand, it is also interesting to discuss what form the CR model for rational communication at both ends of communication might take. In Seo et al. [2023b], drawing inspiration from the RSA model for double-sided CR, the focus was on the problem of minimizing the following objective function:

$$\mathcal{L}_{DS} = \gamma \left[H(q_s, q_m) - \frac{H(q_s)}{\alpha} \right] + (1-\gamma) \left[H(q_r, q_m) - \frac{H(q_r)}{\beta} \right], \tag{9.4}$$

for an arbitrary context $q_m(x,z)$ and given parameters $0 < \gamma < 1$ and $\alpha, \beta \geq 1$. It is noteworthy that the sender and receiver will utilize the converged version of context q_m to make their semantic encoder and decoder. In (9.4), the term $H(q_s) = -\sum_{(x,z)} q_s(x,z) \log q_s(x,z)$ and $H(q_r) = -\sum_{(x,z)} q_r(x,z) \log q_r(x,z)$ are the joint entropies of X and Z concerning the sender and receiver's contexts, respectively. Maximizing $H(q_s)$ increases the uncertainty of semantic encoding, which reduces communication efficiency. Maximizing $H(q_r)$ increases the uncertainty of semantic decoding, which reduces information transmission accuracy. To handle those, hyper-parameters α and β, which are the rationality degrees of the sender and receiver, suppress the maximization of $H(q_s)$ and $H(q_r)$ to promote representational efficiency and decoding accuracy, respectively. The term $H(q_s, q_m) = -\sum_{(x,z)} q_s(x,z) \log q_m(x,z)$ is the cross-entropy of the sender's and arbitrary contexts, and $H(q_r, q_m) = -\sum_{(x,z)} q_r(x,z) \log q_m(x,z)$ is the cross-entropy of the receivers and arbitrary contexts. Since minimizing the cross-entropy results in two distributions being similar, minimizing $H(q_s, q_m)$ and $H(q_r, q_m)$ induces that three contexts of q_s, q_r, and q_m converge to neighbor points.

To get the solution of minimizing (9.4), we can alternately optimize three contexts with the recursion as follows:

$$\begin{cases} q_{m,1}^{[t]}(x,z) = \gamma q_s^{[t-1]}(x,z) + (1-\gamma) q_r^{[t-1]}(x,z), \\ q_s^{[t]}(x,z) = \dfrac{q_{m,1}^{[t]}(x,z)^{\alpha}}{\sum_{(x,z)} q_{m,1}^{[t]}(x,z)^{\alpha}}, \\ q_{m,2}^{[t]}(x,z) = \gamma q_s^{[t]}(x,z) + (1-\gamma) q_r^{[t-1]}(x,z), \\ q_r^{[t]}(x,z) = \dfrac{q_{m,2}^{[t]}(x,z)^{\beta}}{\sum_{(x,z)} q_{m,2}^{[t]}(x,z)^{\beta}}, \end{cases} \tag{9.5}$$

for all $(x,z) \in \mathcal{X} \times \mathcal{Z}$. Then, we can obtain the sender's semantic encoder and the receiver's semantic decoder

$$\begin{cases} p_s^{[*]}(z|x) = \dfrac{q_m^{[*]}(x,z)^{\alpha}}{\sum_z q_m^{[*]}(x,z)^{\alpha}}, \\ p_r^{[*]}(x|z) = \dfrac{q_m^{[*]}(x,z)^{\beta}}{\sum_x q_m^{[*]}(x,z)^{\beta}}, \end{cases} \tag{9.6}$$

for all $(x,z) \in \mathcal{X} \times \mathcal{Z}$, via the contexts that converge to $q_m^{[*]}(x,z) = \lim_{t\to\infty} q_{m,1}^{[t]}(x,z) = \lim_{t\to\infty} q_{m,2}^{[t]}(x,z)$ and $q_s^{[*]}(x,z) = q_r^{[*]}(x,z) = q_m^{[*]}(x,z)$ for all $(x,z) \in \mathcal{X} \times \mathcal{Z}$ as $t \to \infty$, which leads to a local minimum of $\mathcal{L}_{DS}$ [Seo et al., 2023b].

Figure 9.5 shows the impact of CR iteration steps on the reliability of SNC with a single concept delivery. Here the iteration steps are directly related to the amount of computational resource that each agent uses. As shown in Figure 9.5, for a larger number of iteration steps, i.e., with larger computational effort, the reliability of SNC is higher, especially when both parameters α and β approach 1.

9.2.4 Multi-round Contextual Reasoning

The prevailing assumption is that CR typically selects a singular symbolized concept for communication. However, the reliability of semantic transmission can be significantly improved by conveying multiple SRs. To achieve this, a pragmatic approach involves leveraging a greedy algorithm-based method. Despite its optimality, this method entails the joint selection of multiple SRs until communication reliability is assured.

In essence, the greedy algorithm addresses the complexity of determining the necessary representations for reliability and sequentially choosing multiple representations. Its core operation involves updating the sender's priors ($p_s^{[*]}(z|x)$) and the receiver's priors ($p_r^{[*]}(x|z)$) after each SR selection and transmission. For ease

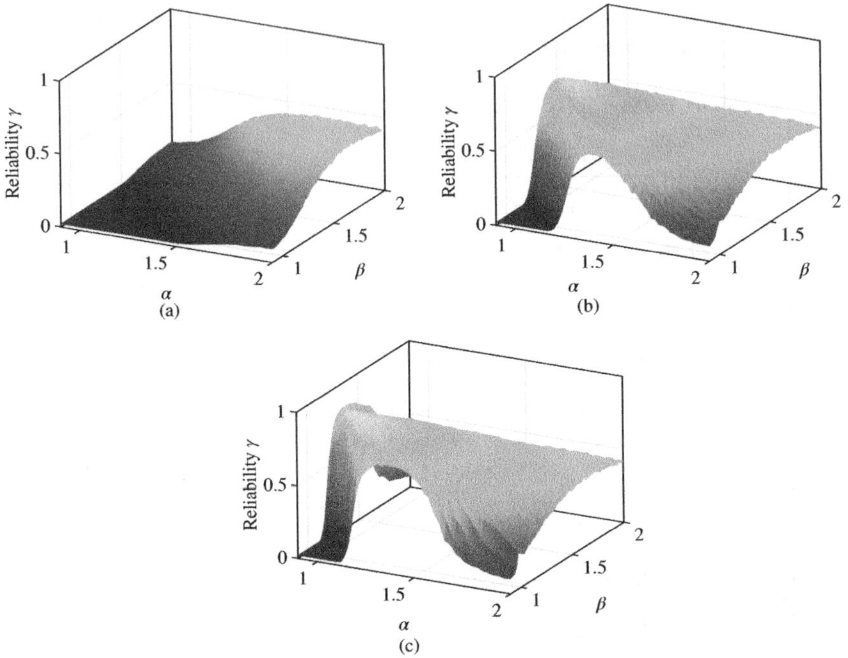

Figure 9.5 Reliability γ with respect to parameters α and β ranging 0.9 to 2 in SNC, for different CR iteration depths $t = 10$, 20, and 100. (a) $t = 20$, (b) $t = 100$, and (c) $t = 200$.

of representation, we denote the sender's encoder and the receiver's decoder at the k-th communication round as $p_s^{(k)}(z|x)$ and $p_r^{(k)}(x|z)$, respectively.

Consider, for instance, a scenario where a sender sequentially selects two SRs, z_1 and z_2, from the set $\mathcal{Z}$ to communicate with a receiver. The initial stationary distribution $p_s^{(1)}(z|x)$ is derived through recursion. The sender then chooses the first SR, $z_1 = \arg\max_z p_s^{(1)}(z|x)$. Upon receiving z_1, the receiver updates the prior distribution $p(x)$ and subsequently adjusts the prior distribution $p(z)$ for the next selection (z_2). This process is iterated, and at each step, $p_s^{(k)}(z|x)$ and $p_r^{(k)}(x|z)$ are updated for the next SR selection. Similarly, the greedy algorithm allows for the sequential selection of meaningful concepts ($k \geq 1$) by obtaining $p_s^{(k)}(z|x)$ and corresponding $p_r^{(k)}(x|z)$. The computational time complexity of the algorithm is asymptotically $\mathcal{O}(KT|\mathcal{X}|^2|\mathcal{Z}|^2)$, where K is the number of selected SRs, and T is the number of recursions until convergence.

Furthermore, the accuracy of inferring the source data by $p_r^{(k)}(x|z)$ improves as the number of communicated SRs increases. This is demonstrated by the inequality $p_r^{(k-1)}(x|z_{k-1}) \leq p_r^{(k)}(x|z_k)$ for $k \geq 2$, under fixed parameters $\alpha, \beta \geq 1$ (excluding $\alpha = \beta = 1$) and $0 < \gamma < 1$, as proven in Seo et al. [2023b]. In practical terms, determining the optimal number of SRs for reliable semantic transmission is challenging. However, the flexibility of the greedy algorithm permits a step-by-step selection of SRs, enabling an early cessation of communication once the desired reliability is achieved.

Figure 9.6 illustrates the impact of the number of communication rounds on the reliability of SNC. Note that a meaningful concept is communicated in each round. The number of communication rounds therefore corresponds to the number of communicating meaningful concepts. Moreover, the larger number of communication rounds indicates higher communication latency. In Figure 9.6a, each communication round runs for 20 iteration steps, and thus the reliability of SNC when $\alpha, \beta = 1.1$ is poor compared to the other settings. However, the reliability γ approaches to 1 after 4 communication rounds. As the iteration steps per communication round increases, the reliability of SNC with smaller α and β enhances owing to their convergence as mentioned earlier. Furthermore, when $\alpha, \beta = 1.1$,

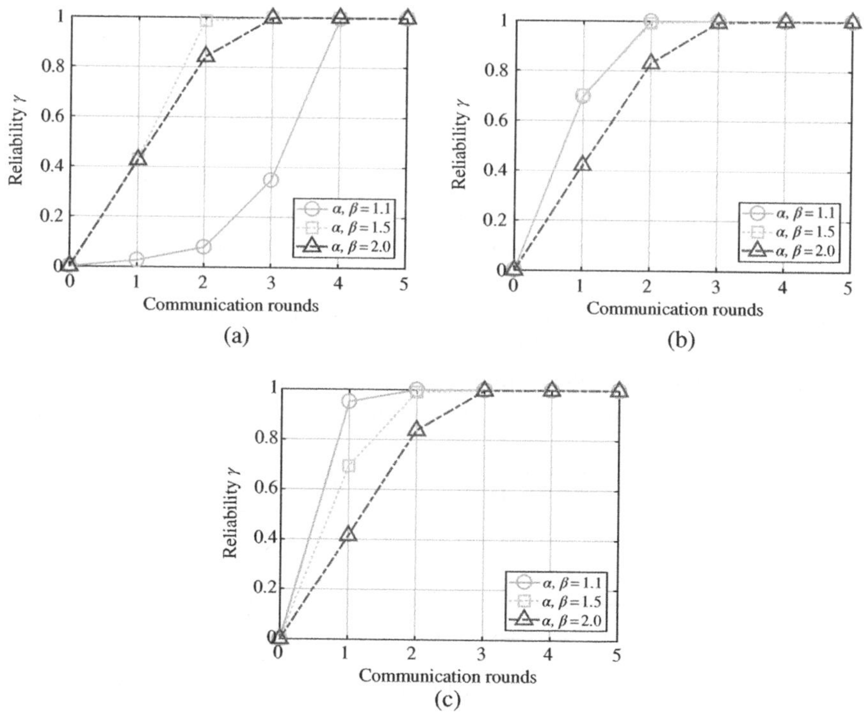

Figure 9.6 Reliability γ versus communication rounds in SNC with different parameters $\alpha, \beta = 1.1$, 1.5, and 2.0, and different CR iteration steps $t = 10$, 20, and 200 in each round. (a) $t = 20$, (b) $t = 100$, and (c) $t = 200$.

the reliability γ approaches 1 within two communication rounds, which means that allocating more computational effort enhances the reliability-latency trade-off in SNC.

9.3 Context Synchronization for Semantics-Native Communication

In a heterogeneous SNC setting devoid of a shared context, the efficacy of CR may diminish, adversely affecting communication efficiency. For instance, as illustrated in Figure 9.7, consider the scenario where Alice and Bob employ CR for task-oriented coordination, while Carol, lacking context awareness, can only observe their communication. To enable effective communication using CR, Carol must infer the context, underscoring the crucial role of synchronization. This process ensures sustained effective communication even amid contextual changes.

To tackle the synchronization challenge among agents, it becomes imperative to address the *inverse contextual reasoning (iCR)* problem. This involves extracting genuine context from noisy communication samples, thereby mitigating communication inefficiencies. Specifically, we present a *Bayesian iCR* method employing a two-staged Metropolis-Hastings (MH) algorithm, aligning with a specified distribution related to inputs, context, and prior distributions.

The elevated computational complexity of Bayesian iCR, primarily arising from CR in calculating acceptance probabilities, prompts the introduction of the *linearized contextual reasoning (LCR)* model. This single-layer linear neural network optimizes inference accuracy, SNC effectiveness, and CR invertibility. The advantages of LCR include expedited computations, transforming the *iCR* problem into an *inverse LCR (iLCR)* problem akin to a compressed sensing issue. The adoption

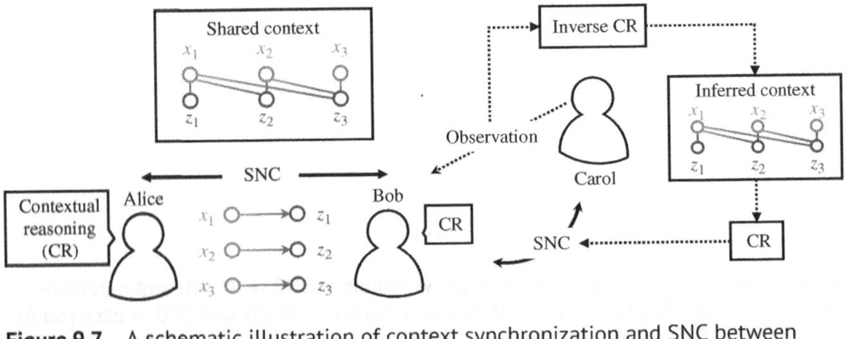

Figure 9.7 A schematic illustration of context synchronization and SNC between heterogeneous agents.

of a *Bayesian iLCR* method with MH significantly enhances computational efficiency, particularly in acceptance probability calculations.

9.3.1 Bayesian Inverse Contextual Reasoning

For easier description, let's redefine the context as a matrix $\mathbf{C} = (c_{z,x}) \in \{0,1\}^{|\mathcal{Z}|\times|\mathcal{X}|}$, where each matrix entry $c_{z,x} \in [0,1]$ reflects the degree of association between action x and concept z across all (x,z) pairs in $\mathcal{X} \times \mathcal{Z}$. Note that if normalization is performed so that the sum of all entries in the matrix becomes 1, it aligns with the definition of the joint distribution-based context previously defined.

Mirroring the recursion expression noted as (9.1), we define $\mathbf{S}^{[t]} = (p_s^{[t]}(z|x))$ and $\mathbf{R}^{[t]} = (p_r^{[t]}(x|z))$, corresponding to the matrices for the rational semantic encoder and decoder at recursion depth $t \in \mathbb{Z}_+$, respectively. The recursion is initialized with the naïve semantic decoder $\mathbf{R}^{[0]}$, given that both $\mathbf{S}^{[t]}$ and $\mathbf{R}^{[t]}$ are stochastic matrices for all $t \in \mathbb{Z}_+$. Moreover, consider a logarithmic utility design to let the recursion (9.1) converge, as previously discussed, and let $\mathbf{S}^*$ and $\mathbf{R}^*$ be the converged encoder and decoder, respectively. Meanwhile, for given context $\mathbf{C}$ and priors $p(x)$ and $p(z)$, since the recursion (9.1) is a deterministic algorithm that gives a unique solution, defined by $\mathcal{S}, \mathcal{R} : \mathbb{C} \times \mathbb{P}_{|\mathcal{X}|-1} \times \mathbb{P}_{|\mathcal{Z}|-1} \to [0,1]^{|\mathcal{Z}|\times|\mathcal{X}|}$ the mapping functions s.t. $\mathbf{S}^* = \mathcal{S}(\mathbf{C}, p(x), p(z))$ and $\mathbf{R}^* = \mathcal{R}(\mathbf{C}, p(x), p(z))$, respectively, where $\mathbb{P}_M$ denotes the standard M-dimensional probability simplex for some positive integer M, i.e., $\sum_{m=1}^{M+1} p_m = 1$ and $p_m \in [0,1]$ for all $m \in \{1, 2, \dots, M+1\}$.

In addressing the context synchronization challenge, our focus is on developing a methodology that empowers Carol to estimate the context, denoted as $\hat{\mathbf{C}}$, and accurately approximate the priors $\hat{p}(x)$ and $\hat{p}(z)$ to closely align with the ground truth distributions $\mathbf{C}$, $p(x)$, and $p(z)$. To tackle this task, we employ the Bayesian iCR method, utilizing Bayesian inference and Markov Chain Monte Carlo (MCMC) sampling to derive context and prior distributions from SNC samples. We treat $\mathbf{C}$, $p(x)$, and $p(z)$ as independent random variables, and to enhance clarity and simplicity, we introduce vector notations, representing $\boldsymbol{x}$ as $p(x)$ and $\boldsymbol{z}$ as $p(z)$, thereby mitigating potential confusion. Additionally, for convenience in our analysis, we define $\mathbf{T} = (\mathbf{C}, \boldsymbol{x}, \boldsymbol{z})$ as the tuple encompassing the random variables $\mathbf{C}$, $\boldsymbol{x}$, and $\boldsymbol{z}$. This systematic approach lays the groundwork for our exploration and resolution of the context synchronization problem.

To begin with, suppose that Carol has obtained a set of noisy data samples $\mathbb{D} = \{\tilde{z}_d, \tilde{x}_d\}_{d=1}^{D}$ from observing what Alice sends and how Bob reacts. Here, "noisy" encompasses all errors that may occur during the communication process between Alice and Bob or during Carol's observation process. By counting the number of each concept–action pair in the samples set, and dividing it by the total number of samples D, Carol obtains an empirical decoder $\bar{\mathbf{R}}$ by aggregating multiple SNC samples collected over time, which is corrupted by a zero-mean additive Gaussian random noise matrix $\mathbf{N} \in \mathbb{R}^{|\mathcal{Z}|\times|\mathcal{X}|}$ with i.i.d. entries of variance σ^2, that is $\bar{\mathbf{R}} = \mathcal{R}(\mathbf{T}) + \mathbf{N}$. This assumption can be supported by considering that $\bar{\mathbf{R}}$ is structured based on some independent samples. In practical terms, to obtain $\bar{\mathbf{R}}$, Carol, in this case, needs exposure to all the various concepts associated with a given action. To accomplish this, a minimum of $|\mathcal{Z}|$ communication samples is required. Unless communication samples are intentionally provided, an average of $|\mathcal{Z}| \sum_{k=1}^{|\mathcal{Z}|-1} \frac{1}{k}$ samples would be necessary.

The likelihood of the empirical decoder $\bar{\mathbf{R}}$ is

$$p(\bar{\mathbf{R}}|\mathbf{T}) = p(\bar{\mathbf{R}}|\mathcal{R}(\mathbf{T}), \mathbf{T}) \tag{9.7}$$

$$= p(\mathbf{N}) \tag{9.8}$$

$$\propto \exp\left(-\frac{1}{2\sigma^2}\|\mathbf{vec}(\mathbf{N})\|_2^2\right), \tag{9.9}$$

where (9.8) holds since the noise is independently generated, and (9.9) is from the probability density function of multivariate normal distribution while $\mathbf{vec}(\mathbf{N})$ is the vectorization of $\mathbf{N}$. Moreover, from Bayes' rule and (9.8), the posterior distribution of $\mathbf{T}$ is $p(\mathbf{T}|\bar{\mathbf{R}}) = \frac{1}{U}\exp\left(-\frac{1}{2\sigma^2}\mathbf{vec}(\bar{\mathbf{R}} - \mathcal{R}(\mathbf{T}))\|_2^2\right) p(\mathbf{T})$, where $U = \int \exp\left(-\frac{1}{2\sigma^2}\|\mathbf{vec}(\bar{\mathbf{R}} - \mathcal{R}(\mathbf{T}))\|_2^2\right) p(\mathbf{T})d\mathbf{T}$ is a normalization term.

Having the posterior distribution, one promising approach for iCR is the maximum a posteriori (MAP) point estimation. In detail, given that the prior $p(\mathbf{T})$ is known and tractable, the MAP method finds the point which maximizes the posterior, i.e., $\hat{\mathbf{T}}_{MAP} = \arg\max_{\mathbf{T}} p(\mathbf{T}|\bar{\mathbf{R}})$. However, in general, it is difficult to obtain the MAP estimate if the prior $p(\mathbf{T})$ is unknown or intractable. As an alternative approach for Bayesian iCR, sampling from the numerator can approximate the posterior distribution and also provide a point estimate. Specifically, we use a two-staged MH algorithm. The algorithm operates in two stages, initially determining the matrix $\mathbf{C}$ using the MH method for each entry, and subsequently finding the vectors $\boldsymbol{x}$ and $\boldsymbol{z}$ collectively in the second stage. It aims to sample from the posterior distribution, utilizing a proposal distribution $q(\mathbf{T}'|\mathbf{T})$ for updates, where $\mathbf{T}' = (\mathbf{C}', \boldsymbol{x}', \boldsymbol{z}')$ represents the proposed updates. The choice of proposal distribution is made from a list of standard, easily samplable distributions. While iteration numbers are typically used in the MH algorithm, they are omitted here for simplicity. A detailed explanation of the two-staged MH algorithm, utilized due to spatial issues, is omitted. For more detailed information, please refer to Seo et al. [2023a].

9.3.2 Inverse Linearized Contextual Reasoning

One critical problem with the SNC-based framework using CR is that running the iterative recursion (9.1) until its convergence is computationally expensive.

This problem becomes more pronounced in the Bayesian iCR introduced in Section 9.3.1 because the CR must be repeatedly done for the accept/reject processes in the MH algorithm. One way to avoid such a problem is to linearize the CR to obtain the rational encoder and decoder without recursive and iterative computation. In this subsection, we propose a *Bayesian iLCR* method based on the LCR model obtained from training a single-layer fully connected neural network and the MH algorithm.

9.3.2.1 Linearizing Contextual Reasoning

Assuming that the mapping function $\mathcal{R}$ is linearized, the rational semantic decoder can be written as $\mathbf{vec}(\tilde{\mathcal{R}}(\mathbf{T})) = \mathbf{\Phi}\mathbf{vec}(\mathbf{T})$ for some $|\mathcal{Z}||\mathcal{X}| \times (|\mathcal{Z}||\mathcal{X}| + |\mathcal{Z}| + |\mathcal{X}|)$ dimensional matrix $\mathbf{\Phi}$, where $\tilde{\mathcal{R}}(\mathbf{T}) = (\tilde{p}_r(x|z)) \in [0,1]^{|\mathcal{Z}|\times|\mathcal{X}|}$ is the matrix that approximates $\mathcal{R}(\mathbf{T})$ incorporating the impact of perturbation that comes from the linearization. In addition, $\mathbf{vec}(\tilde{\mathcal{R}}(\mathbf{T})) \in \mathbb{R}^{|\mathcal{Z}||\mathcal{X}|}$ and $\mathbf{vec}(\mathbf{T}) \in \mathbb{R}^{|\mathcal{Z}||\mathcal{X}|+|\mathcal{Z}|+|\mathcal{X}|}$ are the column vectors obtained by vectorizing $\tilde{\mathcal{R}}(\mathbf{T})$ and $\mathbf{T}$, respectively. The rational semantic encoder obtained through a single iteration of (9.1) upon $\tilde{\mathcal{R}}(\mathbf{T})$ is denoted by $\tilde{\mathcal{S}}(\mathbf{T}) = (\tilde{p}_s(z|x)) \in [0,1]^{|\mathcal{Z}|\times|\mathcal{X}|}$.

One effective way to obtain the matrix $\mathbf{\Phi}$ is to train a single-layer fully connected neural network without non-linear activations which takes $\mathbf{T}$ as input and $\mathcal{R}(\mathbf{T})$ as output. In general, it is difficult to train such a matrix that linearizes a given recursive non-linear evolution with small misfit $\|\mathbf{vec}(\mathcal{R}(\mathbf{T}) - \tilde{\mathcal{R}}(\mathbf{T}))\|_2^2$. However, in the scenario under consideration, it is possible to obtain $\mathbf{\Phi}$ since both $\mathbf{vec}(\mathcal{R}(\mathbf{T}))$ and $\mathbf{vec}(\mathbf{T})$ are sparse vectors from the structure. Later on, we show that whether the recursion is linearizable or not depends on the sparsity of the context $\mathbf{C}$. Moreover, when training $\mathbf{\Phi}$ (or designing loss function for training the linear neural network model), our focus is not only on minimizing the misfit $\|\mathbf{vec}(\mathcal{R}(\mathbf{T}) - \tilde{\mathcal{R}}(\mathbf{T}))\|_2^2$ but also ensuring the communication effectiveness of the approximated rational semantic coding pair $(\tilde{\mathcal{S}}(\mathbf{T}), \tilde{\mathcal{R}}(\mathbf{T}))$, as much as the original pair $(\mathcal{S}(\mathbf{T}), \mathcal{R}(\mathbf{T}))$ provides.

The LCR model is trained as follows. First, generate data samples $(\mathbf{T}_i, \mathcal{R}(\mathbf{T})_i)_{i=1}^N$ for a fixed prior distribution, i.e., $\mathbf{T}_i \sim p(\mathbf{T}), 1 \leq \forall i \leq N$, and then obtain the neural network model weights that minimize the loss function defined by

$$\mathcal{L}_1 = \sum_{i=1}^{N} \left\{ \lambda_{mis}\mathcal{L}_{mis}(\mathbf{T}_i) + \lambda_{eff}\mathcal{L}_{eff}(\mathbf{T}_i) \right\}, \tag{9.10}$$

for some constants $\lambda_{mis}, \lambda_{eff} \geq 0$ such that $\lambda_{mis} + \lambda_{eff} = 1$, where $\mathcal{L}_{mis}(\mathbf{T}) = \|\mathbf{vec}(\mathcal{R}(\mathbf{T}) - \tilde{\mathcal{R}}(\mathbf{T}))\|_2^2$ and $\mathcal{L}_{eff}(\mathbf{T}) = \sum_{x\in\mathcal{X}} p(x) \left(1 - \sum_{z\in\mathcal{Z}} \tilde{p}_s(z|x)\tilde{p}_r(x|z)\right)$ are misfit and effectiveness loss term, respectively. It is worth mentioning that the loss function can be defined in various ways, for example using cross-entropy for each row vector of $\mathcal{R}(\mathbf{T})$ and $\tilde{\mathcal{R}}(\mathbf{T})$, however, we used mean squared error (MSE) misfit function to obtain the linearized form of the CR. The original CR model does not have a one-to-one relationship between input and output. However, the LCR model can be seen as a one-to-one mapping function. Therefore, from the perspective of invertibility, it can be understood that utilizing the LCR model instead of the CR model is more advantageous. On the other hand, in a situation where Carol acts as the eavesdropper and Alice and Bob aim to prevent eavesdropping, it is more advantageous to use CR.

9.3.2.2 Invertible Linearized Contextual Reasoning

Recall that the dimension of $\mathbf{vec}(\mathbf{T})$ is larger than that of $\mathbf{vec}(\tilde{\mathcal{R}}(\mathbf{T}))$, which in turn makes the linear system underdetermined. In addition, note that the entries of $\mathbf{T}$ (strictly saying $\mathbf{C}$) are sparse. In consequence, having the LCR model, the iCR problem is recast as an iLCR problem (or compressed sensing problem), where $\mathbf{vec}(\tilde{\mathcal{R}}(\mathbf{T}))$ is a noiseless observation vector, $\boldsymbol{\Phi}$ is a measurement matrix, and $\mathbf{vec}(\mathbf{T})$ is an unknown vector which requires to be reconstructed. Considering the additive noise in (9.8), a noisy empirical decoder $\bar{\mathbf{R}}$ corrupted by a zero-mean additive Gaussian random noise $\mathbf{N} \in \mathbb{R}^{|\mathcal{Z}|\times|\mathcal{X}|}$ is $\bar{\mathbf{R}} = \tilde{\mathcal{R}}(\mathbf{T}) + \mathbf{N}$, and in a vectorized form $\mathbf{vec}(\bar{\mathbf{R}}) = \boldsymbol{\Phi}\mathbf{vec}(\mathbf{T}) + \mathbf{vec}(\mathbf{N})$.

Now, suppose that $\mathbf{vec}(\mathbf{T})$ is a s-sparse vector, which means that the number of non-zero entries in the vector is (at most) $s \in \mathbb{Z}_+$. Then, according to the compressed sensing theory, the matrix $\boldsymbol{\Phi}$ should satisfy the (s, δ)-restricted isometry property (RIP) condition in order to correctly estimate the unknown vector. The matrix $\boldsymbol{\Phi}$ is said to satisfy the (s, δ)-RIP if there exists $\delta \in (0, \frac{1}{3})$ such that every s-sparse vector $\boldsymbol{v}$ holds

$$(1-\delta)\|\boldsymbol{v}\|_2^2 \leq \|\boldsymbol{\Phi}\boldsymbol{v}\|_2^2 \leq (1+\delta)\|\boldsymbol{v}\|_2^2. \tag{9.11}$$

In normal terms, $\boldsymbol{\Phi}$ behaves like an isometry (orthonormal) matrix, even though it is not a square matrix. As a matter of fact, obtaining $\boldsymbol{\Phi}$ that meets the RIP condition (9.11) is a strongly NP-hard problem [Tillmann and Pfetsch, 2014]. Moreover, the matrix $\boldsymbol{\Phi}$ in our case must also linearize the CR, which makes it harder to obtain such a matrix. However, it is known that some random matrices, for example, the Gaussian or Bernoulli distribution, satisfy the RIP condition with high probability. Considering that the linear neural network for $\boldsymbol{\Phi}$ that minimizes $\mathcal{L}_1$ is trained based on a stochastic optimization such as stochastic gradient descent method under Gaussian random initialization, it is likely that the trained $\boldsymbol{\Phi}$ meets the (s, δ)-RIP condition.

Nevertheless, in order to further ensure an invertible CR, an alternative loss function that incorporates the (s, δ)-RIP constraint is introduced to train the linear neural network. The loss function can be written as

$$\mathcal{L}_2 = \mathcal{L}_1 + \lambda_2 \sum_{i=1}^{N} \mathcal{L}_{RIP}(\mathbf{T}_i), \tag{9.12}$$

for some constant $\lambda_2 \geq 0$, where $\mathcal{L}_{RIP} = \left(\frac{\|\mathbf{\Phi}\mathbf{vec}(\mathbf{T})\|_2^2}{\|\mathbf{vec}(\mathbf{T})\|_2^2} - 1 \right)^2$. When minimizing $\mathcal{L}_2$, the new regularization term forces the $\mathbf{\Phi}$ to satisfy (9.11), for all vectorized data samples $\mathbf{T}_1, \mathbf{T}_2, \dots, \mathbf{T}_N$, which we call a quasi-RIP condition. Since it does not fully characterize the original (s, δ)-RIP condition, in the sense that the condition only holds for the data samples, the regularization term still cannot fully guarantee that the matrix $\mathbf{\Phi}$ trained by using (9.12) holds the (s, δ)-RIP condition. However, compared to that trained with $\mathcal{L}_1$, it is more likely to hold the condition if the constant λ_2 is chosen appropriately.

9.4 Information Bottleneck Contextual Reasoning

The CR methods previously introduced are typically examined under the assumption that, given the source data X, the output Y is deterministic. However, within a broader SC framework, the task executed by the receiver may not always be deterministic in relation to the source. Therefore, it becomes imperative to explore the functionality of SNC within the context of random variables (X, Y, Z), as defined in the earlier system model. Engaging in such discussions requires a comprehensive understanding of the relationships between the random variables (X, Y, Z) crucial to communication systems. Intuitively, for communication to be efficient and task-effective, the semantic representation Z should exhibit minimal mutual information with the source X while maximizing the mutual information between Z and the relevant data Y. From an information-theoretic perspective, the method that precisely captures this balance is the IB method.

9.4.1 Information Bottleneck Method

The IB method revolves around finding a delicate equilibrium between preserving pertinent information and compressing data. Given a random variable X representing the input, another random variable Y representing the output, and an intermediate representation Z, the goal is to determine Z in a manner that preserves relevant information about X while being as independent as possible from Y. The formalization of the IB method involves solving the following optimization problem:

$$\underset{p(z|x)}{\text{minimize}} \; I(X;Z) - \eta I(Z;Y) \;\; \text{subject to} \;\; I(X;Z) \leq \zeta. \tag{9.13}$$

Here, ζ and η function as non-negative Lagrange multipliers. This optimization problem essentially seeks a balance between compression (minimizing $I(X;Z)$) and relevance (minimizing $I(Z;Y)$). The constraint $I(X;Z) \leq \zeta$ governs the level of compression or bottleneck in the system. The IB is widely used in ML for tasks like representation learning, where finding an optimal representation involves balancing between information preservation and compression.

9.4.2 Implementing Information Bottleneck with Contextual Reasoning

Similar to other CR methods, the IB problem, as outlined in (9.13), can also be solved through CR techniques. Specifically, the information bottleneck CR (IB-CR) approach enables the independent development of efficient encoders and decoders via alternating iterations that achieve convergence when both the sender and receiver share an identical context. The IB-CR method draws inspiration from the initial conceptualization of the IB principle, introduced in Tishby et al. [2000], as well as the Blahut–Arimoto algorithm, cited in Blahut [1972] and Arimoto [1972]. Although this document does not delve into the intricate derivation details, the process bears resemblance to that employed by other CR methods. With the encoder and decoder represented as $p_s(z|x)$ and $p_r(y|z)$, respectively, and given fixed $p(x)$ and $p(y|x)$, the algorithm below outlines an alternating iteration process that yields a suboptimal solution to the IB problem (9.13):

$$\begin{cases} p_s^{[t]}(z|x) = \dfrac{p_s^{[t-1]}(z)\exp\left\{-\alpha D_{KL}(p(y|x)||p_r(y|z))\right\}}{\sum_z p_s^{[t-1]}(z)\exp\left\{-\alpha D_{KL}(p(y|x)||p_r(y|z))\right\}}, \\ p_s^{[t]}(z) = \sum_x p(x)p_s^{[t]}(z|x), \\ p_r^{[t]}(y|z) = \sum_x p(y|x)p_s^{[t]}(x|z), \end{cases} \tag{9.14}$$

where $p_s^{[t]}(x|z) = \frac{p_s^{[t]}(z|x)p(x)}{p_s^{[t]}(z)}$ and D_{KL} represents the Kullback–Leibler (KL) divergence.

The encoder and decoder thus created ultimately implement the IB, minimizing the mutual information that Z contains about X, and maximizing the mutual information it contains about Y. This can also be demonstrated through various experiments. Particularly interesting is that the part of the utility function in the RSA model is represented in the IB-CR by the KL divergence between $p(y|x)$ and $p_r(y|z)$. This means that the alternating iteration progresses in such a way that the KL divergence between the two distributions is minimized, intuitively connecting to the philosophy of IB.

9.5 Conclusion

In this chapter, we briefly introduced the SNC model and its core component, CR. SNC facilitates the independent creation of communication-efficient and task-effective encoders and decoders by the sender and receiver when a context is shared between communicating agents. Notably, in SNC, the encoder and decoder are implemented based on the agent's CR, which closely aligns with the concept of pragmatic (contextual) reasoning in linguistics. Specifically, we explored four different methodologies related to CR in this chapter. The first was a single-sided CR method where the sender takes rational communication actions, similar to the linguistically derived RSA model. Its extended concept, double-sided CR, involves both sender and receiver agents in compressing SR to implement efficient encoders and decoders. Additionally, we discussed methods to linearize CR to solve its inverse problem and reduce its computational complexity. Lastly, we briefly looked at how to implement IB utilizing the concept of CR. Beyond the introduced content, future research needs to implement CR in actual ML models. Particularly, CR resembles generative models structurally, as agents create efficient SRs through simulations with a hypothetical counterpart. Therefore, to implement CR-based SNC in more complex and realistic wireless communication environments, a system implementation based on generative models is necessary.

Bibliography

S. Arimoto. An algorithm for computing the capacity of arbitrary discrete memoryless channels. *IEEE Transactions on Information Theory*, 18(1):14–20, 1972. doi: 10.1109/TIT.1972.1054753.

John Bell. *Pragmatic Reasoning A Model-based Theory*. Kluwer Academic Publishers, Amsterdam, 1995.

Yoshua Bengio. From system 1 deep learning to system 2 deep learning. Keynote at *NeurIPS 2019*, 2019.

R. Blahut. Computation of channel capacity and rate-distortion functions. *IEEE Transactions on Information Theory*, 18(4):460–473, 1972. doi: 10.1109/TIT.1972.1054855.

Michael C. Frank and Noah D. Goodman. Predicting pragmatic reasoning in language games. *Science*, 336:998–998, 2012.

Michael C. Frank, Andres Gomez Emilsson, Benjamin Peloquin, Noah D. Goodman, and Christopher Potts. Rational speech act models of pragmatic reasoning in reference games. *PsyArXiv*, 2016. URL http://psyarxiv.com/f9y6b.

Noah Goodman and Michael Frank. Pragmatic language interpretation as probabilistic inference. *Trends in Cognitive Sciences*, 20(11):818–829, Sep. 2016. doi: 10.1016/j.tics.2016.08.005.

Noah D. Goodman and Andreas Stuhlmüller. Knowledge and implicature: Modeling language understanding as social cognition. *Topics in Cognitive Science*, 5(1): 173–184, 2013.

Anirudh Goyal and Yoshua Bengio. Inductive biases for deep learning of higher-level cognition. *CoRR*, abs/2011.15091, 2020. URL https://arxiv.org/abs/2011.15091.

H. Paul Grice. Logic and conversation. In Maite Ezcurdia and Robert J. Stainton, editors, *The Semantics-Pragmatics Boundary in Philosophy*. Broadview Press, 1975.

C. K. Ogden and I. A. Richards. *The Meaning of Meaning A Study of the Influence of Language upon Thought and of the Science of Symbolism*. Routledge & Kegan Paul Ltd., London, UK, 1923.

Bernhard Schölkopf, Francesco Locatello, Stefan Bauer, Nan Rosemary Ke, Nal Kalchbrenner, Anirudh Goyal, and Yoshua Bengio. Toward causal representation learning. *Proceedings of the IEEE*, 109(5):612–634, 2021. doi: 10.1109/JPROC.2021.3058954.

Hyowoon Seo, Yoonseong Kang, Mehdi Bennis, and Wan Choi. Bayesian inverse contextual reasoning for heterogeneous semantics-native communication. *IEEE Transactions on Communications*, 72(2):830–844, 2023a. doi: 10.1109/TCOMM.2023.3327771.

Hyowoon Seo, Jihong Park, Mehdi Bennis, and Mérouane Debbah. Semantics-native communication via contextual reasoning. *IEEE Transactions on Cognitive Communications and Networking*, 9(3):604–617, 2023b. doi: 10.1109/TCCN.2023.3250206.

Richard Sinkhorn. A relationship between arbitrary positive matrices and doubly stochastic matrices. *The Annals of Mathematical Statistics*, 35(2):876–879, 1964. doi: 10.1214/aoms/1177703591.

Andreas M. Tillmann and Marc E. Pfetsch. The computational complexity of the restricted isometry property, the nullspace property, and related concepts in compressed sensing. *IEEE Transactions on Information Theory*, 60(2):1248–1259, 2014. doi: 10.1109/TIT.2013.2290112.

Naftali Tishby, Fernando C. Pereira, and William Bialek. The information bottleneck method. arXiv preprint physics/0004057, 2000.

Pei Wang, Junqi Wang, Pushpi Paranamana, and Patrick Shafto. A mathematical theory of cooperative communication. In *Advances in Neural Information Processing Systems 33*, pages 17582–17593, 2020.

Noga Zaslavsky, Jennifer Hu, and Roger Levy. A Rate—Distortion view of human pragmatic reasoning. *Proceedings of the Society for Computation in Linguistics*, 4(1):347–348, 2021. doi: 10.7275/gc1z-ck09.

10

Interoperable Semantic Communication

Jinhyuk Choi[1], Hyelin Nam[1], Jihong Park[2], Seung-Woo Ko[3], Jinho Choi[4], Mehdi Bennis[5], and Seong-Lyun Kim [1]

[1] *School of Electrical & Electronic Engineering, Yonsei University, Seodaemun-Gu, Seoul, Korea*
[2] *Information Systems Technology and Design Pillar, Singapore University of Technology and Design, Singapore*
[3] *Department of Smart Mobility Engineering, Inha University, Yeonsu-Gu, Incheon, Korea*
[4] *School of Electrical and Mechanical Engineering, the University of Adelaide, North Terrace, Adelaide, SA, Australia*
[5] *Centre for Wireless Communications, University of Oulu, Oulu, Finland*

10.1 Pitfalls of Federated Learning for Semantic Alignment

Federated learning (FL) is a distributed learning framework that constructs a global consensual model by iteratively averaging locally trained modals [McMahan et al., 2017]. In this section, we explore the challenges of aligning semantics through the lens of FL. Precisely, as Figure 10.1 illustrates, we consider a communication scenario involving multiple DeepJSCC transceivers. Each transceiver is structured as a vector-quantized variational autoencoder (VQ-VAE) [Van Den Oord and Vinyals, 2017]. VQ-VAE is a neural network (NN) wherein a vector-quantized codebook bridges its decoder and decoder components. These encoder, decoder, and codebook components are trained concurrently.

For simplicity without loss of generality, consider two pairs of independently trained transceivers: (i) Alice and Bob, trained on the MNIST dataset (hand-written digit images); and (ii) Carol and David, trained on the Fashion-MNIST (FMNIST, clothes images). After training is completed, neither Alice–Bob nor Carol–David can reconstruct images from the FMNIST and MNIST samples, as illustrated in Figure 10.2b,c. Subsequently, we apply FL in the way that the codebooks of Alice-Bob and Carol-David are averaged per each global training iteration, hereafter referred to as federated codebook (FC). Figure 10.3

Foundations of Semantic Communication Networks, First Edition.
Edited by Walid Saad, Christina Chaccour, Christo Kurisummoottil Thomas, and Merouane Debbah.

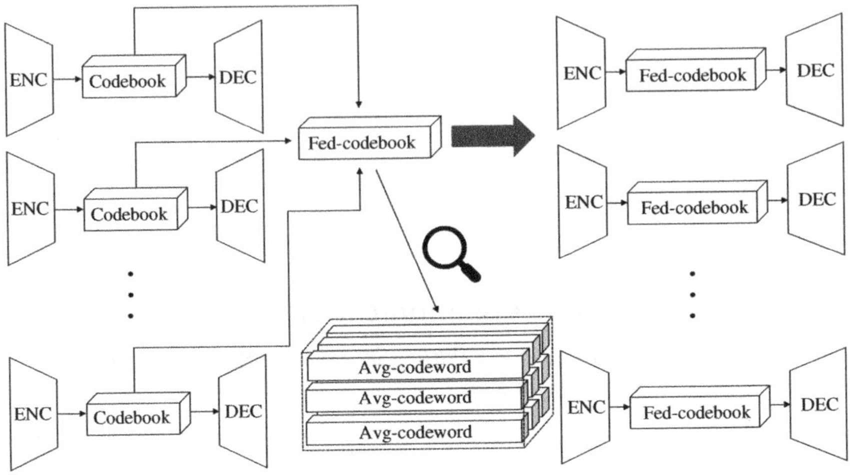

Figure 10.1 A schematic illustration of the federated codebook.

demonstrates that applying FL increases the speed of training convergence. Nevertheless, it does not enhance the quality of image reconstruction, as evidenced by Figure 10.2d,e.

In fact, FL achieves optimal performance when the training data are independent and identically distributed (IID) across data labels and users. From this perspective, the aforementioned scenario involving two distinct datasets (i.e., MNIST and FMNIST) is an extremely non-IID situation where the benefits of federation are negated [Zhu et al., 2021]. This finding suggests that mere model averaging during the training process cannot ensure interoperability across the users associated with different types of source data. Furthermore, the time required for training convergence is significantly greater than the timescale of user mobility, rendering simultaneous training impractical for mobile communication scenarios. These limitations underscore the need for investigating fast and effective fine-tuning techniques for achieving interoperable semantic communication (SC), which will be the focus of subsequent sections.

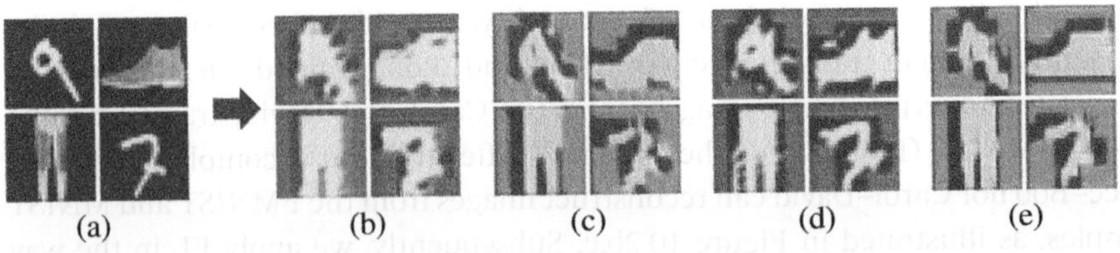

Figure 10.2 A brief illustration of the Federated Codebook: (a) Original, (b) VQ-VAE (MNIST), (c) VQ-VAE (F-MNIST), (d) FC-VQ-VAE (MNIST), and (e) FC-VQ-VAE (F-MNIST).

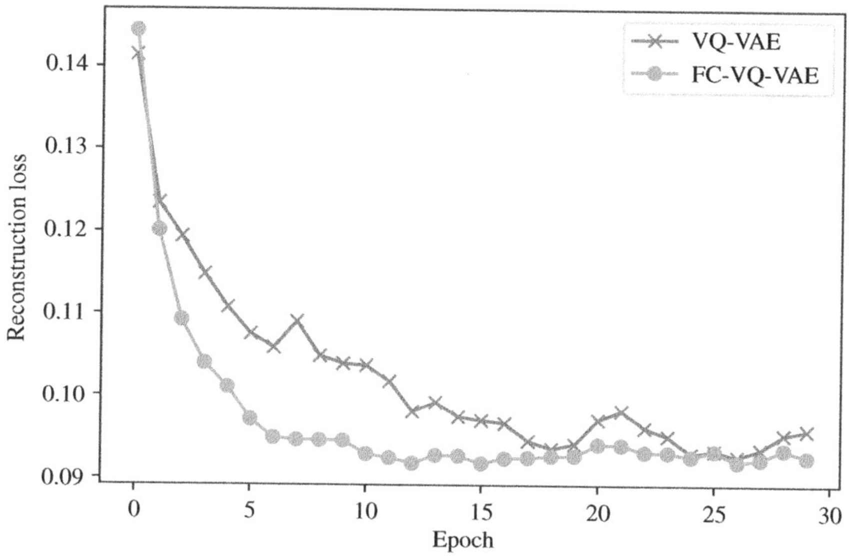

Figure 10.3 Reconstruction loss over the course of training with and without federated codebook method.

10.2 Split Learning for Semantic Alignment

In Section 10.1, we addressed the limitations of model averaging in FL for achieving interoperable SC, primarily attributed to extreme non-IIDness and significant training latency. As an alternative approach, we now explore split learning (SL), which was originally designed to construct a large global model by dividing the model into multiple segments [Vepakomma et al., 2018]. In Vanilla SL, a single NN is divided into its upper and lower segments. The upper segment, stored on a server, is shared among all users, with each possessing a unique lower segment. Each user's lower segment is sequentially connected to the shared upper segment, and both segments are trained until convergence. This method of sequentially transitioning between different users while continuously updating the shared upper segment renders SL particularly effective against non-IID source data while achieving fast convergence, in contrast to FL.

Inspired from this SL architecture and operations, we adapt SL for DeepJSCC transceivers in the following manner. To describe the split architectures of the transceivers, let ENC_1-DEC_1 represent Alice and Bob, trained to communicate semantic representations (SRs) SR_{11}, and the ENC_2-DEC_2 for Carol and David, which are independently trained to communicate SR_{22}. Our focus is on the scenario where Alice transmits SR_{12} of a source data sample for David to reconstruct the original sample. Owing to the independent training processes of ENC_1-DEC_1

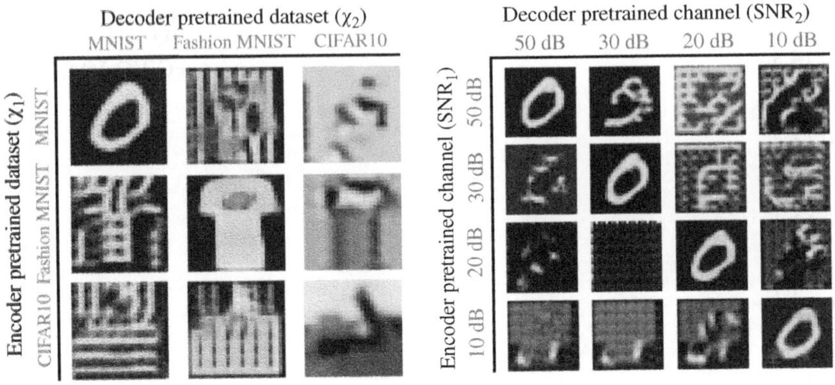

Figure 10.4 Reconstructed image results of encoder–decoder pairs trained in different environments. Source: Park et al. (2022)/IEEE.

and ENC_2-DEC_2, David is unable to decode SR_{12} as intended. For instance, comparison of the same source data and channel characteristics (represented in the diagonal terms of Figure 10.4) with the off-diagonal terms in Figure 10.4 reveals David's failure to decode SR_{12} due to distinct source data and different channel characteristics.

In these transceiver split architectures, as Figure 10.5 illustrates, following SL, we consider that ENC_1 is shared and sequentially connected first to DEC_1 during training and then to DEC_2 for fine-tuning. By downloading DEC_2 from David,

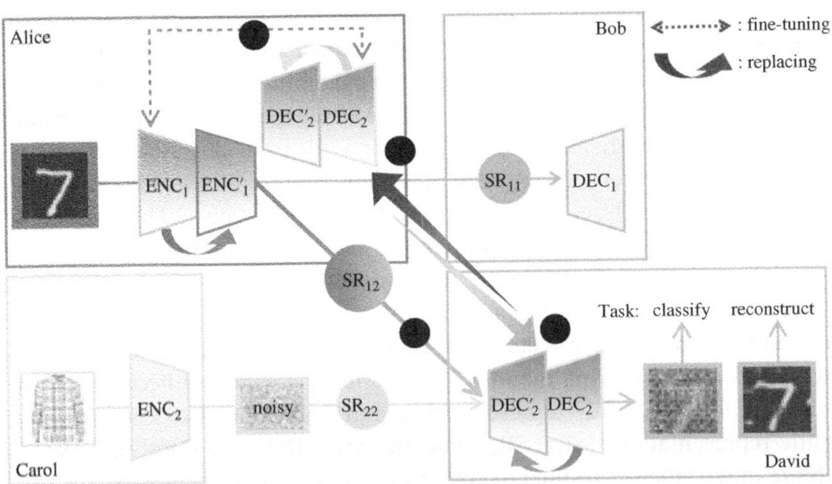

Figure 10.5 Reconstructed image results of encoder–decoder pairs trained in different environments.

Alice can establish the latter connection ENC_1–DEC_2, thereby locally emulating Alice-David communication to align SRs between ENC_1 and DEC_2, as described in the following steps.

1. **Decoder downloading** (Alice ← David): Alice downloads David's decoder DEC_2.
2. **Local fine-tuning** (at Alice): Alice combines the downloaded DEC_2 with its encoder ENC_1, and performs local fine-tuning on the composite NN ENC_1–DEC_2. This fine-tuning utilizes its own source data while assuming the downlink channel statistics are known thanks to the reciprocal downlink–uplink channels between Alice and Bob. The fine-tuned NN is designated as ENC'_1–DEC'_2.
3. **Fine-tuned decoder uploading** (Alice → David): Alice uploads the fine-tuned decoder DEC'_2 back to David.
4. **Aligned SR transmission** (Alice → David): Finally, Alice transmits SRs generated from ENC'_1, which David is now able to decode using its updated decoder DEC'_2.

In Step 2, the extent to which we fine-tune DEC_2 is crucial, impacting both computation and communication overheads. Opting to only fine-tune ENC_1 while keeping DEC_2 frozen can lead to significant savings in terms of not only fine-tuning computation overhead, but also communication costs as it eliminates the need for Step 3. However, this strategy may potentially degrade the quality of the reconstructed data. Therefore, depending on the requirements for specific tasks, the number of fine-tuning layers in DEC_2 should be optimized.

We refer to the aforementioned method as SL with layer freezing (SLF). To demonstrate the effectiveness of SLF, we will present simulations as follows. First, we consider the following three pairs of VQ-VAE structured Deep-JSCC transceivers, trained independently on different datasets and channel characteristics:

- ENC_1–DEC_1: MNIST dataset with $SNR = 5\,\text{dB}$ (Rayleigh),
- ENC_2–DEC_2: MNIST dataset with $SNR = 30\,\text{dB}$ (Rayleigh), and
- ENC_3–DEC_3: CIFAR-10 dataset with $SNR = 30\,\text{dB}$ (Rayleigh).

Here, following Nemati and Choi [2023], we model the channel for each transceiver using a K-ary discrete memory channel (DMC), where K denotes the number of codewords in VQ-VAE. In this model, any transmitted codewords are misdecoded as one of K–1 codewords with probability ϵ. This crossover probability ϵ is linked to the signal-to-noise ratio (SNR) and calculated based on the bit error rate (BER) associated with K-ary quadrature amplitude modulation (QAM).

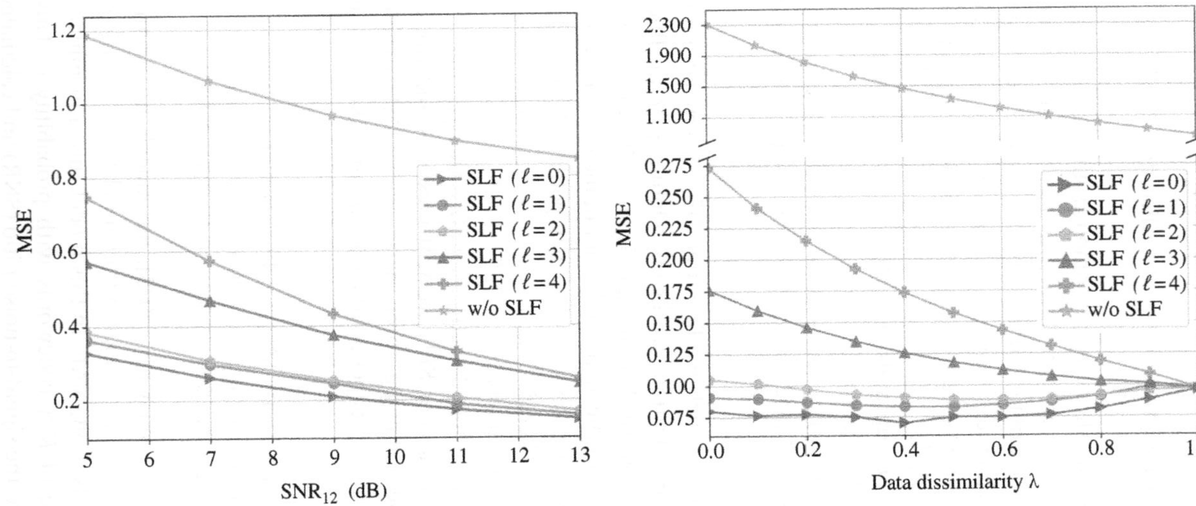

Figure 10.6 SLF result of MSE for reconstruction under semantic misalignment in terms of different channel SNR_{12} and source data dissimilarity λ. The symbol error rates for 16-QAM from 5 to 13 dB are 0.31 through 0.07 for Rayleigh fading channel.

Figure 10.6 shows that SLF yields a lower mean-squared error (MSE) for reconstruction compared to the case not utilizing SLF. Precisely, to assess the impact of different channel characteristics, the left-hand side (LHS) of Figure 10.6 pairs ENC_1–DEC_1 with ENC_2–DEC_2. In this case, ENC_1 trained with DEC_1 is fine-tuned for communicating with DEC_2. As expected, diminishing the number ℓ of frozen layers improves the reconstruction performance, at the cost of incurring additional communication overhead due to the inclusion of Step 3. As channel conditions improve, both scenarios with and without SLF experience-enhanced reconstruction quality, while SLF consistently outperforms the case without SLF.

Next, to assess the impact of different channel characteristics, the right-hand side (RHS) of Figure 10.6 associates ENC_2–DEC_2 with ENC_3–DEC_3 to examine the influence of distinct source data. In this case, ENC_2 trained with ENC_2 is fined-tuned for communicating with ENC_3. In Step 4, the source data of ENC_2' is given by mixing the original datasets $\mathcal{X}_2$ and $\mathcal{X}_3$ of ENC_2–DEC_2 and ENC_3–DEC_3, i.e.,

$$\mathcal{X}_2' = (1 - \lambda)\mathcal{X}_2 + \lambda\mathcal{X}_3. \tag{10.1}$$

Here, the mixing ratio $\lambda \in [0, 1]$ quantifies the data dissimilarity. For all ranges of λ, SLF achieves superior reconstruction results than the case without SLF.

However, fine-tuning Alice to adapt to David may forget their original capability of communicating with Carol. To explore this semantic adaptation-forgetting trade-off, Figure 10.7 illustrates the learning curve of fine-tuning ENC_1 when it continuously alternates between adapting to David (DEC_2) and then going back to Carol (DEC_1), after each fine-tuning convergence. In such a continual learning scenario, ENC_1 may entirely forget the previous communication capability, known as the phenomenon of catastrophic forgetting [De Lange et al., 2021]. The result depicted in Figure 10.7 shows that, indeed, each adaption time Alice partly forgets the previous communication capability, as indicated by the peak at every new encounter. However, interestingly, through repeated encounters, SLF inherently overcomes this issue by significantly reducing not only the peaks but also the peak-to-valley length representing the fine-tuning latency. This indicates that SLF does not suffer from catastrophic forgetting, but instead undergoes a form of meta learning, where ENC_1 continuously adapts to various receivers recognizes common features and is capable of adapting to a specific receiver with few-shot learning [Andrychowicz et al., 2016].

10.3 In-Context Learning for Semantic Alignment

Recent advancements in generative artificial intelligence (GenAI) and large language models (LLMs) have made it possible to convert various types of data into natural language text and to manipulate it for diverse tasks [Grassucci et al., 2024].

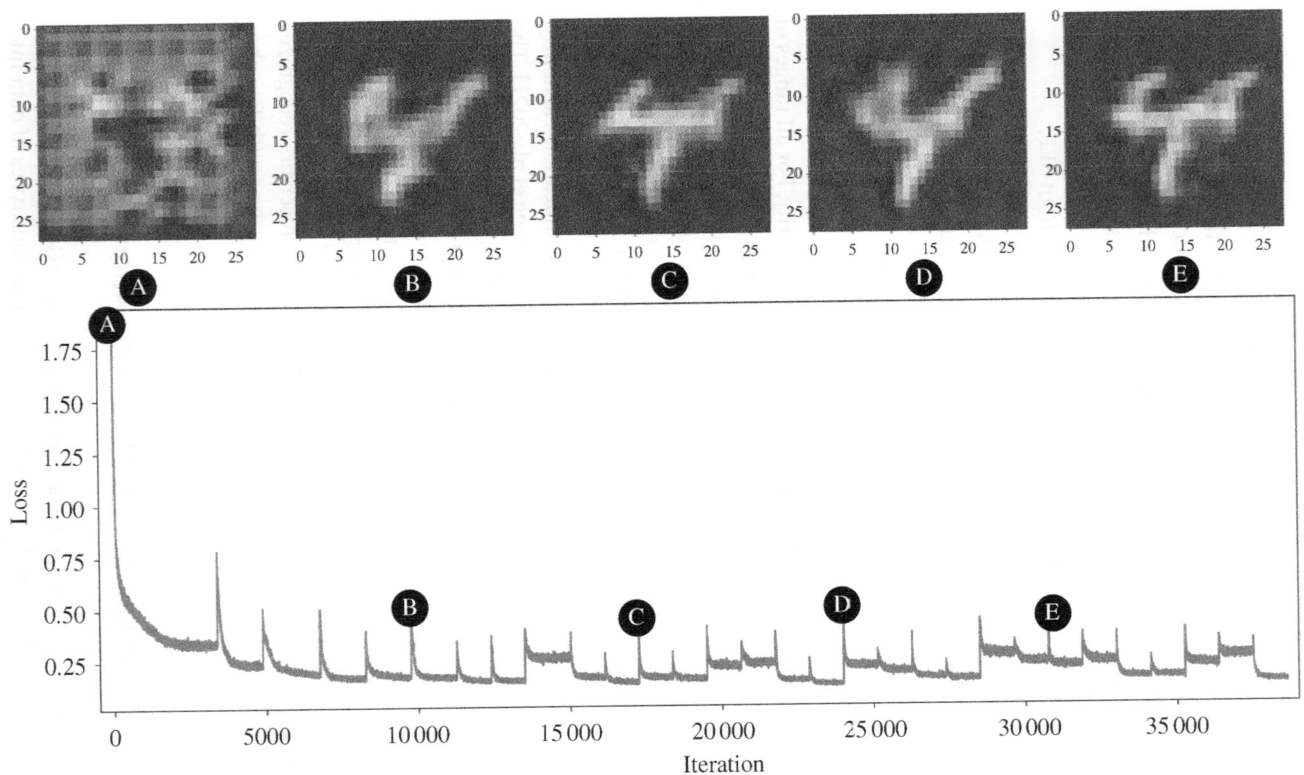

Figure 10.7 Graph of the results of applying SLF when a semantic misalignment situation between ENC–DEC due to different learning channel environments occurred sequentially. Source: Park et al. (2022)/IEEE.

Motivated by these developments, in this section, we consider a language-oriented SC (LSC) framework tailored for a text-to-image (T2I) generation task [Nam et al., 2023]. In this task, Alice and Bob communicate using natural language texts so that Bob can generate an image that is perceptually similar to an intended image located at Alice's end. This process involves Alice converting an image to text (I2T) and Bob subsequently converting text back to an image (T2I). These conversions are facilitated by multimodal NNs, such as bootstrapping language-image pre-training (BLIP) [Li et al., 2022] and contrastive language-image pre-training (CLIP) models [Radford et al., 2021]. These pre-trained NNs are capable of measuring the similarities between text and images through attention scores, thus enabling conversion of text to images and vice versa.

To explore scenarios of semantic misalignment, we suppose that Alice's I2T conversion utilizes BLIP, whereas Bob's T2I conversion is based on CLIP. Since different types of multimodal NNs have distinct interpretations of text-image similarities, the use of an non-identical multimodal NN by both Alice and Bob results in performance degradation. To address such semantic misalignment, as discussed in Section 10.2, we can employ SL to fine-tune I2T by downloading T2I. Now, we turn our attention to the unique ability of LLMs, which offers a pathway to immediate adaptation without additional communication and fine-tuning delays, as elaborated next. Unlike small NNs, LLMs have been trained on a vastly larger dataset with a huge number of model parameters. This extensive training endows LLMs with a broad spectrum of generic knowledge readily adaptable to various tasks. Such knowledge can be immediately tailored to specific tasks by demonstrating a few examples, without adjusting model parameters, which is known as in-context learning (ICL) [Brown et al., 2020].

To facilitate ICL for our task, we explore knowledge distillation (KD), a method for transferring knowledge from a target (teacher) model to another (student) model [Hinton et al., 2015]. KD operates in two major steps. Initially, a common input is fed into both teacher and student models, and their outputs are presented to the student model. Subsequently, the student is learned by minimizing the difference in these outputs for the same input. We hypothesize that LLM's ICL can emulate the second step of KD. Hence, we only follow the first step of KD, in the way that a common image is processed by two I2T models, BLIP and CLIP. The resulting two output text prompts associated with the same image are fed into an LLM as a single example for ICL. This approach is hereafter termed semantic knowledge distillation (SKD) as visualized in Figure 10.8.

Precisely, SKD considers K image samples. The ith image is fed into Alice's BLIP and Bob's CLIP, resulting in the ith text prompts $\mathbf{x}_a^{(i)}$ and $\mathbf{x}_b^{(i)}$, respectively. After presenting a set K of such text prompts, an LLM can play a role as a text-to-text interpreter (T2T) converting Alice's new I2T output $\mathbf{x}_a$ into $\hat{\mathbf{x}}_b$ tailored to Bob, i.e.,

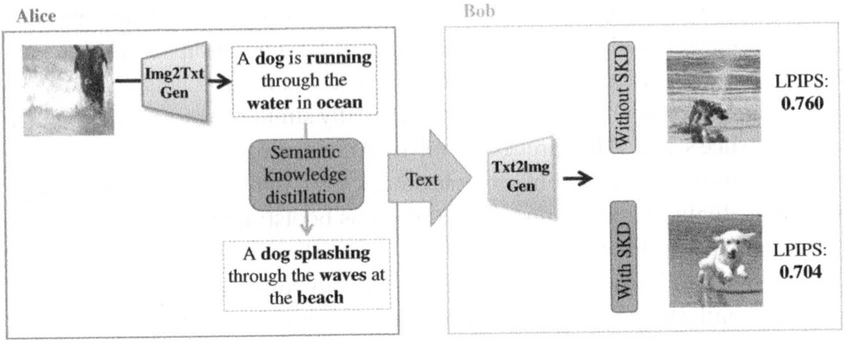

Figure 10.8 Semantic Knowledge Distillation (SKD) in language-oriented semantic communication. Source: Choi et al. (2024)/IEEE/CC BY 4.0.

$$\hat{\mathbf{x}}_b = \mathtt{T2T}\left(\mathbf{x}_a; \{\boldsymbol{v}^{(i)}, \mathbf{x}_a^{(i)}, \mathbf{x}_b^{(i)}\}_{i=1}^{K}\right). \tag{10.2}$$

To demonstrate the effectiveness of SKD, our simulation considers that Alice is equipped with the BLIP image encoder for I2T while Bob is associated with Stable Diffusion v1.5 with 50 denoising steps for T2I generation, conditioned by the CLIP text encoder. Alice transmits images from the Flickr8k dataset [Hodosh et al., 2013], 8092 samples of 256×256 images. Before transmitting text prompts to Bob, Alice runs GPT-4 for SKD with $K = 30$. For each transmission, a text prompt is coded with 8-bit ASCII and modulated using 16-QAM.

The performance, i.e., perceptual similarities between Alice's original images and Bob's generated images, is measured using the learned perceptual image patch similarity (LPIPS) score [Zhang et al., 2018] that calculates the distance at hidden layers of pre-trained AlexNet, i.e.,

$$\mathtt{LPIPS}(\boldsymbol{v}, \hat{\boldsymbol{v}}(t)) = \sum_l \frac{1}{H_l W_l} \sum_{h,w} \|f(\boldsymbol{v}) - f(\hat{\boldsymbol{v}}(t)\|_2^2. \tag{10.3}$$

The term l identifies the lth layer having its width w, height h, and dimension $H_l \times W_l$ with an activation function $f(\cdot)$. The output values of this layer range from 0 to 1, serving as indicators of similarity, where 0 corresponds to identical features and 1 indicates a high degree of inconsistency.

Figure 10.9 shows that in both ideal and noisy channels with various SNRs, SKD consistently reduces LPIPS by an average of 0.008. Figure 10.10 illustrates an exemplary sample of SKD. According to a linguistic style analysis using GPT-4, it reveals that Bob associated with CLIP uses more refined expressions and sophisticated vocabularies compared to Alice equipped with BLIP. After ICL with $K = 30$ examples, Figure 10.10 shows that how Alice's linguistic style becomes aligned with Bob's, showcasing the effectiveness of SKD in semantic alignment.

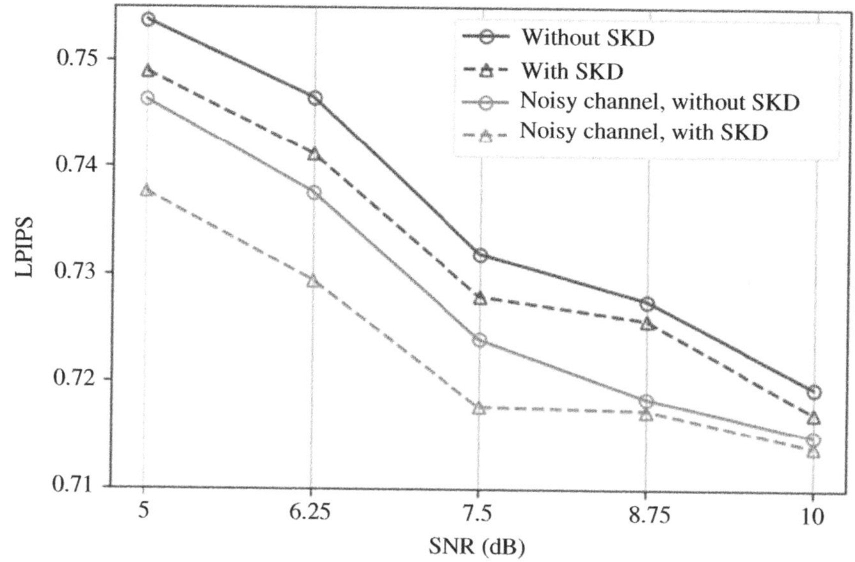

Figure 10.9 Semantic Knowledge Distillation (SKD) performance in various channel conditions.

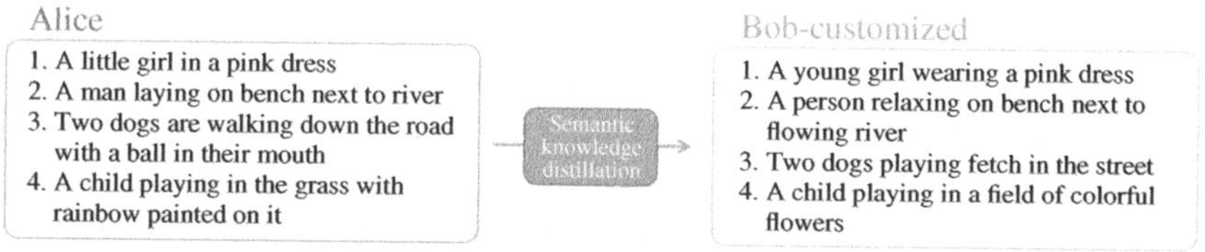

Figure 10.10 Samples of original prompts from Alice and customized prompts through Semantic Knowledge Distillation (SKD).

10.4 Conclusion and Future Directions

In this chapter, we explored the challenge of interoperability in SC, incurred by training biases due to varying source data and channel characteristics prevalent in mobile communication systems. To tackle this issue, we demonstrated that SL offers a fast and efficient fine-tuning solution for achieving semantic alignment in SC. Furthermore, within the context of LSC where communication is confined to text messages, we showed that SKD facilitates immediate adaptation without any fine-tuning delays, thanks to LLM's capability of ICL.

In the SL scenario, we explored, our fcas was on two pairs of transceivers, where the local fine-tuning process adheres to the sequential operations of Vanilla SL.

With many transceivers, an intriguing area for further investigation is the potential impact of parallel SL implementations [Oh et al., 2023] on reducing fine-tuning delays. On the other hand, in our study of LSC, SKD operations require LLM inference that entails huge computing and memory costs. It could be interesting to reduce these costs by exploiting LLM compression and inference acceleration methods for future research.

Bibliography

Marcin Andrychowicz, Misha Denil, Sergio Gomez, Matthew W. Hoffman, David Pfau, Tom Schaul, Brendan Shillingford, and Nando De Freitas. Learning to learn by gradient descent by gradient descent. In *Advances in Neural Information Processing Systems 29*, 2016.

Tom Brown, Benjamin Mann, Nick Ryder, Melanie Subbiah, Jared D. Kaplan, Prafulla Dhariwal, Arvind Neelakantan, Pranav Shyam, Girish Sastry, Amanda Askell, et al. Language models are few-shot learners. In *Advances in Neural Information Processing Systems 33*, pages 1877–1901, 2020.

Matthias De Lange, Rahaf Aljundi, Marc Masana, Sarah Parisot, Xu Jia, Aleš Leonardis, Gregory Slabaugh, and Tinne Tuytelaars. A continual learning survey: Defying forgetting in classification tasks. *IEEE Transactions on Pattern Analysis and Machine Intelligence*, 44(7):3366–3385, 2021.

Eleonora Grassucci, Jihong Park, Sergio Barbarossa, Seong-Lyun Kim, Jinho Choi, and Danilo Comminiello. Generative AI meets semantic communication: Evolution and revolution of communication tasks, 2024.

Geoffrey Hinton, Oriol Vinyals, and Jeff Dean. Distilling the knowledge in a neural network. arXiv preprint arXiv:1503.02531, 2015.

Micah Hodosh, Peter Young, and Julia Hockenmaier. Framing image description as a ranking task: Data, models and evaluation metrics. *Journal of Artificial Intelligence Research*, 47:853–899, 2013.

Junnan Li, Dongxu Li, Caiming Xiong, and Steven Hoi. BLIP: Bootstrapping language-image pre-training for unified vision-language understanding and generation. In *International Conference on Machine Learning*, pages 12888–12900. PMLR, 2022.

Brendan McMahan, Eider Moore, Daniel Ramage, Seth Hampson, and Blaise Aguera y Arcas. Communication-efficient learning of deep networks from decentralized data. In *Artificial Intelligence and Statistics*, pages 1273–1282. PMLR, 2017.

Hyelin Nam, Jihong Park, Jinho Choi, Mehdi Bennis, and Seong-Lyun Kim. Language-oriented communication with semantic coding and knowledge distillation for text-to-image generation. arXiv preprint arXiv:2309.11127, 2023.

Mahyar Nemati and Jinho Choi. All-in-one: VQ-VAE for end-to-end joint source-channel coding. *Authorea Preprints*, 2023.

Seungeun Oh, Hyelin Nam, Jihong Park, Praneeth Vepakomma, Ramesh Raskar, Mehdi Bennis, and Seong-Lyun Kim. Mix2SFL: Two-way mixup for scalable, accurate, and communication-efficient split federated learning. *IEEE Transactions on Big Data*, 10(3):238–248, 2023. doi: 10.1109/TBDATA.2023.3328424.

Alec Radford, Jong Wook Kim, Chris Hallacy, Aditya Ramesh, Gabriel Goh, Sandhini Agarwal, Girish Sastry, Amanda Askell, Pamela Mishkin, Jack Clark, Gretchen Krueger, and Ilya Sutskever. Learning transferable visual models from natural language supervision. In *International Conference on Machine Learning*, pages 8748–8763. PMLR, 2021.

Aaron Van Den Oord and Oriol Vinyals. Neural discrete representation learning. In *Advances in Neural Information Processing Systems 30*, 2017.

Praneeth Vepakomma, Otkrist Gupta, Tristan Swedish, and Ramesh Raskar. Split learning for health: Distributed deep learning without sharing raw patient data. arXiv preprint arXiv:1812.00564, 2018.

Richard Zhang, Phillip Isola, Alexei A. Efros, Eli Shechtman, and Oliver Wang. The unreasonable effectiveness of deep features as a perceptual metric. In *Proceedings of the IEEE Conference on Computer Vision and Pattern Recognition*, pages 586–595, 2018.

Hangyu Zhu, Jinjin Xu, Shiqing Liu, and Yaochu Jin. Federated learning on non-IID data: A survey. *Neurocomputing*, 465:371–390, 2021.

Part III

Machine Reasoning for AI-Native Semantic Communication Networks

Once networking challenges and protocols have been meticulously designed, the logical progression involves delving into the AI and ML foundations that underpin semantic communication systems. Part III of our exploration embarks on this journey, beginning with an in-depth examination of the role deep learning plays in semantic communications.

The inaugural chapter of this section delves into the transformative shift necessary for the design of semantic communication networks. It advocates for a departure from data-driven and information-driven AI-augmented networks, where wireless networks are tethered to raw data, towards knowledge-driven and reasoning-driven AI-native networks. This transition entails scrutinizing the causal foundational aspects of semantic networks and constructing a semantic language grounded in causality, transcending mere associational logic.

Continuing, the second chapter confronts the pressing issue of scalability within semantic communication, confronting the notorious "semantic gap." Through the lens of a lightweight yet potent self-critical supervision mechanism, this analysis endeavors to bridge this gap efficiently during the learning process.

Subsequently, attention shifts to the realm of concealed information, encompassing hidden relationships, conceptual intricacies, and the implicit reasoning mechanisms of users. Herein, the deployment of a semantic-aware communication (iSAC) framework comes into play, employing a graph-inspired model to encapsulate the entirety of a message's semantic spectrum. This approach further harnesses adversarial imitation learning, fostering the development of a reasoning framework within the receiver to discern explicit semantics and their implicit counterparts effectively.

Foundations of Semantic Communication Networks, First Edition.
Edited by Walid Saad, Christina Chaccour, Christo Kurisummoottil Thomas, and Merouane Debbah.

Concluding our exploration, the final chapter adopts a data valuation perspective on semantic and goal-oriented communication. Within this context, we investigate two distinct classes of problems delineated by strategic pull-based and push-based frameworks, shedding light on their unique challenges and potential solutions.

11

Causal Reasoning Foundations of Semantic Communication Systems

Christo Kurisummoottil Thomas[1], *Christina Chaccour*[2], *Walid Saad*[3], *and Merouane Debbah*[4]

[1] *Virginia Tech Research Center, Virginia Tech, Arlington, VA, USA*
[2] *Ericsson, Plano, TX, USA*
[3] *Department of Electrical and Computer Engineering, Virginia Tech, Arlington, VA, USA*
[4] *6G Center, Khalifa University of Science and Technology, Abu Dhabi, United Arab Emirates*

11.1 Introduction

Thus far we have elucidated the information theoretic viewpoints of semantic communication (SC), exploring its potential to revolutionize wireless networks by imbuing communication systems with the ability to perceive and convey meaning. However, to fully realize this vision, it is imperative to recognize the inadequacies of today's information theoretic foundation rooted in classical information theory.

The classical information theory framework, while groundbreaking in its time, falls short in the context of SC. Traditional communication systems, operating within the confines of classical information theory, treat data transmission as a purely technical problem, focusing solely on optimizing the accuracy and efficiency of symbol transmission. However, this approach neglects the semantic dimension of communication, overlooking the crucial aspect of conveying meaning inherent in data.

To bridge this gap, we propose the establishment of a semantic language between transmitters and receivers, transforming their roles into that of a teacher and apprentice. This paradigm shift from transmitter/receiver to teacher/apprentice signifies a departure from conventional data-centric communication toward a more nuanced understanding of communication as a knowledge-driven process. Notably, the transmitter must undergo a transformation from a mere conduit of bits to a facilitator akin to a "teacher," endowed with the capability to discern multiple "semantic content elements" within the source data. These

Foundations of Semantic Communication Networks, First Edition.
Edited by Walid Saad, Christina Chaccour, Christo Kurisummoottil Thomas, and Merouane Debbah.

elements refer to distinct meanings or semantics encapsulated within a message. Subsequently, for each identified semantic content element, the transmitter must construct a "semantic representation" possessing desirable properties. Essentially, the semantic content embodies the meaningful essence of the data, while the semantic representation serves as a concise portrayal of this significance. This process mirrors the way human beings select appropriate words to articulate their observations and thoughts. Furthermore, various semantic content elements may correspond to different modalities within the data. For example, in an audio recording of a person's voice, the tone of voice represents one semantic content element, while the spoken words constitute another. Human cognition effortlessly disentangles and distinguishes between these elements, comprehending the intended meaning of the spoken words. Conversely, contemporary communication system transmitters lack the ability to identify or segregate underlying semantic structures or modalities. Hence, there is a compelling need to re-engineer transmitters to emulate human reasoning capabilities to the best of their ability. This necessitates the incorporation of "reasoning" on the transmission end – a faculty enabling the transmitting agent to recognize semantic content elements, differentiate them from others within the data, and devise efficient representations for each identified content. This stands in stark contrast to the conventional transmitters in today's networks, which treat input data as a random and uncertain string of information, transmitting it as a bit-pipeline that characterizes this uncertainty.

Given this paradigm shift, the smallest meaningful unit is a semantic representation, forming the basis of a semantic language. These languages, while resembling natural languages, prioritize minimalism and generalizability over data rates to enhance autonomy and understanding. Semantic representations within this language must adhere to three key properties:

1. **Minimalism**: They must succinctly characterize the information structure with the fewest language elements possible, thereby minimizing the number of exchanged messages in the long run.
2. **Generalizability**: Semantic representations should be invariant to changes in distribution, domain, and context. This ensures that learned representations can be applied across various domains and contexts, akin to the way words in natural languages generalize to describe diverse events.
3. **Efficiency**: The apprentice must efficiently regenerate information with high fidelity, ensuring that the resolution of generated data matches or exceeds that recoverable by a classical receiver. This implies no trade-off between content quality and minimalism achieved via SC.

Given these properties, it becomes evident that a SC framework necessitates a robust foundation built upon causality. In this chapter, we will explore why integrating causality into the foundational principles of SC in wireless networks is essential. We will argue that causality, through causal discovery, inference, and representation learning, plays a fundamental role in enabling communication systems to effectively understand and convey meaning. By unraveling the causal relationships underlying observed phenomena, communication systems can transcend the limitations of classical information theory and move toward a SC framework capable of comprehending and conveying the significance or meaning embedded within data. Through a comprehensive exploration of causality in this context, our aim is to establish the indispensable role of causality in shaping the fundamentals of SC in wireless networks, paving the way for the development of next-generation communication systems capable of understanding and conveying meaning in a knowledge-driven manner.

11.2 Causality Primer

Before delving into the intricate synergy between causal learning and the development of a semantic language tailored for the teacher–apprentice dynamic within the realm of SC, it is imperative to establish a foundational understanding of the principles underlying causal learning. This preliminary exploration serves as a crucial framework upon which we can build a comprehensive discourse on the interplay between causality and effective communication.

Causal learning lies at the core of our cognitive abilities, enabling us to grasp the intricate web of cause and effect relationships that shape the world around us. It encompasses the cognitive processes through which individuals acquire, process, and utilize information to discern these causal links within their environment. This innate capacity not only deepens our understanding of the world but also serves as the foundation for rational decision-making, problem-solving, and continuous learning. In this section, we delve into the pursuit of three key objectives – minimalism, generalizability, and efficiency – through the fundamental building blocks of causality: causal discovery, causal inference, and causal representation learning (CRL), as illustrated in Figure 11.1.

Definition 11.1 *Causal discovery* employs algorithmic methods to unveil the *underlying causal structure* inherent in observed wireless data, spanning spectrum data, channel information, network topology, or any multimodal sensor-fusion or source data. Unlike statistical AI models that merely infer correlations between entities (features) within the data, causal models accurately depict the directional relationships among these entities, providing explanations for their connections. Consequently, causal discovery naturally offers a pathway to design explainable AI-native wireless networks, enabling clearer insights into the causal mechanisms shaping wireless communication systems.

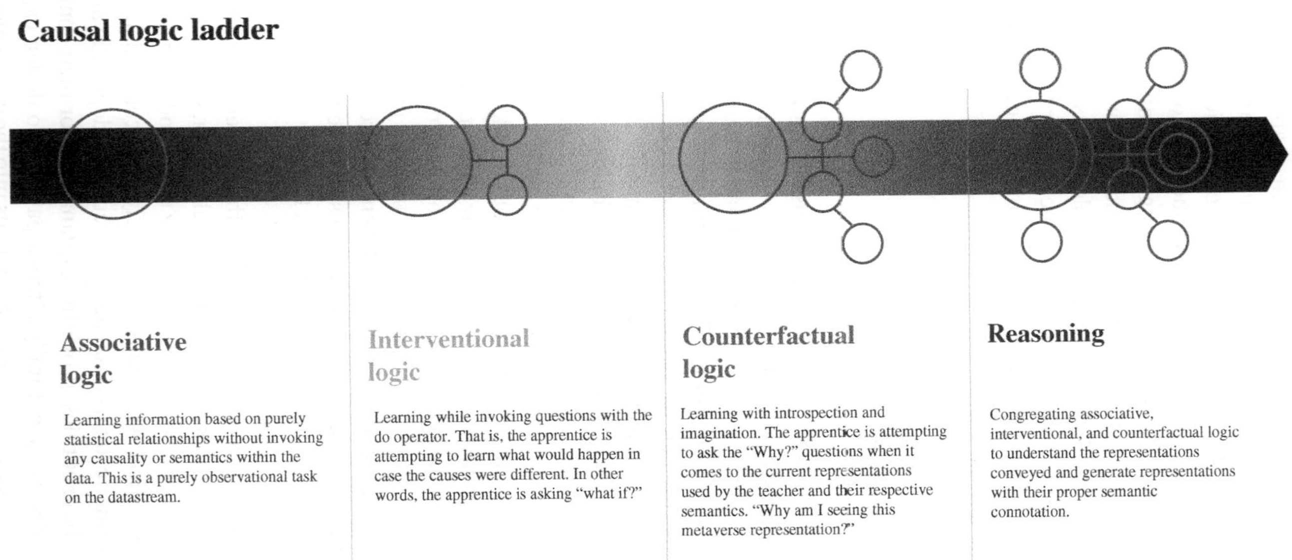

Figure 11.1 Hierarchical levels of causality. Source: Adapted from Chaccour et al. [2024].

An effective and practical approach to implementing causal reasoning via graphical models is through the utilization of structural causal models (SCMs) [Pearl and Mackenzie, 2018]. SCMs offer a structured framework for describing the underlying mechanisms that govern a specific phenomenon, providing a clear and concise language for causal analysis.

Definition 11.2 An *SCM* is a collection of elements $\langle \mathcal{U}, \mathcal{V}, \mathcal{F}, P(\boldsymbol{U})\rangle$, where $\mathcal{V}$ represents endogenous variables (cause and effect variables), and $\mathcal{U}$ represents exogenous variables (random, unknown noise). The structural functions $f_i \in \mathcal{F}$ are designed so that $f_i(PA_i, \boldsymbol{u}_i)$ determines $\boldsymbol{v}_i$, where $PA_i \subset \mathcal{V}$ and $\boldsymbol{u}_i \subset \mathcal{U}$ specify the sets of parents (in the graph) and exogenous inputs, respectively. The exogenous distribution $P(\boldsymbol{u}_i)$ determines the values of $\boldsymbol{u}_i$ and thus the distribution of endogenous variables $\mathcal{V}$.

Causal discovery assumed variables $\boldsymbol{v}_i$ to be random variables connected by a causal graph. However, real-world wireless observations are usually not structured into those variables. For example, high-dimensional channel information or raw source data that is either video or images or holograms. Herein, it is essential to exploit the concept of CRL.

Definition 11.3 *CRL* involves mapping large dimensional observations $\boldsymbol{D} = [d_1, \ldots, d_D]^T$ into a lower dimensional representation involving causal variables $\boldsymbol{v}_1, \ldots, \boldsymbol{v}_n$, captured as:

$$\boldsymbol{v}_i = f_i(PA_i, \boldsymbol{u}_i), \forall i. \tag{11.1}$$

Building a semantic language $\mathcal{L}$ that can be mapped to an SCM model enables disentangling each datastream and its respective representation from other established representations. In other words, the model describing the language can be written as:

$$P(\boldsymbol{X}_l) = P(X_{l,1}, \ldots, X_{l,N}) = \prod P(X_{l,i} \mid PA_i). \tag{11.2}$$

Definition 11.4 *Causal inference* seeks to leverage the knowledge of the causal structure obtained from the causal discovery phase and the learned representations from CRL to make predictions and answer questions about the effects of interventions or counterfactuals over variables of interest. The utilization of causal inference thus empowers wireless systems with both inductive and deductive reasoning capabilities.

Causal inference can be effectively implemented using two key concepts:

- **Interventions**: Interventions can be defined as manipulating any causal variable in the graph using an exogenous random or deterministic process. The do-operator is a mathematical tool used to represent a physical intervention or manipulation of a variable in a causal model [Pearl and Mackenzie, 2018]. In the context of a causal model represented as $Z \rightarrow X \rightarrow Y$, applying the do-operator to X involves removing all incoming edges (that represent the generation of a particular variable from its descendants) to X and setting to a specific value, x_0. This can be represented as $do(X) = x_0$.
- **Counterfactuals**: Counterfactual problems entail the process of reasoning about the causes behind events, envisioning the outcomes of different actions in hindsight, and determining which actions could have led to the desired results. Hence, using counterfactuals, AI models can go beyond just observing the impact of an intervened environment. They can generalize their model behavior to hypothetical scenarios where fundamental changes to the network setup, such as introducing new technologies or altering configurations, could improve performance under various conditions. In traditional wireless problems, the objective is to find the optimal solution for a predefined objective function under specific constraints. In contrast, counterfactuals empower the AI system to adjust these objectives and constraints based on higher-level requirements, offering a more flexible and adaptive approach to problem-solving.

Next, we look at how to define the semantics present in the data using causal reasoning fundamentals discussed above.

11.3 Causal Semantic Communications

11.3.1 System Model

To illustrate the concepts, we consider a point-to-point communication system with a teacher (also called a speaker) who observes a sequence of events that capture the state of the environment, and each event $e \in \mathcal{E}$ leads to the observation $\boldsymbol{x}_e \subset \mathcal{R}^D$. The teacher intends to send a concise description of $\boldsymbol{x}_e$ to a apprentice (also called listener) who wishes to complete a set of sequential tasks drawn from a probability distribution $p(\tau)$. Here, a rigorous approach is to extract the causal structure which represents the *semantics* of information, and communicate it. To compute that, the teacher first extract the different features from $\boldsymbol{x}_e$. These features and their properties are called *entities*. The meaningful content among the entities are defined as *semantic content element* here. We represent the set of N entities as the set of vectors $\{\boldsymbol{s}_0, \dots, \boldsymbol{s}_{N-1}\}$, where each entry $\boldsymbol{s}_i \in \mathcal{R}^M$ (grounded to real logic). If one considers the whole datastream $\boldsymbol{x}_e$ as a bulk, the learning process, i.e., identifying every semantic content element and crafting its corresponding semantic representation, becomes highly inefficient. This is due to the fact that raw data contains a lot of purely random information. This is problematic because such random information increases the complexity of the built language (which we will soon formally define), yet contributes to a language with spurious semantic representations. That is, one might think that the semantic language complexity stems from an inherently complex semantic structure (e.g., a complex high-dimensional hologram that must be described via a semantic language). In reality, the deceiving complexity measure here stems from an abundance of random information in the datastream. Hence, for the sake of efficiency, prior to learning a semantic language, we must first separate two parts of the data, learnable and memorizable, as defined next. An example of communicating using a semantic language is provided in Figure 11.2.

- **The learnable part $\boldsymbol{x}_{e,l}$ of the data**: Is the core of semantic information and captures the inherent causal structure in each semantic content element. Thus, reasoning over $\boldsymbol{x}_{e,l}$ eventually leads to a meaningful semantic language with non-spurious models. One can use $\boldsymbol{x}_{e,l}$ as an input to build a proper semantic language.
- **The memorizable part $\boldsymbol{x}_{e,m}$ of the data**: Is the one that is governed only by pure randomness. For example, these can be details in an image that do not

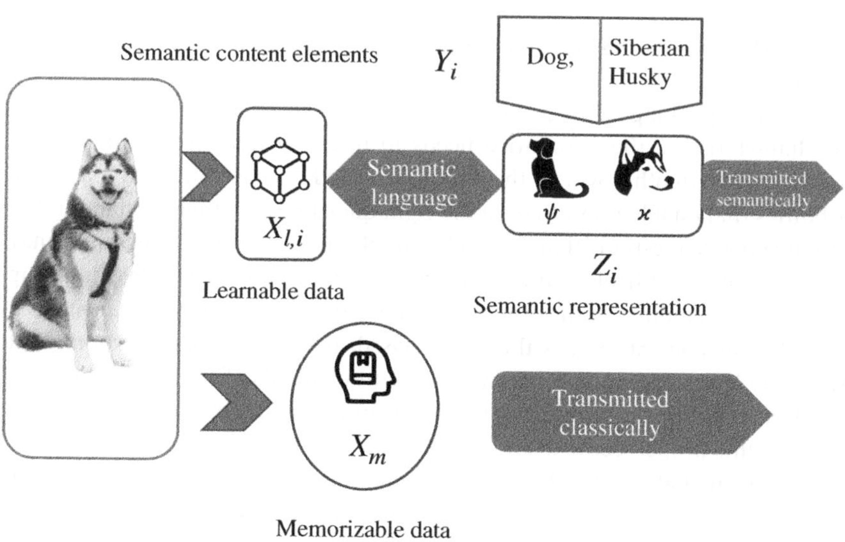

Figure 11.2 Illustrative example showcasing the concept of semantic language. Source: Choi et al., (2024)/IEEE.

contribute to any semantic content element. As such, attributing a semantic representation to $\boldsymbol{x}_{e,m}$ is a highly complex learning process, but a very simple memorization process. In other words, learning the structure of a purely random and structure-less datastream is complex process and does not yield meaningful semantic representation. Thus, allocating computing resources to learning random information is a wasteful process, thus $\boldsymbol{x}_{e,m}$ must be transmitted using classical communication resources.

Next, for the learnable part, our objective is to infer the hidden relations among the entities and compute the causal sequence that best explains the event observed. This component is called the *semantic state descriptor* (a neural component). The semantic state description is defined as the sequence of entities and their relations, $\widetilde{\boldsymbol{z}} = \left(\boldsymbol{s}_0, \boldsymbol{r}_1, \dots, \boldsymbol{r}_{N-1}, \boldsymbol{s}_{N-1}\right)$, of dimension $M(2N-1) \times 1$. Here, the vector $\boldsymbol{r}_i \in \mathcal{R}^M$ denotes the relation between entities $\boldsymbol{s}_i$ and $\boldsymbol{s}_{i+1}$. One way to represent $\boldsymbol{r}_i$ is as a linear relation, $\boldsymbol{s}_{i+1} \approx \boldsymbol{r}_i + \boldsymbol{s}_i$, where $\boldsymbol{r}_i = -\ln p(\boldsymbol{s}_{i+1} \mid \boldsymbol{s}_i)$. If $\boldsymbol{s}_i$ and $\boldsymbol{s}_{i+1}$ are not causally related $p(\boldsymbol{s}_{i+1} \mid \boldsymbol{s}_i) = 0$. Hence, when the entities are far apart in the grounded domain $\mathcal{R}^M$, then $\boldsymbol{r}_i$ has entries ∞. When the $p(\boldsymbol{s}_{i+1} \mid \boldsymbol{s}_i) \approx 1$, then the entities are very close to each other in $\mathcal{R}^M$. A modular approach here to compute *the cause and effect relations* is to utilize the SCM described in Section 11.2. Given the SCM, it is natural to ask how can the apprentice can perform interventions and counterfactuals reasoning.

11.3.1.1 How to Pose the Proper Interventions and Counterfactuals via Queries?

Theoretically speaking, and given a particular datastream that exhibits structural characteristics, the apprentice needs to pose the right interventions and counterfactuals. In other words, the apprentice needs to have minimal reasoning capabilities that enable them to pose the right questions. On the one hand, if the apprentice poses questions that are irrelevant of the context, the transmitter may end up encoding the irrelevant causal structure information, thereby hindering the semantic reliability. On the other hand, if the apprentice has been accustomed to a particular context, and is therefore biased, they might ask questions that only enable learning selective semantic content elements of the data. Here, one aspect that is worth exploring is to send the apprentice information that enable them to build proper context and subsequently pose the proper queries. For instance, given that the teacher uses raw data to complement their semantic representations, the teacher can build on this raw data to guide the apprentice vis-a-vis context and the queries they must ask to understand the language. Here, one can adopt the game-theoretic scheme proposed in Thomas and Saad 2023 to initialize the gradual design of a semantic language.

Definition 11.5 The semantic impact ι_τ generated by a semantic representation $\boldsymbol{u}_i$ during a time τ is the number of packets that would have been needed to be transmitted to regenerate the semantic content element $\boldsymbol{z}_i$.

Lemma 11.1 The communication symmetry index between a teacher b and apprentice d, for a transmission session τ is given by Chaccour et al. 2022:

$$\eta_\tau = \frac{\zeta_{d,\tau}}{\nu_{b,\tau}} \iota_{\tau,\boldsymbol{z}_i}, \tag{11.3}$$

where $\iota_{\tau,\boldsymbol{z}_i}$ is the semantic impact, $\zeta_{d,\tau}$ is the number of query packets (interventions, counterfactuals, etc.) requested by the apprentice to reason over the message transmitted, and $\nu_{b,\tau}$ is the number of raw data packets transmitted by the teacher to accompany the semantic representation sent. We also note that, if the queries are communicated via a semantic representation to the teacher, one can find $\zeta_{d,\tau}$ by applying the concept of semantic impact on the representation used.

Lemma 11.2 The reasoning capacity between a teacher and an apprentice is given by Chaccour et al. 2022:

$$C_R = \Omega \log_2 (1 + \eta), \tag{11.4}$$

where Ω is the maximum computing capability of the server used to represent or generate the semantic representation, and η is the communication symmetry index per second.

Furthermore, given a datastream $\boldsymbol{x}$, with $\boldsymbol{x}_{e,l}$ learnable information and $\boldsymbol{x}_{e,m}$ memorizable ones, one can simply use addition to compute the capacity. That is, the total achievable capacity would be given by:

$$C_T = C_C + C_R = W \log_2 (1 + \gamma) + \Omega \log_2 (1 + \eta), \tag{11.5}$$

where W is the bandwidth and γ is the signal-to-interference-plus-noise-ratio (SINR). One important thing to note here is that while the C_C is limited by Shannon's bound, and the bandwidth of operation; C_R is essentially bounded by the reasoning bound and the computing resources. Notably, claiming an end-to-end (E2E) capacity that might achieve an impact that is higher than what Shannon could have is possible because we are alternatively relying on computing. In other words, if the apprentice is capable of regenerating the message (via a semantic representation) faster than it could be classically transmitted, the E2E capacity could asymptotically reach beyond Shannon's limit. To perform these reasoning computations, it is essential that the apprentice reaches a minimum level of understanding about the data distribution (via causal structure) as well as how the transmitter encodes this causal structure. This can be achieved by

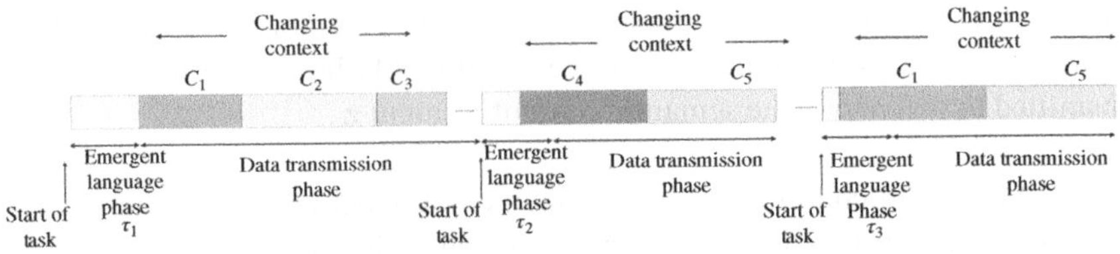

Figure 11.3 Proposed ESC system time split between emergent language and data transmission phases.

emergent language training before the actual data transmission phase. As the apprentice acquires more knowledge over time, the training duration of this emergent language phase should reduce to zero, as shown in Figure 11.3. The emergent language should be built upon such generalization goals in mind, as discussed next.

11.3.2 Emergent Language Model

After causal structure extraction, we need to map $\boldsymbol{z}$ ($\in \mathcal{W}$, the *syntactic space*) to the encoded representation $\boldsymbol{u}$ ($\in \mathcal{U}$, a finite vocabulary) that gets transmitted. For an SC system in which the intentions of both nodes are oblivious to one another, one approach for the teacher to learn semantic awareness is via an interactive conversation. Hence, we propose a novel two-player contextual signaling game [Thomas and Saad, 2023] to learn the teacher transmit distribution $\pi_s = \pi(\boldsymbol{u} \mid \boldsymbol{z})$, the apprentice inference distribution $\pi_l = \pi(\boldsymbol{z} \mid \boldsymbol{u}, c)\pi(c \mid \boldsymbol{u})$, where the context c belongs to a finite set, $C \subset \mathcal{Z}$. *Communication context* here represents the set of all possible state descriptions and the valid logical interpretations defined for a particular scenario corresponding to a task. The communication context, denoted by $c \in C$, captures the richness of the syntactic space in our proposed approach. Specifically, the set of numerical features $\boldsymbol{s}_i$ represents a causal state description, where $\boldsymbol{s}_i \in \mathcal{S}_i$ for each context c_i of a given task τ. There may be possible overlap between different $\mathcal{S}_i$. c_i can be represented as an integer value, indicating which $\mathcal{S}_i$ corresponds to that particular context. Given the syntactic space $\mathcal{W}$, context set C, the representation space $\mathcal{U}$, and the set of all π_s and π_l which are $\mathcal{M}_s$ and $\mathcal{M}_l$, respectively, we define the *signaling game* as the tuple $(C, \mathcal{W}, \mathcal{U}, \mathcal{M}_s, \mathcal{M}_l)$. The use of a game-theoretic approach here is apropos because, at equilibrium, the transmitter will obtain a semantically aware encoder that transmits minimal information such that the apprentice recovers the state description closest semantically to the transmitted version. Using encoded representations for each entity in an event and letting the apprentice extract semantics from them is not communication efficient due to the large number of bits needed to represent all entities.

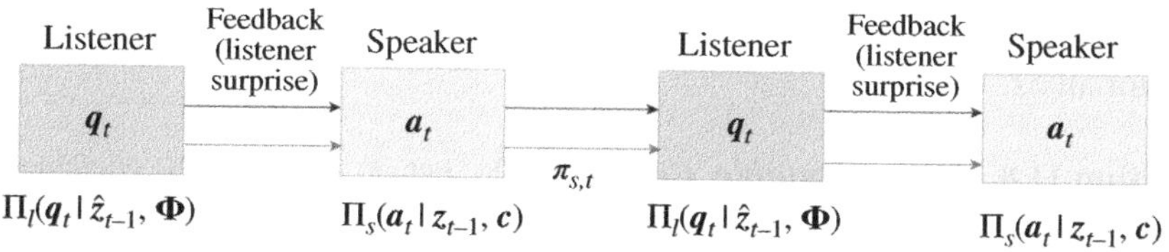

Figure 11.4 Illustration of emergent language.

We define a signaling game [Fudenberg and Tirole, 1991] between the teacher and apprentice as follows. The language, as shown in Figure 11.4, evolves via a series of questions $\boldsymbol{q}_t \in \mathcal{W}$ (question) by the apprentice and answers provided by the teacher, $\boldsymbol{a}_t \in \mathcal{W}$ (answer), which help the apprentice reconstruct the description of the perceived event or partial set of features to achieve its intent. $\boldsymbol{q}_t$ and $\boldsymbol{a}_t$ are vectors of dimensions $M \times 1$. The Q/A-scheme proceeds over multiple communication rounds wherein the answer $\boldsymbol{a}_t$ is encoded using the transmit distribution $\pi_{s,t}$ which is learned. Each question $\boldsymbol{q}_t$ transmitted by the apprentice belongs to a small finite set. Hence, it can be encoded using fewer bits (for example, using an M-QAM, where $M < 100$, assuming number of questions are as low).

To compute the signaling game Nash equilibrium strategies, we first need to define utility functions based on semantic information. Therefore, we will now examine the definition of semantic information.

11.3.3 Semantic Information Measure

To capture the semantic information, we can utilize a category theory based method that is more comprehensive than previous approaches in the field based on propositional logic [Carnap and Bar-Hillel, 1952]. First, we define certain fundamental category theory concepts critical for subsequent discussions.

Definition 11.6 ***(Category)*** A category $\mathcal{C}$ can be represented by: (a) A collection of objects, $Ob(\mathcal{C})$; and (b) a collection of morphisms, $\mathcal{C}(c, d)$ (also denoted as hom(c, d)), which represents the possible relations (represented by arrows in a graph) from one object to another. (c) For every object c, we define the identity morphism, $id_c = \mathcal{C}(c, c) = I$. (d) For every three objects $c, d, e \in Ob(\mathcal{C})$ and morphisms $f \in \mathcal{C}(c, d)$ and $g \in \mathcal{C}(d, e)$, one specifies a morphism $f \circ g \in \mathcal{C}(c, e)$, called the composite of f and g. A $\mathcal{V}$-enriched category $\mathcal{C}$ can be defined as follows: Let $\mathcal{V}$ be a symmetric monoidal preorder (see Fong and Spivak 2018 for more details on preorders), $(\mathcal{V}, \leq, I, \circ)$. A $\mathcal{V}$-enriched category, apart from the properties above for a category satisfies the following too. For every object $c \in Ob(\mathcal{C})$, it satisfies $I \leq C(c, c)$, and for every three objects $c, d, e \in Ob(\mathcal{C})$, it obeys $\mathcal{C}(c, d) \circ \mathcal{C}(d, e) \leq \mathcal{C}(c, e)$.

Definition 11.7 ***(Functor)*** A functor $F : \mathcal{L} \to \hat{\mathcal{L}}$ takes $c \in Ob(\mathcal{L})$ to $F(c) \in Ob(\hat{\mathcal{L}})$ for all $c \in Ob(\mathcal{L})$.

Definition 11.8 ***(Representable Copresheaf)*** Let $\mathcal{L}$ be a $[0, 1]$ enriched category. For every object x, the function $h^x = \mathcal{L}(x, -)$ is a $[0, 1]$ functor and it represents the object x, through its possible morphisms.

Apart from the category theory concepts above, to derive the semantic information, we need the following definitions that are fundamental to deductive reasoning.

Definition 11.9 We define a *deductive logic* as the following conjunction of relations (if involving multiple entities), $r^{(N)} = (r_1 \wedge r_2 \cdots \wedge r_N)$. Here, $r^{(N)}$ holds iff all the relations r_i are true.

Definition 11.10 We define *chain-like rules* as follows, $(e_0, r^{(N)}, e_N) = (e_0, r_1, e_1) \wedge (e_1, r_2, e_2) \wedge \cdots \wedge (e_{N-1}, r_N, e_N)$. The chain rule will be essential for proving analytical or logical theorems using the emergent language. This follows also from the associative property of the relation triplets.

Hence, if the teacher or apprentice is to derive logical conclusions, from the existing or learned relations between the entities, the above properties hold the key. To this end, category theory has excellent algebraic structural properties that allows the grouping of all the syntactic and semantic objects part of the language. Next, we derive here a novel interpretation of the semantic representation of the meaning conveyed by the state $\boldsymbol{z}_t$ using category theory. These novel representations of semantics lead us to define the semantic notion of information. The following definition constructs a category whose objects belong to $\mathcal{W}$.

Definition 11.11 Any entity, represented by $\boldsymbol{a}_t$ or $\boldsymbol{q}_t$ or expressions part of $\mathcal{W}$ form objects in a category $\mathcal{L}$, called the *syntax category*. Expressions are defined as molecular entities that are a combination of multiple entities, e.g., the state $\boldsymbol{z}_t$. A *morphism* (a directed relation between the objects in any category) from object $\boldsymbol{x}$ to object $\boldsymbol{y}$ in $\mathcal{L}$ is formed when expression $\boldsymbol{y}$ extends from the expression $\boldsymbol{x}$.

Moreover, category $\mathcal{L}$ and the morphisms can be represented as a category enriched in $[0, 1]$. This means that we can represent the morphism, $\boldsymbol{x} \to \boldsymbol{y}$ using a probability distribution, $\pi(\boldsymbol{y} \mid \boldsymbol{x}) \in [0, 1]$, that captures the probability that expression $\boldsymbol{x}$ extends to $\boldsymbol{y}$. However, this category $\mathcal{L}$ only represents the language's compositional part (or the syntax part) under consideration. Ideally, we should be able to encode "what goes with what" (*syntactic part*) together with the plausible, logical conclusions one can derive from the expressions (*semantic part*). Further below, we show that *semantic category* $\widehat{\mathcal{L}}$ can be represented as a transformation from the syntax category, represented as, $F : \mathcal{L} \rightarrow \widehat{\mathcal{L}}$.

Lemma 11.3 The semantic category $\widehat{\mathcal{L}} = [0, 1]^L$ within the context c is an $[0, 1]$ enriched category of representable copresheaves from the syntax category $\mathcal{L}$.

Proof: The proof is given in Appendix A.1.

In Lemma 11.3, copresheaves of any object $\boldsymbol{x} \in \mathcal{L}$, represent all the possible morphisms (expressions) from $\boldsymbol{x}$, within context c. The usefulness of the Lemma is two-fold. Firstly, at the apprentice, it helps to reason on any logical formula $\boldsymbol{\phi}_i$ (representing objects in the semantic category) by splitting a complex state description into smaller concepts and evaluating a set of smaller logical formulas $\boldsymbol{\phi}_j$. Further, evaluate $\boldsymbol{\phi}_i$ using the chain-like rule defined previously with $\boldsymbol{\phi}_j$'s as the components. This is only possible if we show that the semantic space is also a category. Secondly, leveraging Lemma 11.3, we provide a novel definition of the semantic information conveyed by any expression $\boldsymbol{x} \in \mathcal{L}$ as follows.

Lemma 11.4 The *semantic information* conveyed by any expression $\boldsymbol{x} \in \mathcal{L}$ can be written as the average across the information carried by the representable copresheaf that follow from $\boldsymbol{x}$. Further, mathematically, we can write the semantic information conveyed by $\boldsymbol{x}$ as

$$S(\boldsymbol{x} \mid c) = \sum_{\boldsymbol{y} \in \mathcal{H}^{\boldsymbol{x}}} \pi(\boldsymbol{y} \mid \boldsymbol{x}, c) \log \frac{\pi(\boldsymbol{y} \mid \boldsymbol{x}, c)}{\pi(\boldsymbol{y})}, \tag{11.6}$$

such that the support of $\boldsymbol{y}$ is defined by the representable copresheaves $\mathcal{H}^{\boldsymbol{x}}$.

Proof: Lemma 11.3 provides a way to categorize the possible logical conclusions that entail from any $\boldsymbol{x}$. The resulting distributional representation provides a rigorous way to formulate the semantic information conveyed by any expression $\boldsymbol{x}$ about $\boldsymbol{y}$, as follows. Inspired by Carnap and Bar-Hillel 1952 and Lindley 1956, we write the information carried by $\boldsymbol{x}$ about $\boldsymbol{y}$ as

$$S(\boldsymbol{y}; \boldsymbol{x} \mid c) = \begin{cases} \log \dfrac{\pi(\boldsymbol{y} \mid \boldsymbol{x}, c)}{\pi(\boldsymbol{y})}, & \text{if } \boldsymbol{y} \geq \boldsymbol{x}, \\ 0, & \text{otherwise.} \end{cases} \tag{11.7}$$

Further, we can obtain the semantic information captured by $\boldsymbol{x}$ as the average of the semantic information conveyed by possible worlds that contain $\boldsymbol{x}$ which is (11.6). Intuitively, (11.7) represents the degree of confidence in a possible logical entailment $\boldsymbol{y}$ after having observed $\boldsymbol{x}$. This means that if $\pi(\boldsymbol{y} \mid \boldsymbol{x}, c) = \pi(\boldsymbol{y})$, then it means $\boldsymbol{y}$ does not entail from $\boldsymbol{x}$. Hence, the corresponding $S(\boldsymbol{y};\boldsymbol{x} \mid c) = 0$; otherwise, it will be a quantity greater than 0, representing non-zero semantic information. Lemma 11.4 provides a way to compute the semantic information conveyed by $\boldsymbol{z}_t$ as $S_s\left(\boldsymbol{z}_t \mid \pi_{s,t}[\boldsymbol{z}_{t-1}], c\right) = \sum_{\boldsymbol{y}\in\mathcal{H}^{\boldsymbol{z}_t}} \pi(\boldsymbol{y} \mid [\boldsymbol{z}_t], c) \log \frac{\pi(\boldsymbol{y}|[\boldsymbol{z}_t],c)}{\pi(\boldsymbol{y}|[\boldsymbol{z}_{t-1}],c)}$. Further, we average across $\boldsymbol{u}_t$ to arrive at the average semantic information conveyed by $\boldsymbol{z}_t$ for a particular policy $\pi_{s,t}$ and context c as in (11.8).

$$\mathbb{E}_{\boldsymbol{u}_t}\left[S_s(\boldsymbol{z}_t;\boldsymbol{u}_t \mid \pi_{s,t}, [\boldsymbol{z}_{t-1}], c)\right] = \sum_{\boldsymbol{z}_t} \pi\left(\boldsymbol{z}_t \mid c, [\boldsymbol{z}_{t-1}]\right) \underbrace{\left[\sum_{\boldsymbol{u}_t} \pi_{s,t} \log \frac{\pi_{s,t}}{\pi(\boldsymbol{u}_t)}\right]}_{\text{avg. number of bits/state description}} \underbrace{S_s(\boldsymbol{z}_t \mid [\boldsymbol{z}_{t-1}], c)}_{\text{semantics}}, \tag{11.8}$$

where $\pi(\boldsymbol{u}_t) = \sum_{\boldsymbol{z}_t} \pi_{s,t}\pi(\boldsymbol{z}_t \mid c, [\boldsymbol{z}_{t-1}])$, and $\pi_{s,t} = \pi(\boldsymbol{u}_t \mid [\boldsymbol{z}_t])$.

An appropriate unit to represent $\frac{S_s(\boldsymbol{z}_t;\boldsymbol{u}_t|\pi_{s,t},[\boldsymbol{z}_{t-1}],c)}{\left[\sum_{\boldsymbol{u}_t}\pi_{s,t}\log\frac{\pi_{s,t}}{\pi(\boldsymbol{u}_t)}\right]}$ will be average semantics/s/Hz.

Corollary 11.1 The similarity metric Z_{xy} between two objects, $\boldsymbol{x}$ and $\boldsymbol{y}$, can be defined as the intersection of the representable copresheaf between them.

$$Z_{xy} = \sqrt{\frac{\sum_{\boldsymbol{z}\in\mathcal{H}^{\boldsymbol{x}}\cap\mathcal{H}^{\boldsymbol{y}}} h^{\boldsymbol{x}}(\boldsymbol{z})}{\sum_{\boldsymbol{z}\in\mathcal{H}^{\boldsymbol{x}}} h^{\boldsymbol{x}}(\boldsymbol{z})} \frac{\sum_{\boldsymbol{z}\in\mathcal{H}^{\boldsymbol{x}}\cap\mathcal{H}^{\boldsymbol{y}}} h^{\boldsymbol{y}}(\boldsymbol{z})}{\sum_{\boldsymbol{z}\in\mathcal{H}^{\boldsymbol{y}}} h^{\boldsymbol{y}}(\boldsymbol{z})}}. \tag{11.9}$$

Whenever, $\mathcal{H}^{\boldsymbol{x}} = \mathcal{H}^{\boldsymbol{y}}$, then $Z_{ij} = 1$ and when $\mathcal{H}^{\boldsymbol{x}} \cap \mathcal{H}^{\boldsymbol{y}} = \emptyset$, $Z_{ij} = 0$.

Leveraging Corollary 11.1, we obtain the average semantic information extracted by apprentice as in (11.10).

$$\mathbb{E}_{\boldsymbol{u}_t}\left[S_l(\widehat{\boldsymbol{z}}_t;\boldsymbol{u}_t \mid \boldsymbol{z}_t, [\widehat{\boldsymbol{z}}_{t-1}], c, \pi_{l,t})\right] = \sum_{\boldsymbol{u}_t} \pi\left(\boldsymbol{u}_t \mid [\boldsymbol{z}_t]\right) S_s(\widehat{\boldsymbol{z}}_t \mid [\widehat{\boldsymbol{z}}_{t-1}], c) \times \left[\sum_{\widehat{\boldsymbol{z}}_t,\widehat{c}} \pi_{l,t} \log \frac{\pi_{l,t}}{\pi(\widehat{\boldsymbol{z}}_t \mid c, [\widehat{\boldsymbol{z}}_{t-1}])} Z_{\widehat{\boldsymbol{z}}_t\boldsymbol{z}_t}\right]. \tag{11.10}$$

In (11.10), $[\sum_{\widehat{\boldsymbol{z}}_t,c}\pi_{l,t}\log\frac{\pi_{l,t}}{\pi(\widehat{\boldsymbol{z}}_t|c,[\widehat{\boldsymbol{z}}_{t-1}])}Z_{\widehat{\boldsymbol{z}}_t\boldsymbol{z}_t}]$ represents the probability that $\boldsymbol{z}_t = \widehat{\boldsymbol{z}}_t$, semantically. We now have the necessary ingredients to define the two-player signaling game NE problem.

11.3.4 Signaling Game Model and Generalized Nash Equilibrium Problem

Using the metrics above, we further formulate the solutions of the signaling game defined using the tuple $(\mathcal{C}, \mathcal{W}, \mathcal{U}, \mathcal{M}_s, \mathcal{M}_l)$ as a *Bayesian generalized NE* Facchinei and Kanzow 2010 and Fudenberg and Tirole 1991 (generalized due to the constraints part of the problem formulation below). The NE strategies are such that for all $\boldsymbol{z}_t \in \mathcal{W}$ and $c \in C$

$$\pi^*_{s,t} \in \arg\max_{\pi_{s,t}} -\mathbb{E}_{\boldsymbol{u}_t}\left[S_s(\boldsymbol{z}_t; \boldsymbol{u}_t \mid \pi_{s,t}, [\boldsymbol{z}_{t-1}], c)\right], \tag{11.11a}$$

$$\text{s.t.} \qquad \mathbb{E}[V(\pi_{s,t}, \pi_{l,t})] \leq D \tag{11.11b}$$

$$\pi^*_{l,t} \in \arg\max_{\pi_{l,t}} \mathbb{E}_{\boldsymbol{u}_t}\left[S_l(\widehat{\boldsymbol{z}}_t; \boldsymbol{u}_t \mid \pi_{l,t}, \pi^*_{s,t}, [\widehat{\boldsymbol{z}}_{t-1}])\right], \tag{11.11c}$$

where $V(\pi_{s,t}, \pi_{l,t}) = c(\boldsymbol{u}_t) - \log \pi_{l,t}$, and the expectation of V is over the policies $\pi_{s,t}, \pi_{l,t}$. $c(\boldsymbol{u}_t)$ represents the cost of producing $\boldsymbol{u}_t$ and $-\log \pi_{l,t}$ represents the apprentice's surprisal about inferring $\boldsymbol{z}_t$ given $\boldsymbol{u}_t$. Equation (11.11a) states that the strategy space $\mathcal{M}_{s,t}$ consists of signals $\boldsymbol{u}_t \in \mathcal{U}$ that minimizes the expected semantic information at teacher's end. Constraint (11.11b) guarantees that the teacher responds optimally to the apprentice's strategy. Equation (11.11c) states that the strategy space $\mathcal{M}_{l,t}$ consists of states $\widehat{\boldsymbol{z}}_t \in \mathcal{W}$ that maximize the expected semantic information at the apprentice and responds optimally to the teacher strategy. Next, we formally define the equilibrium points.

Definition 11.12 As defined in Ratliff et al. 2016, we define a strategy $(\pi^*_{s,t}, \pi^*_{l,t})$ as a *local NE* for (11.11) if there exist open sets $\mathcal{M}_{g_r} \subset \mathcal{M}_{r,t}$ such that $\pi^*_{r,t} \in \mathcal{M}_{g_r}$, where $r \in \{s, l\}$, and satisfies

$$-\mathbb{E}_{\boldsymbol{u}_t}\left[S_s(\boldsymbol{z}_t; \boldsymbol{u}_t \mid \pi^*_{s,t}, [\boldsymbol{z}_{t-1}], c)\right] \tag{11.12a}$$

$$\geq -\mathbb{E}_{\boldsymbol{u}_t}\left[S_s(\boldsymbol{z}_t; \boldsymbol{u}_t \mid \pi_{s,t}, [\boldsymbol{z}_{t-1}], c)\right], \ \forall \pi_{s,t} \in \mathcal{M}_{g_s}(\pi^*_{l,t}), \tag{11.12b}$$

$$\mathbb{E}_{\boldsymbol{u}_t}\left[S_l(\widehat{\boldsymbol{z}}_t; \boldsymbol{u}_t \mid \pi^*_{l,t}, \pi^*_{s,t}, [\widehat{\boldsymbol{z}}_{t-1}], c)\right] \tag{11.12c}$$

$$\geq \mathbb{E}_{\boldsymbol{u}_t}\left[S_l(\widehat{\boldsymbol{z}}_t; \boldsymbol{u}_t \mid \pi_{l,t}, \pi^*_{s,t}, [\widehat{\boldsymbol{z}}_{t-1}], c)\right], \ \forall \pi_{l,t} \in \mathcal{M}_{g_l}(\pi^*_{s,t}). \tag{11.12d}$$

When $\mathcal{M}_{g_r} = \mathcal{M}_r, \forall r$, then the equilibrium is unique.

Next, we evaluate the equilibrium points corresponding to (11.11). The teacher in this communication instance aims to minimize the average amount of transmitted semantic information carried by $\boldsymbol{u}_t$, represented by $S_s(\boldsymbol{z}_t; \boldsymbol{u}_t \mid \pi_{s,t}, [\boldsymbol{z}_{t-1}], c)$. However, the teacher cannot independently optimize this since its strategy space $\mathcal{M}_{s,t}$ depends on the apprentice inference distribution via the constraint (11.11b). The mutual understanding evolves via the two-player game proposed in (11.11), where the apprentice communicates about the logical formulas it wants to evaluate, and the teacher responds by providing an encoded state description $\boldsymbol{u}_t$ corresponding to the question. We assume that $\boldsymbol{q}_t$ and $\boldsymbol{u}_t$ are recovered without any error at the teacher and apprentice, respectively (reliable wireless links during the emergent language phase).

11.3.5 Characterization of the Generalized Local NE

To find the generalized local NE, one conventional approach is alternating maximization between speaker and listener objectives. Alternating maximization is an iterative procedure for solving the NE objectives jointly over all variables by alternating maximizations over the individual subsets of variables (two NE distributions and Lagrange multipliers). The estimated NE distributions using alternating maximization can be computed as follows.

Lemma 11.5 The NE strategies in (11.11) can be obtained as the following alternating updates. The speaker's strategy is obtained as (11.13)

$$\pi^*_{s,t} = \pi(\boldsymbol{u}_{t-1}) \exp\left(\frac{-1}{S_s(\boldsymbol{z}_t|[\boldsymbol{z}_{t-1}],c)\pi(\boldsymbol{z}_t|c)}(\lambda_s(c(\boldsymbol{u}_t)) - \log \pi_{l,t}) - \alpha_{\boldsymbol{z}_t}) - 1\right). \quad (11.13)$$

and the listener's inference distribution is (11.14).

$$\pi^*_{l,t} = \frac{1}{\pi(\boldsymbol{u}_t)} \exp\left(\frac{-1}{\pi\left(\boldsymbol{u}_t \mid [\boldsymbol{z}_t]\right) S_s(\widehat{\boldsymbol{z}}_t \mid [\widehat{\boldsymbol{z}}_{t-1}], c) Z_{\widehat{\boldsymbol{z}}_t \boldsymbol{z}_t}} \times \left(\alpha_{\boldsymbol{u}_t} - \pi(\boldsymbol{u}_t \mid [\boldsymbol{z}_t]) S_s(\widehat{\boldsymbol{z}}_t \mid [\widehat{\boldsymbol{z}}_{t-1}], c) Z_{\widehat{\boldsymbol{z}}_t \boldsymbol{z}_t} \log \pi_{l,t-1}\right)\right). \quad (11.14)$$

Proof: The proof is provided in Appendix A.3.

The node policies (11.13) and (11.14) suggest that the encoded signal depends on the semantics conveyed by the state description and not on the syntactic representation. Intuitively, the emergent language can represent two state descriptions $\boldsymbol{z}_1, \boldsymbol{z}_2$ leading to the same logical conclusions using the same $\boldsymbol{u}$. Moreover, the speaker transmit policy (11.13) tends to choose a transmit message $\boldsymbol{u}$ that requires a smaller transmission cost and causes the listener to be less surprised. If the semantic information in $\boldsymbol{z}$ is high, then the slope parameter $\frac{1}{S_s(\boldsymbol{z}_t|[\boldsymbol{z}_{t-1}],c)\pi(\boldsymbol{z}_t|c)}$ becomes low. This also suggests that $\pi_{s,t}$ will be very low for short-length codewords, due to higher listener ambiguity. Hence more bits are required to represent $\boldsymbol{z}_t$ with high semantic information. When $\boldsymbol{z}_t$ provides no new semantic information, the speaker policy's slope becomes infinite due to considering past state descriptions $[\boldsymbol{z}_{t-1}]$ in player strategies. This causes $\pi_{s,t}$ to converge to a Dirac delta distribution, resulting in zero-cost transmission of $\boldsymbol{u}_t$, and no signal is sent. As a result, there is reduced transmission compared to a classical system that doesn't account for semantic history. Equation (11.14) can be intuitively interpreted as choosing the $\pi_{l,t}$ with its peak close to $\hat{\boldsymbol{z}}_t$ that maximizes the average semantic information extracted by the listener for any $\boldsymbol{u}_t$. Next, we prove local convergence of the resulting alternating maximization updates.

Lemma 11.6 The alternating maximization updates obtained in (11.13) and (11.14) for the two-player signaling game proposed converges to a local Nash equilibrium as defined in Definition 11.12.

Proof: The objective functions in (11.11) are jointly non-concave w.r.t the combined strategy $(\pi_{s,t}, \pi_{l,t})$. Hence, the resulting NE cannot be unique. However, for a fixed listener policy, the speaker objective function (11.11a) is concave and hence leads to the global optimum, thus satisfying (11.12b) for the strategy $\pi^*_{s,t}$. The listener objective function in (11.11c) is non-concave w.r.t $\pi_{l,t}$ and can be written as summation of a concave and convex part. To solve this, we computed an approximate function $\hat{F}_l(\pi_{l,t}, \pi_{s,t})$ (defined in (A.8)) to the original listener objective function (11.11c) by approximating the convex part using the first order Taylor series expansion. It can be shown that the approximate function (A.8) satisfies $F_l(\pi_{l,t}, \pi_{s,t}) \geq \hat{F}_l(\pi_{l,t}, \pi_{s,t})$. The resulting lower bound is concave [Hiriart-Urruty, 1985]; hence, the achieved solution (11.14) is a local optimum of the original listener objective function. Hence, an alternating maximization update between the speaker and listener policy here leads to local NE. The proof for convergence of the alternating maximization updates is classical [Hong et al., 2015].

11.3.6 Analysis of the Signaling Game Equilibria for Emergent Language

We have so far looked at computing the encoder and decoder probability distributions using the signaling game. Next, we look at how the teacher chooses the transmit signal and the apprentice maps the received signal to the state description given these distributions. Naive approach would be to consider the signals $(\boldsymbol{u}, \hat{\boldsymbol{z}})$ as those which maximize the respective probabilities. However, as we show here, that need not be the optimal solution (in terms of the teacher transmitting less and the apprentice extracting maximum). For notational simplicity, we remove the subscript t.

Definition 11.13 An equilibrium (π^*_s, π^*_l) is a *separating equilibrium* [Ratliff et al., 2016] if each state $\boldsymbol{z}$ is encoded as different $\boldsymbol{u}$. That is, $\mathcal{W}$ can be partitioned into regions $\mathcal{W}_k$ such that $\sum_{z \in \mathcal{W}_k} \pi(\boldsymbol{u}_k \mid \boldsymbol{z}) = 1$. A *pooling equilibrium* occurs if the same $\boldsymbol{u}$ is sent for all $\boldsymbol{z}$. When $\|\mathcal{U}\| < \|\mathcal{W}\|$, with the same signal transmitted for multiple states but not for all, the resulting equilibrium is a *partial pooling equilibrium*.

Using these definitions, we analyze the teacher and apprentice strategies for different separating and pooling signaling equilibria, as a function of system parameters λ_s and $\|\mathcal{U}\|$.

Theorem 11.1 When the language emerges at the convergence of the signaling game, the following equilibria occurs.

1. For $\lambda_s \in [0, 1]$, the minimum semantic information at teacher side can be achieved by choosing uniformly random the transmit message. Hence, the optimal policies are $\pi(\boldsymbol{u} \mid \boldsymbol{z}) = \frac{1}{K}, \pi(\boldsymbol{z} \mid \boldsymbol{u}) = \pi(\boldsymbol{z})$ and $H_s(\boldsymbol{u}) = \log K$, where K denotes the cardinality of $\mathcal{U}$. $H(\boldsymbol{x})$ is the Shannon entropy here.
2. For $\lambda_s > 1$, the teacher objective can be maximized by choosing a bijective transformation from $\boldsymbol{z}$ to $\boldsymbol{u}$ for the transmit policy (means $|\mathcal{U}| = |\mathcal{W}|$). This is the perfect signaling or separating equilibrium.
3. For the case of pooling signaling equilibrium, the teacher may send the same $\boldsymbol{u}$ for all $\boldsymbol{z}$ ($|\mathcal{U}| = 1$). While this scheme may achieve the minimum semantic information at the teacher, it does not provide any semantic information at the apprentice and hence it is not suitable for emergent SC (ESC).
4. In partial pooling, the apprentice extracts maximum semantic information when the teacher partition its semantic category space into a Voronoi tessellation, where each $\boldsymbol{u}$ corresponds to a distinct partition.

Proof: Proof is provided in Appendix A.2. From Theorem 11.1, for an ESC system where $|\mathcal{U}| < |\mathcal{W}|$, the partial pooling equilibrium is more realistic. An optimal partition at the teacher is such that $\boldsymbol{u}_k$ is transmitted whenever the average semantic information extracted at the apprentice is the highest among all possible partitions$\mathcal{W}_k$. Analytically, this can be represented as choosing$\boldsymbol{u}_t = \arg\max_{\boldsymbol{u}_t'} S_l(\hat{\boldsymbol{z}}_t; \boldsymbol{u}_t' \mid \boldsymbol{z}_t, [\hat{\boldsymbol{z}}_{t-1}], c, \pi_{l,t})$. The proposed ESC system can compute $\pi^*_{l,t}$ using the apprentice's surprise $(-\log \pi^*_{l,t})$ fed back during the emergent language. Hence, the average semantic information extracted by the apprentice for different possible partitions can be evaluated. This makes it possible for the teacher to choose the Voronoi tessellation for the encoded signal. In this case, given a Voronoi tessellation of $\mathcal{W}$, the apprentice in turn can choose an $\hat{\boldsymbol{z}} \in \mathcal{W}$ for each signal $\boldsymbol{u}_k$ that describes the average state in $\mathcal{W}_k$ optimally (in semantic sense and given the environment or prior for $\boldsymbol{z}$). An optimal interpretation consists thus of Bayesian estimators for each cell$\mathcal{W}_k$, that is, arg min $_{z\in\hat{\mathcal{W}}} \int_{\hat{\mathcal{W}}} \|z - \hat{z}\|_S \pi(d\hat{z} \mid \boldsymbol{u})$, where $\hat{\mathcal{W}}$ represents the space over which $\pi(dz \mid \boldsymbol{u})$ is defined.

11.3.7 Average Semantic Representation Length for Classical and Emergent Language Based ESC

Let c be the context, drawn from $p_c(c)$. Depending on c, it is possible that a certain entity relation pair (or entity relation sequence of pairs) is not valid. Hence, the possible state sequences that generate the data will be a subset of $\mathcal{W}$. Depending on this subset, the transmitted vocabulary $\mathcal{U}$ size can be dynamically adjusted if the emergent language is aware of the context c. Motivated by the importance of communication context significance in the optimal vocabulary design, we derive the following result.

Theorem 11.2 For a particular syntactic space, $\mathcal{W}$ and context distribution $p(c)$ over C, the average number of bits to represent the state description in an ESC system can be bounded as (11.15).

$$\sum_{c\in C} \pi(c)H(\boldsymbol{z}_i \mid c) \leq \sum_{\boldsymbol{u}_i\in\mathcal{U}} \pi(\boldsymbol{u}_i)l_i \leq -\sum_{c} \pi(c) \sum_{\boldsymbol{z}_i} \pi(\boldsymbol{z}_i \mid c) \lceil \log \pi(\boldsymbol{z}_i \mid c) \rceil, \quad (11.15)$$

where l_i represents the number of bits to represent $\boldsymbol{u}_i$. For a classical system, the bounds are (11.16).

$$H_c(\boldsymbol{z}_i) \leq \sum_{\boldsymbol{u}_i\in\mathcal{U}} \pi(\boldsymbol{u}_i)l_i \leq \max_{c} \sum_{\boldsymbol{z}_i} \pi(\boldsymbol{z}_i \mid c) \left[\sum_{\boldsymbol{s}_i\in\boldsymbol{z}_i} \lceil \log \pi(\boldsymbol{s}_i \mid c) \rceil \right]. \quad (11.16)$$

Proof: See Appendix A.5. From Theorem 11.2, as explained in Appendix A.5, it can be shown that for an ESC system, the lower and upper bounds for a physical representation of the semantics are smaller to those in a classical system (transmitting a compressed version of the entities), justifying its transmission efficiency. Next, we look at whether the apprentice's semantic reliability improves by incorporating reasoning at the nodes.

Theorem 11.3 For a given representation space $\mathcal{U}$, the lower bound on the semantic error probability (S_e-representing reliability) is always less than or equal to the lower bound on the probability of bit error (P_e) measure achieved using classical communication system.

$$P_e \geq \frac{H(\hat{z}_c|z_c)-1}{\log \|\mathcal{W}\|}, \quad S_e \geq \frac{H(z|\hat{z})-H(e|\hat{z})}{\log \|\mathcal{W}\|}, \quad (11.17)$$

where $\frac{H(\hat{z}_c|z_c)-1}{\log \|\mathcal{W}\|} \geq \frac{H(z|\hat{z})-H(e|\hat{z})}{\log \|\mathcal{W}\|}$.

Proof: See Appendix A.4. Here, $\boldsymbol{e}$ represents the error in semantics recovered at the apprentice, and $\boldsymbol{z}_c, \hat{\boldsymbol{z}}_c$ means the state descriptions for the classical system. Theorem 11.3 shows that inducing reasoning at teacher and apprentice can improve semantic reliability compared to a classical system that uses the same number of bits to communicate. In other words, the ESC system could choose to transmit at a higher rate than a classical system to achieve similar semantic reliability. Next, we look at the structure of reasoning components at the teacher and apprentice that can help the ESC system achieve gains discussed above.

11.4 Numerical Results

In this section, we first start with a simple illustrative scenario to demonstrate the gains from a causal reasoning based ESC system as derived in Theorems 11.2 and 11.3.

11.4.1 Illustrative Example for NeSy AI's Potential in Wireless Versus Classical AI Based Wireless

Our proposed ESC framework can support several wireless applications. Nevertheless, for an illustrative example that is simple to understand, we consider the events as the images observed from a camera at the speaker. Assume that the image's background is irrelevant, and the particular entity present (with the possible attributes as in Figure 11.6a is of interest to the listener. We use SHAPES which is a synthetic dataset created by Andreas et al. 2016. In this dataset, each image is a 30×30 RGB image showing a 3×3 grid of objects. Each object is characterized by shape, colour and style as shown in Figure 11.6a. This collection comprises complex questions pertaining to simple arrangements of colored shapes. Each question involves two to four characteristics, object types, or relationships. For example: Is there a red shape above a circle? Overall, the dataset contains 244 questions and 15, 616 images. During training, for every minibatch of the dataset, the listener samples a random question from the dataset that acts as the logical conclusion here. Upon receiving the responses from the speaker, listener performs logical reasoning by combining the individual semantic concepts as shown in Figure 11.5. During the communication phase, the listener employs the decoder distribution $\pi_{l,t}$ to reconstruct the sequence of entities $\hat{\boldsymbol{z}}_t$, with each entity corresponding to a specific object. The relations among these entities are then determined using GFlowNet, and logical formulas are evaluated through a combination of GFlowNet and logical neural networks, as discussed in Section IV of Thomas and Saad 2023. As an illustrative example, consider that, in task τ_1, learning can occur over contexts described by state descriptions of the form (a, b, c), where the entities

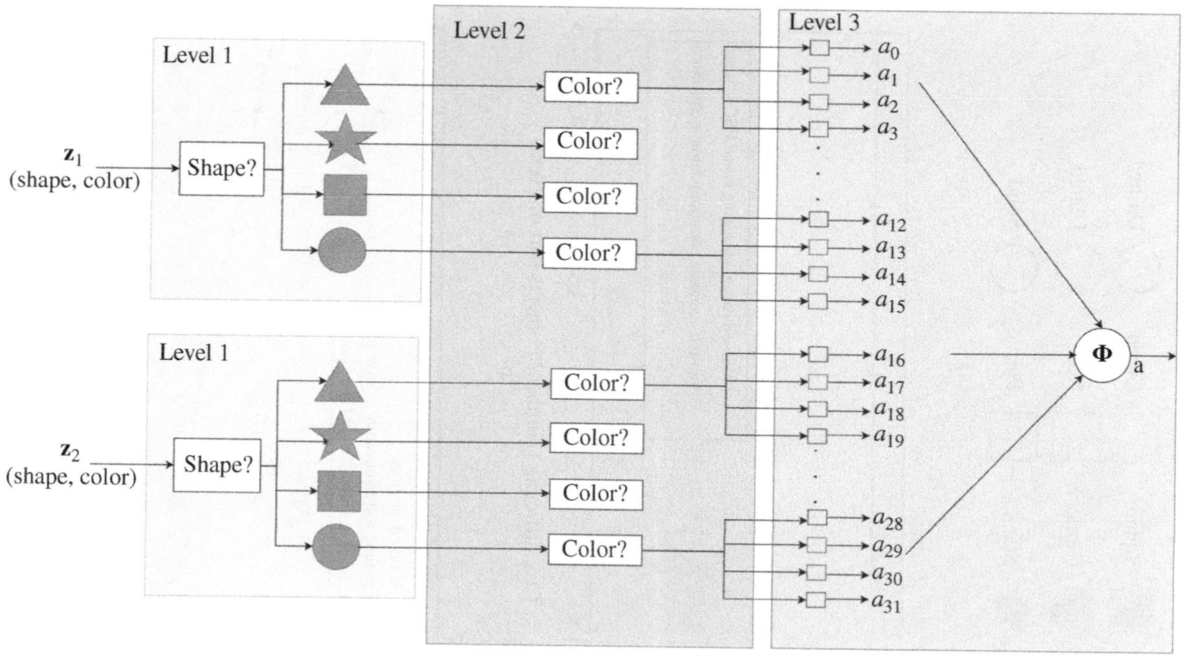

Figure 11.5 Example on how logical reasoning (task with two objects present) is performed.

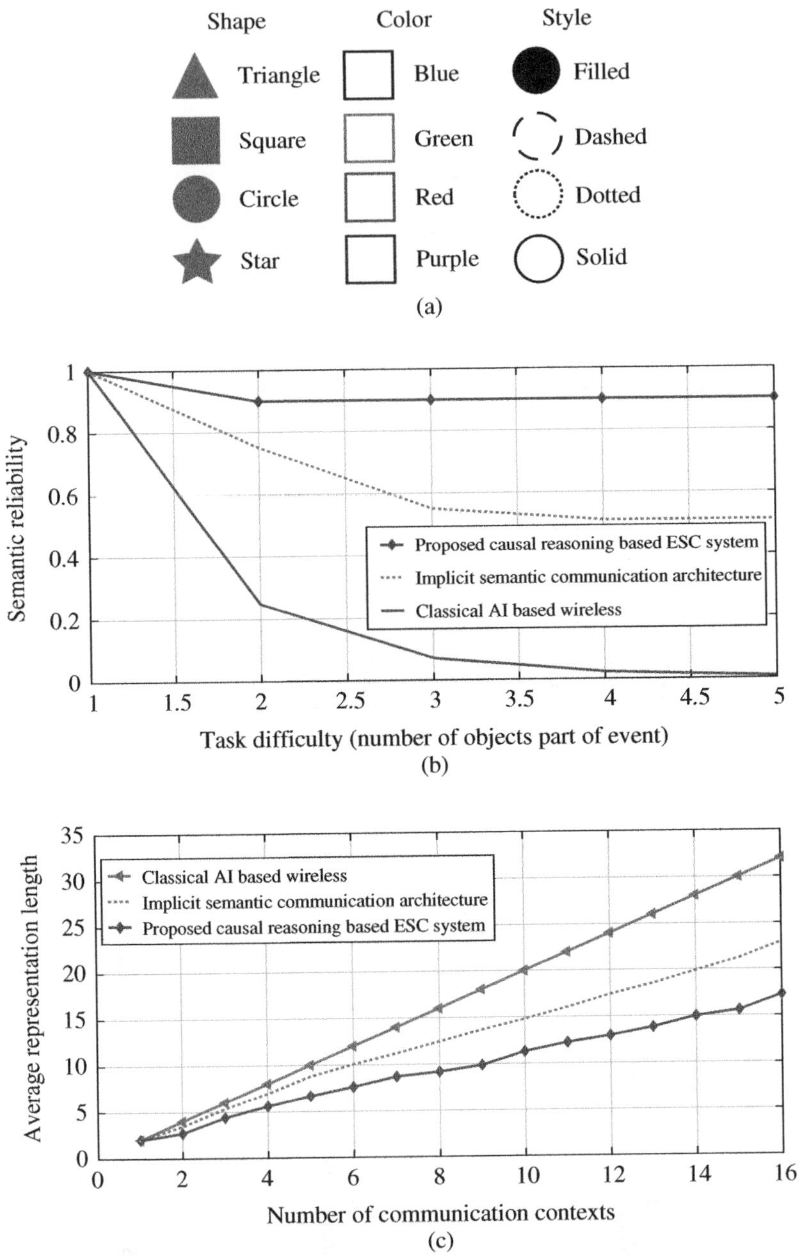

Figure 11.6 (a) The testbed for the illustrative example is the state description grounded in a synthetic world of objects with 4 shapes × 4 colors × 4 styles. (b) Classical AI based system fails with task difficulty (increase in number of objects describing an event). (c) Classical wireless system is designed based on worst case communication context while contextual aware ESC has a dynamically changing representation length, resulting in an efficient transmission.

a, b, and c can be one of three distinct values: {"shape," "color," "style"}. However, "style" is irrelevant to the task, leaving only 16 relevant state descriptions. Among these state descriptions, from the listener's perspective, suppose there exist four unique conclusions that can be drawn from a subset of four states. Therefore, a speaker aware of the listener's semantics can communicate using only two bits, as states with the same logical conclusions can be encoded as the same transmit signal. In contrast, a naive wireless system unaware of the semantics and redundancy may require up to six bits to communicate all 64 possible state descriptions. Considering training complexity, for an ESC system, task τ_1 just requires 8 communication rounds, wherein each round, the information about a single attribute is communicated from speaker to listener. However, a classical AI system would require to be trained with enough data that involves possible 16 different objects. As the complexity of the task increases, which implies each context involving 3 attributes and multiple objects, the training complexity (to achieve required reliability) of the classical AI system increases exponentially. Figure 11.6b compares the semantic reliability for classical and ESC systems with the same amount of training data used, where the ESC is superior due to its generalization capability (with fewer training data). Additionally, the ESC system proposed in this study exhibits a 63% improvement in semantic reliability compared to implicit semantic communication architecture (at task difficulty = 3) based on reasoning. This improvement can be attributed to extracting causal reasoning and emergent language components. Figure 11.6c conveys the potential gain by using a causal reasoning based ESC system in terms of the average representation length (bits). Figure 11.6c shows a 23% reduction in semantic representation length for the ESC system compared to the implicit semantic communication architecture. Here, we used a binary symmetric channel (BSC) to represent the underlying wireless channel for communication. Having understood the fundamental performance gains from a reasoning based system, we further look at evaluating our algorithms.

11.5 Conclusion

In this chapter, we first outlined the fundamental principles of causal reasoning such as causal discovery, causal representation learning, and causal inference. Further, we have presented a new vision of SC systems entitled ESC that relies upon causal reasoning which allows it to be task agnostic over time with less training data. We have formulated a two-player contextual signaling game to learn an optimal representation space for transmission and extracting the semantic state. Simulation results show that the ESC system achieves better reliability with lesser number of bits communicated, thus achieving the transmission efficiency as envisioned in semantic communications. Furthermore, the proposed scheme performs superior state of the art techniques that does not incorporate reasoning into semantic representation and extraction.

Bibliography

J. Andreas, M. Rohrbach T. Darrell, and D. Klein. Deep compositional question answering with neural module networks. In *Proceedings of the IEEE Conference on Computer Vision and Pattern Recognition*, pages 39–48, Las Vegas, NV, USA, Jun. 2016.

R. Carnap and Y. Bar-Hillel. An Outline of a Theory of Semantic Information. *Technical Report No. 247*, Oct. 1952.

C. Chaccour, W. Saad, M. Debbah, Z. Han, and H. V. Poor. Less data, more knowledge: Building next generation semantic communication networks. arXiv preprint arXiv:2211.14343, Nov. 2022.

Christina Chaccour, Walid Saad, Merouane Debbah, Zhu Han, and H. Vincent Poor. Less data, more knowledge: Building next generation semantic communication networks. *To appear in IEEE Communication Surveys and Tutorials*, 2024. https://ieeexplore.ieee.org/document/10554663

F. Facchinei and C. Kanzow. Generalized Nash equilibrium problems. *Annals of Operations Research*, 175(1):177–2011, Sep. 2010.

B. Fong and D. I. Spivak. Seven sketches in compositionality: An invitation to applied category theory. arXiv preprint arXiv:1803.05316, Oct. 2018.

D. Fudenberg and J. Tirole. *Game Theory*. MIT Press, Cambridge, MA, 1991.

J. B. Hiriart-Urruty. Generalized differentiability, duality and optimization for problems dealing with differences of convex functions. In *Convexity and Duality in Optimization*. Springer, 1985.

M. Hong, M. Razaviyayn, Z. Q. Luo, and J. S. Pang. A unified algorithmic framework for block-structured optimization involving big data: With applications in machine learning and signal processing. *IEEE Signal Processing Magazine*, 33(1):57–77, Dec. 2015.

D. V. Lindley. On a measure of the information provided by an experiment. *The Annals of Mathematical Statistics*, 27(4):986–1005, Jul. 1956.

J. Pearl and D. Mackenzie. *The Book of Why*. Basic Books, New York, 2018.

L. J. Ratliff, S. A. Burden, and S. S. Shastry. On the characterization of local Nash equilibria in continuous games. *IEEE Transactions on Automatic Control*, 61(8): 2301–2307, Jun. 2016.

C. K. Thomas and W. Saad. Neuro-symbolic causal reasoning meets signaling game for emergent semantic communications. *IEEE Transactions on Wireless Communications*, 23(5):4546–4563, 2023.

12

Reinforcement Learning-Based Unicast and Broadcast Wireless Semantic Communications

Zhilin Lu[1], Rongpeng Li[1], Ekram Hossain[2], Zhifeng Zhao[1,3], and Honggang Zhang[1,3]

[1] *College of Information Science and Electronic Engineering, Zhejiang University, Hangzhou, China*
[2] *Department of Electrical and Computer Engineering, University of Manitoba, Winnipeg, Manitoba, Canada*
[3] *Zhejiang Lab, Hangzhou, China*

12.1 Introduction

With the rapid development of communication technologies and the rise of artificial intelligence (AI), tremendous intelligent applications, such as extended reality (XR), holographic communication, and autonomous driving flourish and impose stringent communication requirements [Wang et al., 2023] which are beyond the capabilities of 5G. In order to deal with this performance dilemma, researchers are resorting to reap the merits of AI, and deep learning (DL)-enabled semantic communication (SC) emerges as one of the promising solutions. In particular, targeted at the semantic and/or effectiveness level communications [Weaver, 1949], SC has attained intense research interest [Lu et al., 2023b] with encouraging results in text transmission [Farsad et al., 2018; Lu et al., 2021; Xie et al., 2021; Zhou et al., 2021; Sana and Strinati, 2022], speech/audio transmission [Tong et al., 2021; Weng and Qin, 2021; Han et al., 2023], and image/video transmission [Bourtsoulatze et al., 2019; Kurka and Gündüz, 2020; Jiang et al., 2022; Tung and Gündüz, 2022; Zhang et al., 2022b].

SC primarily concentrates on the source information reconstruction based on the received information. For text-based SC systems, Farsad et al. [2018] propose the initial joint source-channel coding (JSCC) based on the recurrent neural network (RNN) codecs for fixed length sentences, which shows that the DL-based codecs are capable of achieving lower word error rate and preserve semantic information by embedding sentences into a semantic space. Inspired by this success,

Foundations of Semantic Communication Networks, First Edition.
Edited by Walid Saad, Christina Chaccour, Christo Kurisummoottil Thomas, and Merouane Debbah.

Xie et al. [2021] designed a transformer-based SC system, named DeepSC, which aims to recover the semantics for arbitrary lengths of sentences under varying channels. Since then, SC has attracted a lot of attention and many works utilize JSCC frameworks to transmit sentence semantics [Xie and Qin, 2020; Zhou et al., 2021; Sana and Strinati, 2022], such as UT [Dehghani et al., 2018] based semantic coding system [Zhou et al., 2021; Xie et al., 2023], and mutual information (MI) based performance optimization between semantic compression and semantic fidelity. Moreover, Wang et al. [2022] and Seo et al. [2023] adopt knowledge graph (KG) and probability distribution respectively to infer semantics information explicitly. In addition to text, Weng and Qin [2021] designed an attention-based end-to-end (E2E) SC system for speech transmission. Tong et al. [2021] developed a federated learning (FL) trained model to deliver audio semantics between multiple devices and the server. Besides, Bourtsoulatze et al. [2019] initially extended the JSCC framework and applied it to the deep image SC system, which provided graceful degradation as signal-to-noise ratio (SNR) changes. Moreover, Tung and Gündüz [2022] and Jiang et al. [2022] introduced the semantics to maximize overall visual quality by transmitting some key motion points, so as to further improve transmission efficiency.

Apart from the achievements in semantic level SC, some results are achieved at the effectiveness level as well, so as to transmit essential semantics timely for successful task execution. For example, Jankowski et al. [2020a,2020b] target the identification of individuals or vehicles and design two alternative (digital and analog communications-based) schemes to enhance retrieval accuracy. Kountouris and Pappas [2021] developed a communication paradigm, in which smart devices sampled the "semantics of information" to steer their traffic for the purpose of real-time remote actuation. Additionally, in Xie et al. [2022], a transformer-based framework was designed to unify the structure of the transmitter (TX) for different tasks with different types of data (single modal data and multimodal data), wherein many-to-many and many-to-one scenarios are exemplified. Similarly, Zhang et al. [2022a] also presents a unified E2E framework that can serve different tasks with multiple modalities by dynamically adjusting feature dimensions and deep neural network (DNN) layers.

However, there are still several underlying problems that cannot be properly solved with existing SC approaches. First and foremost, in existing works, the objective function commonly leverages a differentiable function, such as cross-entropy (CE) or mean square error (MSE), so as to facilitate direct gradient back-propagation. This kind of bit-level metric, despite being feasible in most cases, inevitably introduces a "semantic gap" [Lu et al., 2023b] due to inconsistencies between evaluation metrics and optimization objectives, i.e., the training stage is optimized by the differentiable CE or MSE function, while the test stage is evaluated by nondifferentiable semantic metrics that lack a well-defined or continuous derivative across their domain, such as BLEU, CIDEr [Lu et al., 2023b].

Meanwhile, the difficulty of dealing with the nondifferentiability of semantic similarity metrics could be exaggerated by the large dimension of sequence decoding space, which demands a simple algorithm. It remains challenging to solve the nondifferentiability issue in SC. Besides, almost all existing works are built on the JSCC framework with a strong assumption of the differentiable channel model or are designed to transmit only a few semantic tokens, which can hardly be satisfied or scaled in reality.

In addition, it is noteworthy that most of the literature predominantly focuses on point-to-point communication scenarios, while shedding little light on exploring one-to-many semantic broadcast communication (BC). For example, Ding et al. [2021] proposed to estimate the SNR at the decoder before adaptively decoding transmitted images. Liu and Chen [2022] extended the unidirectional and bidirectional decoders to a degraded broadcast channel, and receivers (RXs) therein decoded the text in different languages yet with the same meanings as needed. Similarly, Hu et al. [2022] designed a semantic recognizer DistilBERT to distinguish different receivers according to emotional properties (i.e., positive or negative). Different from Ding et al. [2021], Liu and Chen [2022], and Hu et al. [2022] that broadcast all extracted features, the semantic encoder in Ma et al. [2023] extracts disentangled semantic features and only broadcasts the receiver's intended semantic information to further improve transmission efficiency. Apparently, these efforts on semantic BC lay the very foundation for resource-efficient transmission to multiple receivers requiring the same content [Shrivastava et al., 2022].

Besides, existing semantic BC schemes [Ding et al., 2021; Hu et al., 2022; Liu and Chen, 2022; Shrivastava et al., 2022; Ma et al., 2023] continuously adopt JSCC to optimize the end-to-end (E2E) DNNs encompassing multiple receivers. Hence, it is cost-ineffective to directly scale the JSCC framework to multiuser communication systems (e.g., semantic BC), due to the awful size of parameters corresponding to exponentially increased receivers. Instead, an alternative framework, which is built on top of a state-of-the-art transformer DNN structure (e.g., universal transformer (UT) [Dehghani et al., 2018]) and capably separates the training of TX and RX, is highly essential. In this regard, alternate learning between TX and RX sounds intuitive, but its convergence could be questionable.

In summary, the design of semantic unicast and broadcast communications is still hindered by the following critical issues:

Q1: *(Compatibility) How to compatibly deal with the nondifferentiability of semantic similarity metrics as well as practical nondifferentiable channel, so as to develop a learning-efficient algorithm?*

Q2: *(Scalability) How to design a cost-effective semantic BC framework?*

Q3: *(Convergence) How to establish a theoretical guidance for alternate learning and calibrate the convergent hyperparameters correspondingly?*

Since unicast is a special case of broadcast, we have unified unicast and multicast to jointly address these critical yet challenging issues and propose a convergent semantic BC framework based on self-critical alternate learning (SCAL), named SemanticBC-SCAL [Lu et al., 2023a]. In particular, SemanticBC-SCAL capably adapts to different channel environments from one TX to multiple RXs and competently solves the nondifferentiability issue in the objective function (i.e., semantic similarity metrics) by incorporating reinforcement learning (RL) techniques that regard the semantic similarity as the reward. Meanwhile, considering the difficulty of handling large semantic decision space, SemanticBC-SCAL utilizes a lightweight self-critical supervision [Luo, 2020], which provides a simple and expedited baseline estimation with low variance, to yield a stable and efficient gradient policy. Furthermore, SemanticBC-SCAL is equipped with an asynchronous alternate learning mechanism to unify TX and RX, swiftly adapting to different numbers of RXs, and also addresses the nondifferentiable channel. Additionally, to verify its applicability in a point-to-point scenario, we conducted experiments in both UT and long short-term memory (LSTM) architectures, named SemanticBC-SCAL and SemanticRL [Lu et al., 2021], respectively. The primary contributions of our work are summarized as follows:

- In order to enable stable optimization via a nondifferentiable semantic similarity metric, we regard semantic similarity as a reward and formulate it as an RL problem, which is compliant with any semantic metric. Meanwhile, we adopt self-critical supervision to provide simple and efficient gradient estimation. This becomes especially suitable for complex SC systems with large semantic decision space, and partially addresses the mentioned *Q1*.
- Within the abovementioned optimization, built upon the UT architecture [Dehghani et al., 2018], we propose a semantic BC framework, named SemanticBC-SCAL, comprising one TX and multiple RXs, the superiority of which has been widely validated in Xie et al. [2023] and Zhou et al. [2021]. Notably, the decoders in SemanticBC-SCAL share a consistent structure with trained parameters, which could be leveraged for further adaptive learning. Thus, it allows the trained decoder model to readily expand to additional RXs at a reasonable cost, effectively addressing the *Q2*.
- Combining the above approaches, a self-critical based alternate learning mechanism is devised to alternately train the encoder at TX and multiple decoders at RXs, as the conventional supervised training can not directly perform back-propagation through nondifferentiable channels. In particular, under self-critical supervision, when the TX remains fixed, multiple RXs locally update their parameters according to their unique channel properties. After certain iterations, the RXs are fixed, and the TX updates itself by aggregating the information of multiple RXs uniformly. This alternate learning mechanism just

addresses the nondifferentiable channel in *Q1*. Moreover, we carefully validate the performance. The results show that the RXs can adapt to different channels, and yield superior performance compared to traditional communication method [Reed and Solomon, 1960] and JSCC-based semantic model [Xie et al., 2021]. This addresses the mentioned *Q3*.

12.2 System Model And Problem Formulation

12.2.1 Unicast Model

First, we consider a unicast communication system with three major parts: TX side (parameterized by $\mathcal{F}_{TX}(\cdot)$), random channel ($\mathcal{H}(\cdot)$), and RX side ($\mathcal{F}_{RX}(\cdot)$), where TX and RX can contain multiple learnable modules and are tacitly assumed to be abstract in this chapter. Given a source message $\boldsymbol{m}$, we aim to recover it from the observed degraded d-dimensional representation $\mathcal{F}'_{TX}(\boldsymbol{m}) = \mathcal{H}(\mathcal{F}_{TX}(\boldsymbol{m})) \in \mathbb{R}^d$. The paradigm of such a transmission system is illustrated in Figure 12.1a. Also note that in this chapter, we always parameterize the abstract encoder with θ, and the abstract decoder with ϕ.

To enable a reliable transmission, mainstream schemes adopt pixel-level or word-level supervision like MSE and CE, and train the whole system in a deep learning-based JSCC manner (see Figure 12.1b). To further facilitate the gradient back-propagation, the channel function $\mathcal{H}(\cdot)$ is usually modeled as additive or multiplicative ones [O'shea and Hoydis, 2017; Bourtsoulatze et al., 2019]. Different

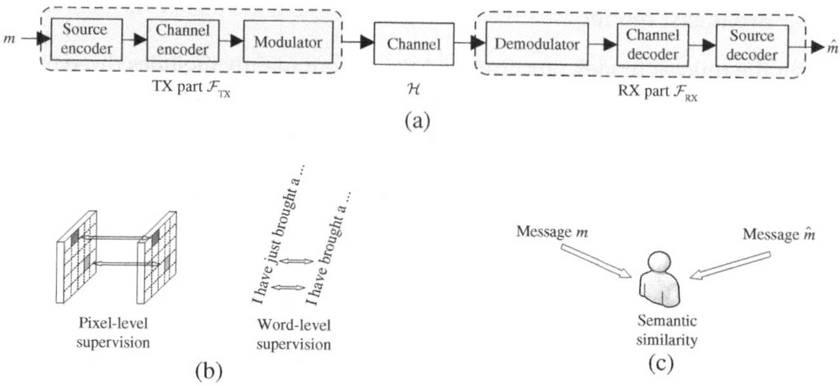

Figure 12.1 System model and comparisons of varied communication schemes [Lu et al., 2021]. (a) The framework of a neural-network enabled communication system, where the TX part and RX part are modeled as abstract learnable modules. (b) The commonly used bit-level supervisions in existing joint source-channel coding schemes. (c) Our proposal. We develop a paradigm shift that communicates by learning from semantic similarity.

from the aforementioned popular schemes, a SC system should concentrate more on the understanding and expression of semantics, instead of merely pursuing a simple bit-level recurrence. To formalize this task, we denote a semantic similarity metric by Θ and encourage the whole learning system $\{\mathcal{F}_{TX}, \mathcal{F}_{RX}\}$ to achieve higher semantic scores, as illustrated in Figure 12.1c. This is systematically different from conventional JSCC-based transmission pipelines or popular SC schemes [O'shea and Hoydis, 2017; Farsad et al., 2018; Bourtsoulatze et al., 2019; Weng and Qin, 2021] that interoperate the semantics from a model level but with bit-level supervisions. Our formulation, instead, evaluates how the semantics are represented and concentrates on the accuracy of semantic expression.

As another key factor, the impact of random channel is a long-discussed yet troublesome problem. Although recent progress in machine learning has provided plenty of state-of-theart noise elimination schemes, like alternative training [Aoudia and Hoydis, 2019], generative adversarial network (GAN)-based estimation [Ye et al., 2020], and RL-based policy solution [Aoudia and Hoydis, 2018], these methods are usually computational intensive, or much sensitive to the parametric or environmental settings, which makes them hard to be integrated directly. Existing works on JSCC or semantic coding either make a strong hypothesis of an additive white Gaussian noise (AWGN) channel or directly sort to the off-the-shelf denoising techniques, with this fundamental problem only touched upon but not typically analyzed. In this chapter, we take a step further by no longer assuming its differentiability, and aim to simultaneously tackle this problem with a unified learning framework.

12.2.2 Broadcast Model

Based on the unicast model, our proposed BC system mainly includes three parts as illustrated in Figure 12.2, that is, one TX, N random channels, and N RXs. Generally speaking, at TX side, the message $\boldsymbol{m}$ is first converted to an embedding by the embedding layer. Then the embedding is sent to implement semantic encoding, including a semantic encoder for features extraction and a dense layer for channel encoding. Then, the quantization layer quantifies $\boldsymbol{x}$ as $\boldsymbol{b}$ to facilitate the subsequent transmission and this extracted valuable information is broadcasted through different physical channels to reach RX_n $(n \sim N)$.[1] Likewise, at RX_n side, the dequantization layer detects and reconstructs the received symbol $\hat{\boldsymbol{b}}_n$ from the received noisy signals, and the semantic decoding, i.e., including channel decoder and semantic decoder, estimate and recover the source as $\hat{\boldsymbol{m}}_n$ based on the knowledge base (KB). Different from the classic SC [Xie et al., 2021],

1 The $\sim$ operator implies $n \in \{1, 2, \dots, N\}$.

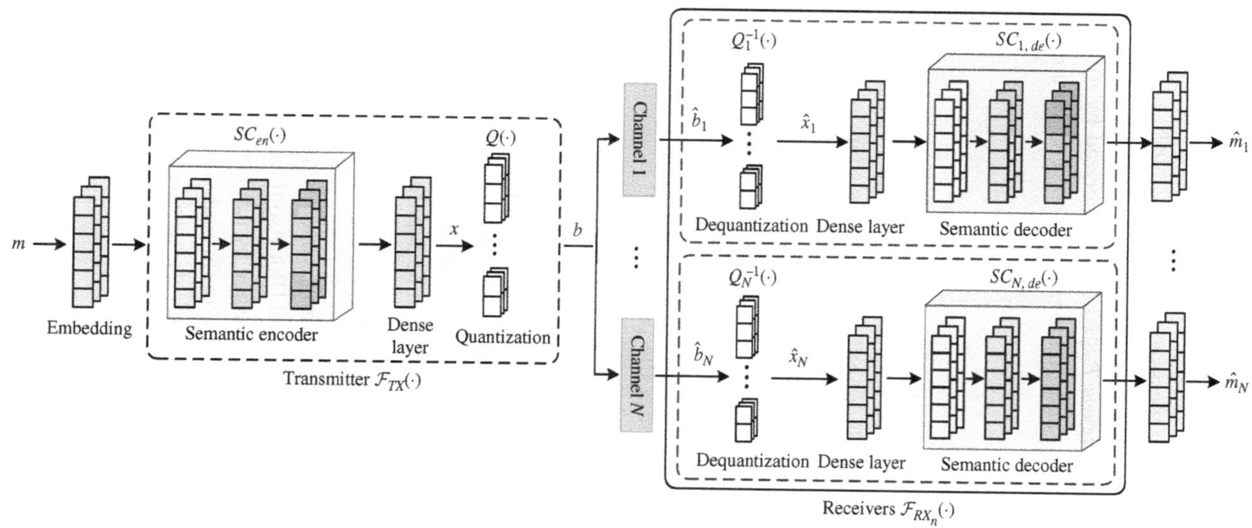

Figure 12.2 The system model for semantic BC system.

we adopt the UT [Dehghani et al., 2018] based semantic encoder/decoders to achieve semantic coding/decoding. Besides, RXs share the same structure but have individual DNN parameters accommodating to channel properties, so as to recover the original information as accurately as possible.

Mathematically, for a source message $\boldsymbol{m} = \{w^{(1)}, \dots, w^{(i)}, \dots, w^{(T)}\}$, where T is the sequence length, and $w^{(i)}$ can be regarded as a word from dictionary W_m (i.e., $w^{(i)} \in W_m$). The encoding process $\mathcal{F}_{TX}(\cdot)$ (parameterized by θ) for TX consists of semantic encoding $SC_{en}(\cdot)$ and quantization $Q(\cdot)$. The semantic encoding is given by

$$\boldsymbol{x} = \mathrm{SC}_{en}(\boldsymbol{m}; \theta_{en}), \tag{12.1}$$

where $\boldsymbol{x} \in \mathbb{R}^{T \times D}$, D is the dimension of each symbol in $\boldsymbol{x}$, $SC_{en}(\cdot)$ is the combination of semantic encoder and channel encoder parameterized by θ_{en}.

After that, the embedding $\boldsymbol{x}$ is quantified into discrete bits by a θ_q-induced quantization $Q(\cdot)$ as

$$\boldsymbol{b} = Q(\boldsymbol{x}; \theta_q), \tag{12.2}$$

where $\boldsymbol{b} \in \mathbb{R}^{T \times B}$, B is the number of bits for each quantified symbol of $\boldsymbol{x}$. Then TX broadcasts $\boldsymbol{b}$ to multiple physical channels.

At the RX_n side, different from the point-to-point transmission with a single receiver, the transmitted data in semantic BC experience heterogeneous channel conditions corresponding to different receivers. Analogously, the decoding process $\mathcal{F}_{RX_n}(\cdot)$ (parameterized by ϕ_n) for RX_n involves the reverse steps, including dequantization and semantic decoding, which are represented as $Q_n^{-1}(\cdot)$ and $SC_{n,de}(\cdot)$, respectively. With the aid of channel state information (CSI), the semantic decoder can be dynamically adjusted to achieve adaptive decoding. As such, the decoding message $\hat{\boldsymbol{m}}_n$ of RX_n can be denoted as:

$$\hat{\boldsymbol{m}}_n = SC_{n,de}\left(Q_n^{-1}(\mathcal{H}_n(\boldsymbol{b}); \phi_{n,dq}); \phi_{n,de}\right), \tag{12.3}$$

where $\mathcal{H}_n(\cdot)$ $(n \sim N)$ is the channel layer and is usually modeled as AWGN channel or multiplicative Rayleigh fading channel [Bourtsoulatze et al., 2019].

As such, RX_n recovers the source message by predicting the probability distribution of the next word in the dictionary. In the end, a complete decoded message can be denoted as $\hat{\boldsymbol{m}}_n = \{\hat{w}_n^{(1)}, \hat{w}_n^{(2)}, \dots, \hat{w}_n^{(\hat{T}_n)}\}$, where $\hat{T}_n$ is the length of decoded message for RX_n.

12.2.3 Problem Formulation

As mentioned above, unicast is a special case of broadcast, that is, $N = 1$. Hence, we have unified the above two models within the broadcast model. Given the described broadcast model, our goal is to maximize semantic similarity for RX_n by learning the optimal parameters for the system $\langle \mathcal{F}_{TX}, \mathcal{F}_{RX_n} \rangle$, that is,

$$\left\langle \theta^*, \phi_n^* \right\rangle = \underset{\mathcal{F}_{TX}, \mathcal{F}_{RX_n}}{\arg\max}\ \Theta(\boldsymbol{m}, \hat{\boldsymbol{m}}_n), \tag{12.4}$$

where $\Theta(\cdot)$ can be any reasonable semantic similarity metric, and we do not assume its differentiability. Therefore, in this chapter, bilingual evaluation understudy (BLEU) scores [Papineni et al., 2002], BERT (bidirectional encoder representation from transformers) [Devlin et al., 2019] based similarity metric (i.e., BERT-SIM), word accuracy rate (WAR), and CIDEr [Vedantam et al., 2015] are adopted to measure the degree of consistency between the input $\boldsymbol{m}$ and output $\hat{\boldsymbol{m}}_n$ [Xie et al., 2021; Lu et al., 2023b].

Nevertheless, the problem defined in (12.4) for semantic BC is challenging to solve by traditional JSCC frameworks. The difficulties are twofold. On one hand, in order to address the inherent "semantic gap," the commonly used differentiable objective will no longer be applicable since the gradient cannot be directly propagated backward through the channel. On the other hand, as the number of RX increases, the scalability issue emerges, primarily as the commonly used point-to-point optimization method becomes invalid and untenable. Furthermore, to alleviate the "semantic gap," we formulate (12.4) as an RL problem by regarding nondifferentiable semantic similarity as a reward. Besides, we adopt an alternate learning mechanism to cope with the scalability issue by iteratively updating the encoders and decoders, in which the encoders and decoders are all conceptualized as independent agents.

Alternate learning mechanism: Especially, upon freezing the parameters of $\mathcal{F}_{TX}$ (i.e., θ), $\mathcal{F}_{RX_n}$ is locally updated κ iterations. Conversely, when $\mathcal{F}_{RX_n}$ (i.e., ϕ_n) is held constant, $\mathcal{F}_{TX}$ is updated once independently by averaging the decoders' results. It is worth emphasizing that the encoder updates once, while each decoder locally updates κ iterations, collectively forming a single *update cycle*. Corresponding to this alternate learning mechanism, the objective functions for the TX and RX_n diverge intrinsically. In this regard, the overall optimization objective outlined in (12.4) is formulated into the following two parts

$$\theta^* = \underset{\mathcal{F}_{TX}(\cdot)}{\arg\max}\ \Theta(\boldsymbol{m}, avg(\underbrace{\mathcal{F}_{RX_n}}_{no\ grad}(\mathcal{H}_n(\mathcal{F}_{TX}(\boldsymbol{m}))))) \tag{12.5a}$$

$$\phi_n^* = \underset{\mathcal{F}_{RX_n}(\cdot)}{\arg\max}\ \Theta(\boldsymbol{m}, \mathcal{F}_{RX_n}(\mathcal{H}_n(\underbrace{\mathcal{F}_{TX}(\boldsymbol{m})}_{no\ grad}))), \tag{12.5b}$$

where $avg(\cdot)$ is to get the mean value.

In this chapter, we discuss about how to alternately optimize (12.5a) and (12.5b) on top of RL, the details of which shall be presented in Section 12.3.

12.3 SemanticBC-SCAL Schemes with Alternating Learning Mechanism

In this section, we first elaborate on how to formulate a Markov decision process (MDP) to facilitate the application of RL, optimizing the semantic BC system with semantic level supervision. Afterwards, we discuss about the SCAL based algorithm, SemanticBC-SCAL, to produce a solution.

12.3.1 The Markov Decision Process (MDP) Framework

Following this multiuser semantic BC system, where the encoder can be formulated as one agent at TX and the decoders can be modeled as N independent agents at RXs, which interacts with the environment (i.e., the text dictionary W_m) by taking actions on the basis of states and receiving the rewards from the environment.

Without loss of generality, the MDP setting can be denoted as $(\mathcal{S}, \mathcal{A}, P, R, \gamma)$, where $\mathcal{S}$ and $\mathcal{A}$ is the state space and the action space respectively, P is the state transition probability, and R denotes reward function with discounted factor $\gamma \in (0, 1)$ [Lu et al., 2021]. Since the actions (i.e., word generation process) are implemented at the decoders, the state space and action space are determined by RX_n and shared for the encoder. As such, we define the state space as $\mathcal{S} = \{s_1, \dots, s_n, \dots, s_N\}, n \sim N$ and action space (i.e., vocabulary) as $\mathcal{A} = \{w_1, \dots, w_n, \dots, w_N\}, w_n \in W_m$ (where the dimension of dictionary W_m is V). Moreover, the state transition function as $P = \{P_{\pi_\theta}, P_{\pi_{\phi_n}}\}$, and the reward for TX and RX_n are distinctly parameterized by π_θ and π_{ϕ_n}, respectively, which are further delineated as $R = \{R_{\pi_\theta}, R_{\pi_{\phi_n}}\}$.

In particular, the MDP can be detailed as follows:

- **Action**: The definition of action is to generate the next token $a_n^{(t)} = \hat{w}_n^{(t)}$ from the dictionary W_m at RX_n, where the sequence generation process starts with token "SOS" and ends with "EOS." In contrast to traditional communication where there exists a one-to-one relationship among the source message $\boldsymbol{m}$, SC introduces a scenario where multiple pairs of interpretations can exist between them. Hence, the state-action space can be larger than traditional methods.
- **State**: We define the state as the sequential state of decoder n at step t and decoded tokens (i.e., words in the text tasks) so far. Furthermore, for each RX_n, $s_{n,de}^{(t)}$ can be viewed as a combination of previous decoder output and buffer generation (words) so far, i.e., $s_n^{(t)} = \{s_{n,de}^{(t)}, \hat{w}_n^{(1)}, \hat{w}_n^{(2)}, \dots, \hat{w}_n^{(t)}\} \in \mathcal{S}$. It merits emphasis that our goal is to interpret messages at the semantic level, which may lead to variations in length between the decoded sentences and their original counterparts. Here, for different RX_n, the maximum length of decoded sentence is $\hat{T}_n$. Meanwhile, once the next token $\hat{w}_n^{(t+1)}$ is generated, the state transition is deterministic between any adjacent state.

- **Policy**: Notably, the policies for TX and RX shall be independently calibrated, since the semantic encoder at TX aims to optimize the encoding process in (12.5a), while the semantic decoder at RX_n is responsible for generating semantically accurate messages by maximizing (12.5b).
  Specifically, for TX side, the semantic encoding of the encoder is a continuous one and only needs to be implemented once for each sample. In this sense, we model this process as a continuous Gaussian distribution with the mean value $\boldsymbol{\mu} = \mathcal{F}_{TX}(\boldsymbol{m}) \in \mathbb{R}^{1\times D}$, and the covariance matrix $\Sigma = (\sigma \boldsymbol{I})^2 \in \mathbb{R}^{D\times D}$ (where σ is typically set to 0.1). In other words, some Gaussian noise is intentionally added to the semantic encoding process, to encourage a certain exploration. Mathematically, the encoder policy can be written as
  $$\pi_\theta := \text{Sample}\left(\mathcal{N}\left(\boldsymbol{\mu} = \mathcal{F}_{TX}(\boldsymbol{m}), \Sigma = (\sigma \boldsymbol{I})^2\right)\right), \tag{12.6}$$
  where $\mathcal{N}$ indicates a Gaussian distribution.
  For RX_n, instead of one deterministic distribution, the input of the decoder is decoded by K times and the output likelihood is modeled as a probabilistic multinomial distribution (termed as $\mathcal{MD}$) under self-critical supervision (it will be detailed in Section 12.3.2), which means RX_n can obtain a group of different output from K views, so as to better reap the potentially diversified semantics of one same sentence. As such, sampling from this kind of multinomial distribution ensures the collection of samples with the maximum likelihood from multiview. For each multinomial distribution, it is modeled by the "softmax" function that is parameterized on the ϕ_n. Then for step t, one has
  $$\pi^t_{\phi_n} := \text{Sample}(\mathcal{MD}([p_n(\hat{w}^{(t)}_1), p_n(\hat{w}^{(t)}_2), \dots, p_n(\hat{w}^{(t)}_V)])), \tag{12.7}$$
  where the superscript t means the token time step for the sampling operation.
- **State transition probability**: According to the abovementioned definitions, the state transition function P is deterministic between two adjacent states. Formally, we have $P(s^{(t+1)}_n | s^{(t)}_n, \hat{w}^{(t)}_n) = 1$.
- **Reward**: We use the semantic similarity as a reward and it is denoted as $\Theta(\boldsymbol{m}, \hat{\boldsymbol{m}}_n)$. In the sentence generation model, the reward can only be obtained until the last token is generated. Therefore, the reward is sparse in SemanticBC-SCAL. In other words, the intermediate reward is always zero except for the last time step $\hat{T}_n$, that is,
  $$r^{(t)}_n = \begin{cases} 0, & \text{if } t \neq \hat{T}_n, \\ \Theta(\boldsymbol{m}, \hat{\boldsymbol{m}}_n), & \text{if } t = \hat{T}_n. \end{cases} \tag{12.8}$$

For the sake of simplification, we assume the discounted factor $\gamma = 1$, so without loss of generality, the *return* for our system is formulated as:

$$G^{(t)}_n = \sum_{k=t+1}^{\hat{T}_n} \gamma^{k-t-1} r^{(k)}_n = \sum_{t=1}^{\hat{T}_n} r^{(t)}_n. \tag{12.9}$$

Accordingly, the state-value function is given by $V(s_n^{(t)}) = \mathbb{E}\left[G_n^{(t)} | s_n^{(t)}\right]$. Consistent with the reformulated problem in (12.5), at decoder sides, the objective function for RX_n is to maximize the value function, which can be viewed as maximizing *return* for a complete trajectory $\{\hat{w}_n^{(t)}, s_n^{(t+1)}, \hat{w}_n^{(t+1)}, \ldots\}$ under decoder policy π_{ϕ_n} starting from $s_n^{(0)}$, which is given by

$$J(\phi_n) = V_{\pi_{\phi_n}}(s_n^{(0)}) = \mathbb{E}_{\pi_{\phi_n}}\left[G_n^0 | s_n^0\right] = \mathbb{E}_{\pi_{\phi_n}}\left[\sum_{t=1}^{\hat{T}_n} r_n^{(t)}\right]. \tag{12.10}$$

On the other hand, for encoder optimization, TX aims to find an optimal encoding to maximize the semantic accuracy for all decoders. Under the encoder policy π_θ, we regard the maximum average *return* of all decoders as the objective function, that is

$$J(\theta) = \frac{1}{N}\sum_{n=1}^{N} V_{\pi_\theta}(s_n^{(0)}) = \frac{1}{N}\sum_{n=1}^{N} \mathbb{E}_{\pi_\theta}\left[\sum_{t=1}^{\hat{T}_n} r_n^{(t)}\right]. \tag{12.11}$$

12.3.2 Self-Critical Optimization Under Alternate Learning Mechanism

In order to find a proper solution to maximize the objectives functions (12.10) and (12.11), one most straightforward solution is to simulate the environment to obtain Monte-Carlo trajectory $\{(s_n^{(t)}, \hat{w}_n^{(t)}, r_n^{(t)})\}_{t=1}^{\hat{T}_n}$ for each time step t [Gao et al., 2019]. Though Monte-Carlo rollout provides an unbiased estimation for expected *return*, large state-action space unavoidably leads to high variance [Dubey et al., 2016]. Another alternative solution is the "actor-critic" method and its variants [Dogru et al., 2021], in which the baseline is established by constructing another *value network* to criticize the policy [Lu et al., 2021]. However, this way will introduce extra parameters and estimation bias. Meanwhile, due to the extensive number of tokens included in the dictionary W_m, the size of state-action space exceeds 10^4. In this sense, merely the Monte-Carlo trajectory or actor-critic methods are impractical to implement for sentence generation in the SC system.

Based on the aforementioned discussions, to handle this large-scale sentence generation, we develop the SemanticBC-SCAL algorithm to combine a simper and practical RL algorithm, i.e., self-critical learning [Luo, 2020], with the abovementioned alternate learning mechanism in Section 12.2.3. Especially, SCAL samples a group of parallel Monte-Carlo trajectories and utilizes the mean reward from the rest of $K-1$ parallel samples, that is, $\hat{\boldsymbol{m}}_n = \{\hat{\boldsymbol{m}}_{n,1}, \ldots, \hat{\boldsymbol{m}}_{n,K-1}\}$ (where K is the number of samples), to act as the baseline function [Lu et al., 2021]. To generate a complete trajectory by introducing a certain exploration, the semantic encoder/decoders could employ the policy π_θ and π_{ϕ_n}, respectively to

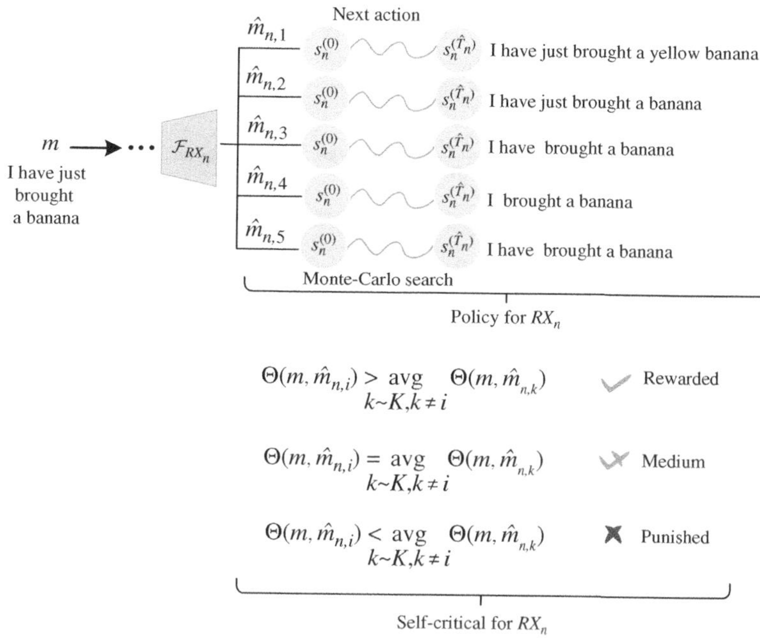

Figure 12.3 Illustration of the self-critic optimization for decoder side at RX_n.

guide the system in accommodating multiview semantic representations. Since the baseline comes from itself, i.e., its other $K - 1$ decoding results, this kind of supervision can be regarded as "self-critical" manner [Rennie et al., 2017; Gao et al., 2019; Luo, 2020], and the empirical results show that larger K will have a larger absolute value of the mean and introduce lower variance [Gao et al., 2019]. Taking an example of decoder optimization in Figure 12.3, the TX sends the message ("I have just brought a banana") to the broadcast channel, and the RX_n first samples $K = 5$ parallel samples before channel decoding, in which the decoded parallel sentences could be different. Then, RX_n can be viewed as an actor responsible for taking actions by Monte-Carlo sampling to generate parallel decoding results. As such, the advantage of one decoding result will be evaluated by the other four parallel samples.

Furthermore, combining the self-critical learning and alternate learning mechanism, a learning-efficient method is devised to asynchronously update the one TX and multiple RXs to adapt to its own channel environment. To be specific, in the decoder training stage, we freeze the encoder parameters θ and locally update the decoder parameters ϕ_n. Since the semantic similarity score can only be obtained via the self-critical manner after decoding a complete sentence, the decoder updating occurs only after a complete sentence transmission. The details of gradients and flow in SemanticBC-SCAL are illustrated in Figure 12.4a.

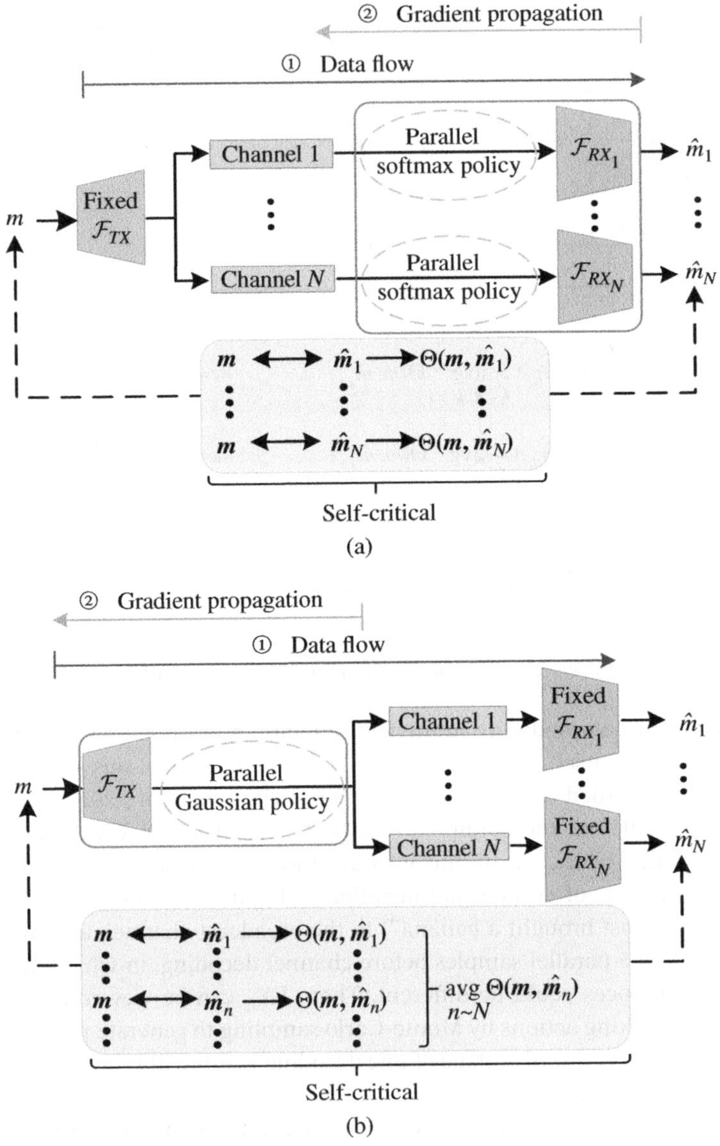

Figure 12.4 Optimization process under the alternate learning mechanism. (a) The training process of RX_n. (b) The training process of TX. (The black solid line is the direction of data flow, while the gray solid line denotes the direction of gradient propagation.)

Moreover, we provide how to compute the gradient of (12.10) by the following theorem.

Theorem 12.1 ***(Self-critical semantic policy gradient for RX_n)*** For any semantic decoder n optimized by the alternate learning mechanism with K parallel sampling, suppose that the decoder is updated by the policy $\pi_{\phi_n}^{(t)}$ to generate a complete sequence $\{\hat{\omega}_n^{(1)}, \hat{\omega}_n^{(2)}, \dots, \hat{\omega}_n^{(\hat{T}_n)}\}$, the gradient can be approximated by

$$\nabla_{\phi_n} J(\phi_n) \approx \frac{1}{K}\sum_{i=1}^{K}\left[\sum_{t=1}^{\hat{T}_n} \nabla_{\phi_n} \log \pi_{\phi_{n,i}}(\hat{\omega}_n^{(t)}|s_n^{(t)}) \cdot \left(\Theta_{n,i} - \underset{k\sim K, k\neq i}{\text{avg}} \Theta_{n,k}\right)\right], \quad (12.12)$$

where the subscript i in $\pi_{\phi_n,i}$ denotes the ith parallel sampling policy of sample i, while $avg(\cdot)$ gets the mean value. Θ_i is the return of current sample i, and Θ_k represents any one of parallel samples.

Proof: During the decoder optimization process, a complete trajectory $\hat{\boldsymbol{m}}_n = \{\hat{\omega}_n^{(1)}, \hat{\omega}_n^{(2)}, \dots, \hat{\omega}_n^{(\hat{T}_n)}\}$ has the probability of $P(\hat{\boldsymbol{m}}_n|\phi_n)$, and the gradient of RX_n is

$$\begin{aligned} J(\phi_n) =& \nabla_{\phi_n}\left[\sum_{\hat{\boldsymbol{m}}_n} P(\hat{\boldsymbol{m}}_n|\phi_n) \sum_{t=1}^{\hat{T}_n} r_n^{(t)}\right] \\ =& \sum_{\hat{\boldsymbol{m}}_n}\left[\nabla_{\phi_n} P(\hat{\boldsymbol{m}}_n|\phi_n) \cdot \sum_{t=1}^{\hat{T}_n} r_n^{(t)}\right] \\ \overset{(a)}{=}& \sum_{\hat{\boldsymbol{m}}_n}\left[P(\hat{\boldsymbol{m}}_n|\phi_n)\nabla_{\phi_n} \log\left(P(\hat{\boldsymbol{m}}_n|\phi_n)\right) \cdot \sum_{t=1}^{\hat{T}_n} r_n^{(t)}\right] \\ =& \mathbb{E}_{\hat{\boldsymbol{m}}_n}\left[\nabla_{\phi_n} \log\left(P(\hat{\boldsymbol{m}}_n|\phi_n)\right) \cdot \sum_{t=1}^{\hat{T}_n} r_n^{(t)}\right], \end{aligned} \quad (12.13)$$

where the equality (a) is derived by using the log-trick $\nabla \log x = \nabla x / x$. Since we sample one Monte-Carlo trajectory to estimate the $\hat{\boldsymbol{m}}_n$, then one obtains

$$\nabla_{\phi_n} J(\phi_n) \approx \nabla_{\phi_n} \log\left(P(\hat{\boldsymbol{m}}_n|\phi_n)\right) \cdot \sum_{t=1}^{\hat{T}_n} r_n^{(t)}. \quad (12.14)$$

As the sentence is generated step by step through $\hat{T}_n + 1$ actions, $P(\hat{\boldsymbol{m}}_n|\phi_n)$ is further written as $P(\hat{\boldsymbol{m}}_n|\phi_n) = p(s_n^0)\prod_{t=1}^{\hat{T}_n}\left[\pi_{\phi_n}(\hat{\omega}_n^{(t)}|s_n^{(t)})P_{\pi_{\phi_n}}(s_n^{(t)}|s_n^{(t-1)}, \hat{\omega}_n^{(t-1)})\right]$. Meanwhile, the reward function is sparse, i.e., $\sum_{t=1}^{\hat{T}_n} r_n^{(t)} = \Theta(\boldsymbol{m}, \hat{\boldsymbol{m}}_n)$, and the state transition probability $P_{\pi_{\phi_n}}(s_n^{(t)}|s_n^{(t-1)}, \hat{w}_n^{(t-1)})$ is deterministic between two adjacent states. Therefore, one has

$$\nabla_{\phi_n} J(\theta, \phi_n) \approx \sum_{t=1}^{\hat{T}_n} \nabla_{\phi_n} \log \pi_{\phi_n}(\hat{w}_n^{(t)}|s_n^{(t)})\Theta(\boldsymbol{m}, \hat{\boldsymbol{m}}_n). \tag{12.15}$$

Finally, the self-critical baseline function is implemented by utilizing the mean reward from the rest of $K-1$ parallel samples, thus, the proof ends with (12.12).

Remark 12.1 It is worth mentioning that the term $\left(\Theta_{n,i} - \text{avg}_{k\sim K,k\neq i}\Theta_{n,k}\right)$ is regarded as baseline or advantage of policy $\pi_{\phi_{n,i}}$, which is a success of SCAL because the critical samples come from its own model and also obtain a gradient variance reduction [Luo, 2020]. Furthermore, in semantics representation, the one with higher semantic similarity gives a larger positive reward, while those with lower advantage will be punished.

As for the encoder, recalling that as in Figure 12.4b, the encoder output is converted to K parallel samples by Gaussian policy, which are then sent to broadcast channels. Therefore, the objective of the encoder is to maximize the average reward of all decoders, the average decoded semantic similarity of different decoders is conversely used to optimize the $\mathcal{F}_{TX}$. Analogously, we give the optimization direction of (12.11) by the following theorem.

Theorem 12.2 ***(Self-critical semantic policy gradient for*** *TX)* For semantic encoder optimized by the alternate learning mechanism, suppose the encoder is updated by the parallel Gaussian strategy π_θ, the gradient can be approximated by the Monte-Carlo sampling as

$$\nabla_\theta J(\theta) \approx \frac{1}{N} \cdot \frac{1}{K} \sum_{n=1}^{N} \sum_{i=1}^{K} \left[\left[\widetilde{\mathcal{F}_{TX}}(\mathbf{m}) - \mathcal{F}_{TX}(\mathbf{m}) \right]^T \right. \tag{12.16}$$
$$\left. \cdot \boldsymbol{\Sigma}^{-1} \left[\nabla_\theta \mathcal{F}_{TX}(\mathbf{m}) \right] \cdot \left(\Theta_{n,i} - \underset{k\sim K,k\neq i}{\text{avg}} \Theta_{n,k} \right) \right],$$

where $\widetilde{\mathcal{F}_{TX}}$ is the Gaussian sampling for $\mathbf{m}$, T is the length of source message $\boldsymbol{m}$.

Proof: The proof is similar to that for Theorem 12.1. Thus, we only provide a sketch here.

Basically, for a Gaussian distribution

$$\pi_\theta(\boldsymbol{x}; \boldsymbol{\mu}, \boldsymbol{\Sigma}) = \frac{1}{(2\pi)^{D/2}\sqrt{\det\{\boldsymbol{\Sigma}\}}} \exp\left(-\frac{1}{2}(\boldsymbol{x}-\boldsymbol{\mu})^T\boldsymbol{\Sigma}^{-1}(\boldsymbol{x}-\boldsymbol{\mu})\right), \tag{12.17}$$

we have

$$\nabla_\theta \log \pi_\theta(\boldsymbol{x}; \boldsymbol{\mu}, \boldsymbol{\Sigma}) = [\boldsymbol{x}-\boldsymbol{\mu}]^T \boldsymbol{\Sigma}^{-1} \left[\nabla_\phi \boldsymbol{\mu}\right]. \tag{12.18}$$

Taking $\boldsymbol{x} = \widetilde{\mathcal{F}_{TX}}$ and $\boldsymbol{\mu} = \mathcal{F}_{TX}(\boldsymbol{m})$, as well as following the proof for Theorem 12.1, we have the theorem.

Remark 12.2 On the basis of this stable and reasonable learning policy, our SCAL-based semantic BC system can adaptively encode/decode by introducing as few parameters and costs as possible. To obtain a stable and expected reward (semantic similarity score) in the training process, we adopt the aforementioned alternate learning mechanism at both *TX* and RX_n to implement a Monte-Carlo simulation. Specifically, in an *update cycle*, the RXs update κ iterations and then

---

Algorithm 12.1 The training process of SemanticBC-SCAL.

Input: Batch size, learning rate lr, parallel samples K, input sequence $\boldsymbol{m}$, local iterations κ, number of pretraining end epochs E_p, number of end epochs E_e, number of end batches E_b, and number of RXs N.

Output: Encoder parameter θ, decoder parameter ϕ_n.

// **Pretraining stage**
for epoch=1 : E_p **do**
  Sample a batch of data to train parameters of encoder θ and decoder ϕ optimized by CE loss [Zhou et al., 2021].
  Update θ and ϕ to obtain the pretrained parameters θ_{pre} and ϕ_{pre}.
end for
// **RL-based alternate learning**
$\theta \leftarrow \theta_{pre}$, $\phi_n \leftarrow \phi_{pre}$
for epoch=$E_p + 1$: E_e **do**
  for $i = 1$: E_b **do**
    Sample a batch of data
    // **Training Decoder**
    while $(i \mod \kappa) \neq 0$ **do**
      for $n = 1$: N **do**
        Freeze θ, RX_n samples K parallel trajectories, and updates $\phi_n \leftarrow \phi_n + lr \cdot \nabla_{\phi_n} J(\phi_n)$ as in (12.12).
      end for
    end while
    // **Training Encoder**
    For each RX_n, freeze ϕ_n and sample K parallel trajectories.
    Update $\theta \leftarrow \theta + lr \cdot \nabla_\theta J(\theta)$ by (12.16).
  end for
end for
// **Finish training**
Return θ and ϕ_n.

TX updates once. Therefore, this process is implemented iteratively until both the encoders and decoders converge. With this updating mechanism through self-critical training, we provide a stable and precise gradient surrogate, in which we neither need an additional module to estimate the gradient, nor suffer from a poor estimation of expected reward.

Finally, the detailed procedures for training SemanticBC-SCAL are summarized in Algorithm 12.1.

12.4 Performance Evaluation

In this section, we compare the proposed SemanticBC-SCAL with the traditional reliable communication scheme and typical SC algorithms under both AWGN and Rayleigh fading channels, respectively. In addition, we also give an ablation study to comprehensively evaluate and analyze the performance.

12.4.1 Simulation Settings

We adopt the standard English version of the proceedings of the European Parliament [Koehn, 2005] as the dataset, which consists of more than 2.0 million sentences and 53 million words. In our experiment, the dataset is preprocessed into sentences with the length of $4 \sim 30$ words, those shorter than 30 words being padded with zeros and the number of units D in the hidden layer behind each word equals 128. Furthermore, each word is converted to 30 bits (i.e., $B = 30$) by the quantization layer for further transmission. The preproposed dataset contains a total of $1,136,816$ sentences and $32,478$ words (dictionary size), which are then divided into training and testing with a ratio of $4:1$. Furthermore, the backbone of semantic encoder/decoders is composed of UT [Dehghani et al., 2018] while the channel encoder and decoder are based on dense layers. The detailed DNN structure and hyperparameters are summarized in Table 12.1.

As for the wireless channel, we consider the commonly used AWGN and Rayleigh fading channel as discussed in Section 12.2.2. As a comparison, we utilize the following baselines.

1. **Huffman-RS**: A traditional communication system where source coding and channel coding are implemented as Huffman coding and Reed–Solomon (RS) [Reed and Solomon, 1960] coding, respectively. Meanwhile, the coding length with RS is set to (30×42) in 64-QAM.
2. **SemanticRL**: It optimizes the entire semantic system on top of RL but its backbone is built around LSTM networks [Lu et al., 2021].

Table 12.1 Network structure and hyperparameters used in SemanitcBC-SCAL.

	Modules	Layer	Dimensions	Activation
	Input	$\boldsymbol{m}$	30	—
TX	$SC_{en}(\cdot)$	Embedding UT encoder Dense layer Gaussian policy	(30 × 128) (30 × 128) (30 × 16) (30 × 16)	— — ReLU —
	Q	Dense layer Quantization	(30 × 30) (30 × 30)	ReLU —
Channel	AWGN	—	—	—
$3 \times \boldsymbol{RX}_n$	$Q_n^{-1}(\cdot)$	Dense layer	(30 × 16)	ReLU
	$SC_{n,de}(\cdot)$	Dense layer UT decoder Dense layer Softmax	(30 × 128) (30 × 128) (30 × 32,478) $(32{,}478 \times \hat{T}_n)$	ReLU ReLU Dictionary Softmax

3. **SemanticBC-SCAL(point-to-point):** It is the point-to-point case of our proposed SemanticBC-SCAL.
4. **CE-baseline**: It refers to a degenerative point-to-point model that shares the same network structure as our proposed model. Put differently, a special case arises when SemanticRL-SCAL-UT is optimized by the CE function.
5. **DeepSC**: It belongs to the pioneering work in SC [Xie et al., 2021], and leverages the transformer architecture optimized by CE function and mutual information.

Moreover, for SemanticBC-SCAL, the pretraining process is implemented by utilizing the CE function to accelerate the training. Afterward, all pretrained RXs undergo the described training procedure in SemanticBC-SCAL until convergence. Besides, in order to characterize the inherently noisy channel environment, we have implemented two types of SNR simulation methods. First, we set deterministic channel SNR i.e., μ_{SNR} for each receiver to indicate practical channel conditions. Secondly, in broadcast cases, we add different noise variances δ^2_{SNR} of SNR for different RXs to reflect their heterogeneous estimation capabilities and adaptability. Especially, for both AWGN and Rayleigh fading channels, we set different mean values μ_{SNR} and noise variance δ^2_{SNR} during the training stage respectively, including $\mu_{SNR,1} = 6\,\text{dB}$, $\delta_{SNR,1} = 1\,\text{dB}$; $\mu_{SNR,2} = 10\,\text{dB}$, $\delta_{SNR,2} = 1\,\text{dB}$; $\mu_{SNR,3} = 10\,\text{dB}$, $\delta_{SNR,3} = 2\,\text{dB}$. While for SemanticBC-SCAL in the point-to-point scenario, we set the SNR of the training process by default with $\mu_{SNR} = 10\,\text{dB}$ and $\delta_{SNR} = 1\,\text{dB}$. The related parameters are listed in Table 12.2.

Table 12.2 The default parameters in SemanitcBC-SCAL.

Parameters	Description	Value
—	Batch size	64
lr	Learning rate	1e−5
K	Number of parallel samples	5
κ	Local iterations for decoders in each *update cycle*	1000
E_p	Number of pretraining end epochs	50
E_e	Number of end epochs	180

12.4.2 Numerical Results And Analysis

12.4.2.1 Performance In Point-to-Point SC

We start with the performance comparison of the SemanticBC-SCAL with other baselines in the point-to-point case. In particular, we provide the performance in terms of BLEU scores and BERT-SIM under AWGN and Rayleigh fading channel in Figures 12.5 and 12.6, respectively. As shown in Figure 12.5, our proposed SemanticBC-SCAL outperforms CE-baseline and SemanticRL, especially in terms of BLEU-4. Since a larger n-gram phrase carries more contextual information, the result demonstrates that our approach is capable of preserving semantics and prone to provide a semantic level transmission. Furthermore, our approach also outperforms baselines in terms of BERT-SIM, allowing for accommodating some synonyms. Meanwhile, as illustrated in Figure 12.6, attributed to the volatile channel gain, the Rayleigh fading channel imposes a stronger detrimental impact compared to the AWGN channel on both the traditional method and SC models, leading to a more pronounced attenuation during the signal transmission.

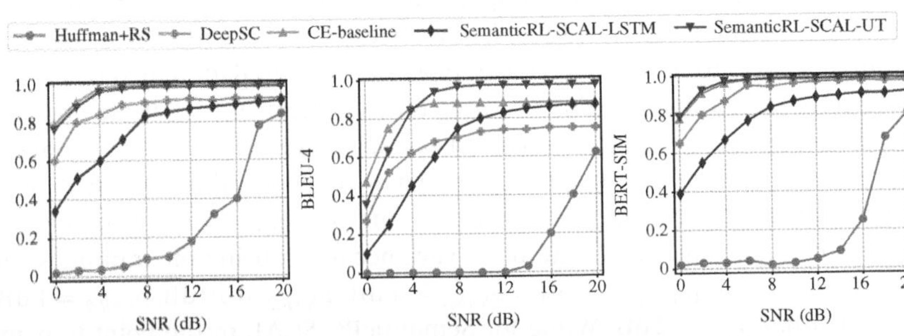

Figure 12.5 BLEU score and BERT-SIM versus SNR in AWGN channel for the point-to-point case.

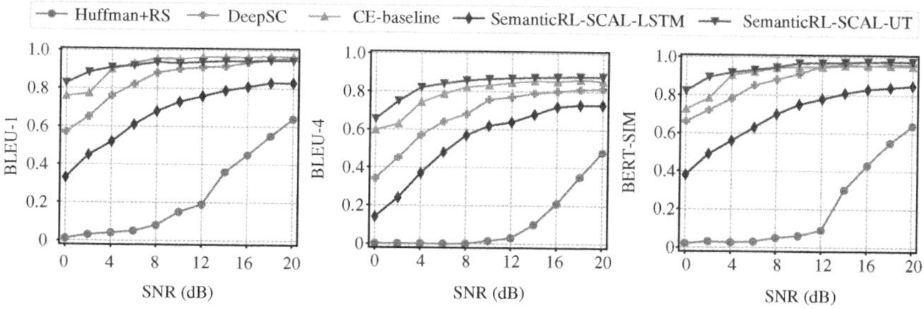

Figure 12.6 BLEU score and BERT-SIM versus SNR in Rayleigh fading channel for the point-to-point case.

Furthermore, our method exhibits significant superiority over the traditional Huffman-RS method, and stands out among other DNN-based SC approaches. Compared to the traditional Huffman-RS method, SC can better utilize background knowledge to enhance transmission quality, especially in low SNRs. Most impressively, our approach obtains a decoding accuracy that closely approximates 100% at around $SNR = 8\,\text{dB}$ in AWGN channels. This improved performance can be primarily attributed to the adaptive transformer layer in the UT, which has the ability to reconstruct partial semantic information in a poor channel environment by leveraging shared background knowledge, thereby mitigating the distortion introduced by channel noise. Meanwhile, by adopting semantic similarity scores as rewards, our approach fosters an environment conducive to the system's learning and understanding of semantic representations at a semantic level, which addresses the $Q1$.

Notably, even if our approach has a lower BLEU-1 compared to the CE-baseline in Figures 12.5 and 12.6, the existence of synonyms in the recovered sentences actually lead to a slightly higher sentence similarity in terms of BLEU-4. More importantly, although BERT-SIM is not directly adopted as a reward function, the semantic level supervision under BLEU score also leads to the superiority of SemanticBC-SCAL in the point-to-point case and manifests its effectiveness.

Additionally, considering a broader case of SC, we claim that by slightly modifying the decision process and learning details, it is convenient to introduce the proposed SemanticRL paradigm [Lu et al., 2021] on other media (e.g., pictures), so long as one can provide a meaningful semantic measurement. To better demonstrate this viewpoint, we provide an example where we show how to transmit images in a SemanticRL manner.

By formulating the decoding process as progressively adding or subtracting a small number for each pixel value, we can simply define the intermediate reward as the MSE gain between the previous image and the current one. This leads to a new parametric setting: the intermediate reward function is given in (12.19);

and $G^{(t)}$ is then defined in a more general way by setting the discount factor $\gamma = 0.99$ [Lu et al., 2021].

$$r_i^{(t)} = (I_i - \hat{I}_i^{(t)})^2 - (I_i - \hat{I}_i^{(t+1)})^2 \quad I_i \in I, \tag{12.19}$$

where I_i indicates any single pixel in an image I. This reward function can be interpreted as encouraging the decisions that lead to a lower MSE. The major difference as compared to the above mentioned sentence-transmission task is to use a globally shared policy for all pixels, and turns the single-agent RL learning into a multiagent RL learning task, similar to PixelRL [Furuta et al., 2019].

We illustrate how this image-targeted model learns and behaves in Figure 12.7 [Lu et al., 2021]. The model is trained on MNIST dataset[2] with the weight of encoder and the first layer of decoder initialized from a regularly pretrained one. All the images are quantized every 25.5 pixels for simplicity (so that after normalization, one pixel has 10 values $\{0, 0.1, 0.2, \dots, 0.9\}$ to take). We set the initially decoded image $I^{(0)}$ as one filled with pixel value 0.5 and let each action adjust the pixel value by increasing 0.1 or decreasing 0.1, or keeping it unchanged, so that all images could be recovered in five time steps. Note that here we attempt to provide a simple yet proof-of-the-concept example; one may further conduct more complex tasks such as directly optimizing the semantic segmentation or detection/reconstruction process with task-specific nondifferentiable reward functions like intersection over union (IOU), semantic classification accuracy, etc.

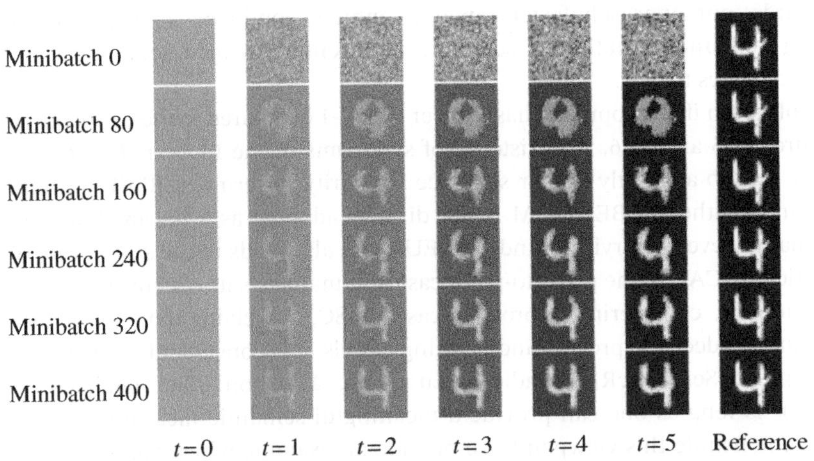

Figure 12.7 An example on image-transmission extension of SemanticRL [Lu et al., 2021], where the model is transformed into a semantic agent and gradually learns to recover input semantics by acting progressively. Source: Choi et al., (2024)/IEEE/CC BY 4.0.

2 http://yann.lecun.com/exdb/mnist.

12.4.2.2 Performance In Semantic BC

For semantic BC, in addition to the inherent channel noise as discussed in (12.3), we also consider the estimation variance related to SNRs to better adapt to the varying conditions. Consequently, during the training process, we employ lower SNR to acclimate to these demanding channel conditions, i.e., exemplified by scenarios, such as the AWGN and Rayleigh channel under $\mu_{SNR,1} = 6\,\text{dB}$, $\delta_{SNR,1} = 1\,\text{dB}$. Conversely, in favorable channel environment, we frequently employ higher SNR settings during training to guarantee robust performance across a substantial portion of the channel environment while maintaining a competitive level of resilience. Figure 12.8 visualizes the corresponding BLEU-1, BLEU-4, and BERT-SIM in semantic BC with three different receivers under both AWGN and Rayleigh channels.

From Figure 12.8, it can be observed that our SemanticBC-SCAL demonstrates adaptability to the varying SNR, particularly excelling in low SNR regime. Notably, when the test SNR ranges from 0 to 6 dB, the RX_n trained under low SNR (i.e., $\mu_{SNR,1} = 6\,\text{dB}$, $\delta_{SNR,1} = 1\,\text{dB}$) yields superior results than others for both AWGN and fading channels, while DeepSC in BC scenarios follows a similar trend but

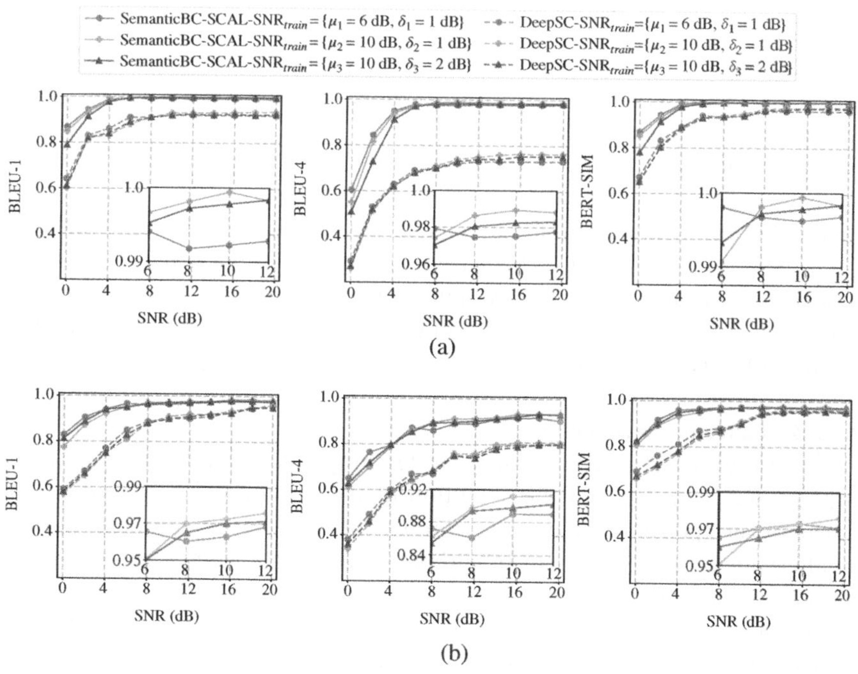

Figure 12.8 Comparison of semantic BC schemes in terms of BLEU-1 score, BLEU-4 score, and BERT-SIM under both AWGN (a) and Rayleigh fading (b) channel.

Table 12.3 Comparisons on the varying number of RXs of SemanticBC-SCAL under AWGN channel, wherein all BLEU scores and WAR denote the averages with SNR ranging from 0 ~ 20 dB.

Number of RXs	BLEU-1	BLEU-2	BLEU-3	BLEU-4	WAR
1	0.9528	0.9262	0.9004	0.8697	0.9272
3	0.9760	0.9583	0.9442	0.9209	0.9614
5	0.9624	0.9338	0.9120	0.8834	0.9466
7	0.9517	0.9288	0.9018	0.8811	0.9358

has an inferior performance. To elaborate, when the mean value of training SNR is configured at $\mu_{SNR,1} = 10\,\text{dB}$, for SemanticBC-SCAL in the AWGN channel, the training SNR with a narrower variance (i.e., $\delta_{SNR,1} = 1\,\text{dB}$) outperforms that with a wider variance (i.e., $\delta_{SNR,1} = 2\,\text{dB}$) in a low SNR regime, while in Rayleigh fading channels the opposite is true, thanks to its generalization capabilities. Meanwhile, in the higher SNR regime, the performance of the channel trained under low SNR exhibits insubstantial improvement and the channel with lower variance estimation performs better.

Additionally, we also validate the scalability of our semantic BC system in scenarios with a varying number of RXs, as depicted in Table 12.3. It is noteworthy that as the number of RX increases, there is a modest decrease in semantic decoding proficiency, but the overall BLEU scores and WAR consistently maintain a high level. Furthermore, when compared to the point-to-point system, our SemanticBC-SCAL demonstrates a notable advantage in utilizing semantics within the broadcast scenario. This advantage can be attributed to the localized adaptive learning and comprehensive integration of all decoders' outcomes. This indicates that our SemanticBC-SCAL can be effectively extended to semantic BC systems with varying numbers of RXs, thereby addressing the Q2.

12.4.2.3 Convergence Analyses

In order to testify whether our alternate learning mechanism guarantees convergence, we have depicted the evolution of the reward for the encoder and decoders over epochs in Figure 12.9. From Figure 12.9a, we can observe that our alternate learning mechanism ultimately converges with a steady learning and low variance (among different receivers) for the different number of RXs, which promises an effective learning approach for the large-scale KB. Furthermore, the convergence of these interleaved iterations between one encoder

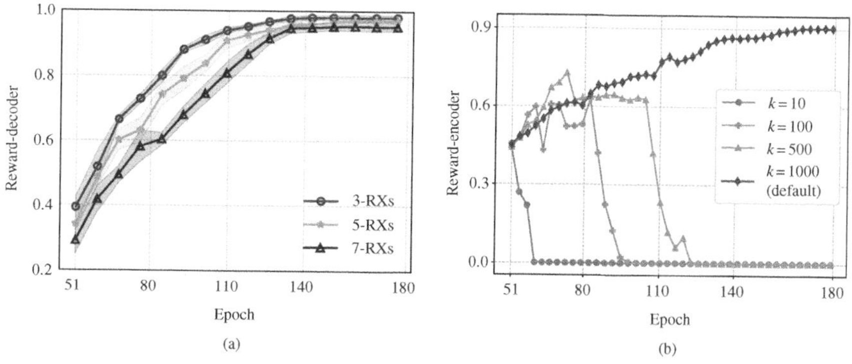

Figure 12.9 Convergence of SemanticBC-SCAL, wherein the reward curves are the average BLEU score for all broadcast receivers. (a) The training reward of the decoder sides with varying numbers of RXs. (b) The training reward of the encoder side with varying local iterations κ under three RXs.

and multiple decoders can be ensured by certain locally updated iterations κ. Consequently, we delve into exploring the influence of κ on the convergence of the encoder, as illustrated in Figure 12.9b. Based on the observations from Figure 12.9b, it becomes apparent that when κ is relatively small, it indicates a low number of local iterations for decoders and frequent updates for the encoder per epoch. In this sense, the encoder fails to learn effectively even after a few training epochs. As κ increases, the encoder can sustain longer training epochs but still struggles to converge, until $\kappa = 1000$. Only at this point, it achieves continuous and stable learning, ultimately reaching the convergence, which aligns with those presented in Corollary 1. These empirical findings clearly address the $Q2$ and $Q3$.

Additionally, we also give the runtime return of SemanticRL-SCAL-LSTM as shown in Figure 12.10 [Lu et al., 2021]. It is vividly shown that both the TX and RX are able to learn stably and tend to converge as the training proceeds. It can be obviously seen that using LSTM as the backbone yields worse results compared to UT, which to some extent reflects the advantage of UT in word generation task.

12.4.2.4 Ablation Study

The ablation study for comparing different numbers of parallel samples K in (12.12) and (12.16) is given in Figure 12.11. According to (12.12) and (12.16), as K increases, the mean reward from K rollouts of state-action value can be estimated more precisely. Consistently, as depicted in Figure 12.11, a larger K obtains a higher semantic similarity, which is consistent with the referred [Gao et al., 2019].

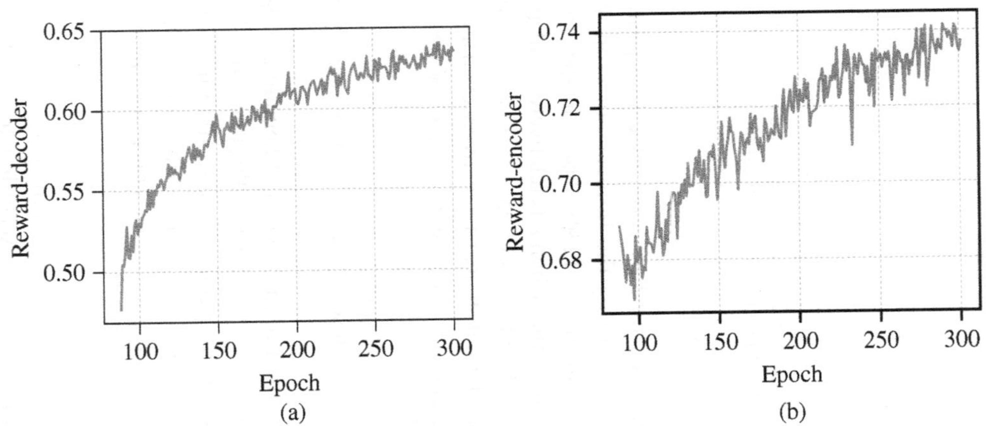

Figure 12.10 Convergence performance of SemanticRL. From (a) to (b), we provide the runtime reward (CIDEr-D score) of the decoder and encoder respectively [Lu et al., 2021].

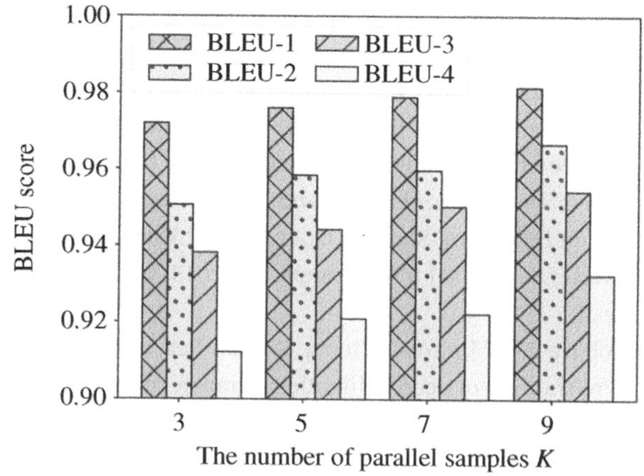

Figure 12.11 The impact of the number of parallel samples K in the SemanticBC-SCAL system.

12.5 Conclusions

In this chapter, in order to cope with compatibility, scalability, and convergence issues in semantic unicast and BCs, we have proposed a semantic BC framework optimized by an RL-based SCAL to unify these two scenarios, called SemanticBC-SCAL, which provides semantic level supervision and improves the robustness under varying numbers of RXs. In particular, we have regarded semantic similarity as a reward and formulated the optimization process as an RL problem. Furthermore, we have adopted self-critical learning to obtain a simple and efficient gradient estimation, especially suitable for complex SC systems with large-scale samples. To accommodate varying numbers of RXs and address nondifferentiable channels, we have adopted a cost-effective alternate training mechanism to asynchronously learn the encoder and multiple decoders with different learning iterations instead of directly back-propagating gradients through the channel. The simulation results have demonstrated that our proposed approach can achieve high semantic accuracy in both unicast and broadcast scenarios. Especially suitable for different numbers of RXs, and has exhibited strong scalability and robustness for different channel environments.

Bibliography

Fayçal Ait Aoudia and Jakob Hoydis. End-to-end learning of communications systems without a channel model. In *2018 52nd Asilomar Conference on Signals, Systems, and Computers (ACSSC)*, pages 298–303, Pacific Grove, CA, USA, Oct. 2018.

Fayçal Ait Aoudia and Jakob Hoydis. Model-free training of end-to-end communication systems. *IEEE Journal on Selected Areas in Communications*, 37(11):2503–2516, Aug. 2019.

Eirina Bourtsoulatze, David Burth Kurka, and Deniz Gündüz. Deep joint source-channel coding for wireless image transmission. *IEEE Transactions on Cognitive Communications and Networking*, 5(3):567–579, Sep. 2019.

Mostafa Dehghani, Stephan Gouws, Oriol Vinyals, Jakob Uszkoreit, and Łukasz Kaiser. Universal transformers, Jul. 2018. URL https://arxiv.org/pdf/1807.03819.pdf.

Jacob Devlin, Ming-Wei Chang, Kenton Lee, and Kristina Toutanova. BERT: Pre-training of deep bidirectional transformers for language understanding. In *Proceedings of the 2019 Conference of the North American Chapter of the Association for Computational Linguistics: Human Language Technologies (NAACL HLT)*, pages 4171–4186, Minneapolis, Minnesota, USA, Jun. 2019.

Mingze Ding, Jiahui Li, Mengyao Ma, and Xiaopeng Fan. SNR-adaptive deep joint source-channel coding for wireless image transmission. In *Proceedings of IEEE International Conference on Acoustics, Speech and Signal Processing (ICASSP)*, pages 1555–1559, Toronto, ON, Canada, Jun. 2021.

Oguzhan Dogru, Kirubakaran Velswamy, and Biao Huang. Actor–critic reinforcement learning and application in developing computer-vision-based interface tracking. *Engineering*, 7(9):1248–1261, Sep. 2021.

Kumar Avinava Dubey, Sashank J. Reddi, Sinead A. Williamson, Barnabas Poczos, Alexander J. Smola, and Eric P. Xing. Variance reduction in stochastic gradient Langevin dynamics. In *Proceedings of Advances in Neural Information Processing Systems (NIPS)*, pages 1162–1170, Barcelona, Spain, Dec. 2016.

Nariman Farsad, Milind Rao, and Andrea Goldsmith. Deep learning for joint source-channel coding of text. In *Proceedings of IEEE International Conference on Acoustics, Speech and Signal Processing (ICASSP)*, pages 2326–2330, Calgary, Canada, Apr. 2018.

Ryosuke Furuta, Naoto Inoue, and Toshihiko Yamasaki. PixelRL: Fully convolutional network with reinforcement learning for image processing. *IEEE Transactions on Multimedia*, 22(7):1704–1719, Jul. 2019.

Junlong Gao, Shiqi Wang, Shanshe Wang, Siwei Ma, and Wen Gao. Self-critical n-step training for image captioning. In *Proceedings of the IEEE/CVF Conference on Computer Vision and Pattern Recognition (CVPR)*, pages 6300–6308, Long Beach, CA, US, Jun. 2019.

Tianxiao Han, Qianqian Yang, Zhiguo Shi, Shibo He, and Zhaoyang Zhang. Semantic-preserved communication system for highly efficient speech transmission. *IEEE Journal on Selected Areas in Communications*, 41(1):245–259, Jan. 2023.

Han Hu, Xingwu Zhu, Fuhui Zhou, Wei Wu, Rose Qingyang Hu, and Hongbo Zhu. One-to-many semantic communication systems: Design, implementation, performance evaluation. *IEEE Communications Letters*, 26(12):2959–2963, Sep. 2022.

Mikolaj Jankowski, Deniz Gündüz, and Krystian Mikolajczyk. Deep joint source-channel coding for wireless image retrieval. In *Proceedings of IEEE International Conference on Acoustics, Speech and Signal Processing (ICASSP)*, pages 5070–5074, Barcelona, Spain, May 2020a.

Mikolaj Jankowski, Deniz Gündüz, and Krystian Mikolajczyk. Wireless image retrieval at the edge. *IEEE Journal on Selected Areas in Communications*, 39(1):89–100, Nov. 2020b.

Peiwen Jiang, Chao-Kai Wen, Shi Jin, and Geoffrey Ye Li. Wireless semantic communications for video conferencing. *IEEE Journal on Selected Areas in Communications*, 41(1):230–244, Nov. 2022.

Philipp Koehn. Europarl: A parallel corpus for statistical machine translation. In *Proceedings of AAMT 10th Machine Translation Summit X: Papers*, pages 79–86, Phuket, Thailand, Sep. 2005.

Marios Kountouris and Nikolaos Pappas. Semantics-empowered communication for networked intelligent systems. *IEEE Communications Magazine*, 59(6):96–102, Jun. 2021.

David Burth Kurka and Deniz Gündüz. Deep joint source-channel coding of images with feedback. In *Proceedings of IEEE International Conference on Acoustics, Speech and Signal Processing (ICASSP)*, pages 5235–5239, Barcelona, Spain, May 2020.

Ting Liu and Xuechen Chen. Attention-based neural joint source-channel coding of text for point to point and broadcast channel. *Artificial Intelligence Review*, 55(3):2379–2407, Sep. 2022.

Kun Lu, Rongpeng Li, Xianfu Chen, Zhifeng Zhao, and Honggang Zhang. Reinforcement learning-powered semantic communication via semantic similarity, Aug. 2021. URL https://arxiv.org/abs/2108.12121.

Zhilin Lu, Rongpeng Li, Ming Lei, Chan Wang, Zhifeng Zhao, and Honggang Zhang. Self-critical alternate learning based semantic broadcast communication, Nov. 2023a. URL https://arxiv.org/abs/2312.01423.

Zhilin Lu, Rongpeng Li, Kun Lu, Xianfu Chen, Ekram Hossain, Zhifeng Zhao, and Honggang Zhang. Semantics-empowered communication: A tutorial-cum-survey. *IEEE Communications Surveys and Tutorials*, Nov. 2023b. URL https://arxiv.org/abs/2212.08487.

Ruotian Luo. A better variant of self-critical sequence training, Mar. 2020. URL https://arxiv.org/pdf/2003.09971.pdf.

Shuai Ma, Weining Qiao, Youlong Wu, Hang Li, Guangming Shi, Dahua Gao, Yuanming Shi, Shiyin Li, and Naofal Al-Dhahir. Features disentangled semantic broadcast communication networks, Mar. 2023. URL https://arxiv.org/pdf/2303.01892.pdf.

Timothy O'shea and Jakob Hoydis. An introduction to deep learning for the physical layer. *IEEE Transactions on Cognitive Communications and Networking*, 3(4):563–575, Dec. 2017.

Kishore Papineni, Salim Roukos, Todd Ward, and Wei-Jing Zhu. BLEU: A method for automatic evaluation of machine translation. In *Proceedings of Annual Meeting of the Association for Computational Linguistics (ACL)*, pages 311–318, Philadelphia, USA, Jul. 2002.

Irving S. Reed and Gustave Solomon. Polynomial codes over certain finite fields. *Journal of the Society for Industrial and Applied Mathematics*, 8(2):300–304, 1960.

Steven J. Rennie, Etienne Marcheret, Youssef Mroueh, Jerret Ross, and Vaibhava Goel. Self-critical sequence training for image captioning. In *Proceedings of IEEE Conference on Computer Vision and Pattern Recognition (CVPR)*, pages 7008–7024, Honolulu, HI, US, Jul. 2017.

Mohamed Sana and Emilio Calvanese Strinati. Learning semantics: An opportunity for effective 6G communications. In *Proceedings of IEEE Annual Consumer Communications & Networking Conference (CCNC)*, pages 631–636, Las Vegas, NV, USA, Jan. 2022.

Hyowoon Seo, Jihong Park, Mehdi Bennis, and Mérouane Debbah. Semantics-native communication with contextual reasoning. *IEEE Transactions on Cognitive Communications and Networking*, 9(3):604–617, Jun. 2023.

Vinay Kumar Shrivastava, Sangkyu Baek, and Youngkyo Baek. 5G evolution for multicast and broadcast services in 3GPP release 17. *IEEE Communications Standards Magazine*, 6(3):70–76, Sep. 2022.

Haonan Tong, Zhaohui Yang, Sihua Wang, Ye Hu, Omid Semiari, Walid Saad, and Changchuan Yin. Federated learning for audio semantic communication. *Frontiers in Communications and Networks*, 2:734402, Sep. 2021.

Tze-Yang Tung and Deniz Gündüz. Deep-learning-aided wireless video transmission. In *Proceedings of IEEE International Workshop on Signal Processing Advances in Wireless Communication (SPAWC)*, pages 1–5, Oulu, Finland, Jul. 2022.

Ramakrishna Vedantam, C. Lawrence Zitnick, and Devi Parikh. CIDEr: Consensus-based image description evaluation. In *Proceedings of IEEE Conference on Computer Vision and Pattern Recognition (CVPR)*, pages 4566–4575, Boston, MA, USA, Jun. 2015.

Yining Wang, Mingzhe Chen, Tao Luo, Walid Saad, Dusit Niyato, H. Vincent Poor, and Shuguang Cui. Performance optimization for semantic communications: An attention-based reinforcement learning approach. *IEEE Journal on Selected Areas in Communications*, 40(9):2598–2613, Jul. 2022.

Cheng-Xiang Wang, Xiaohu You, Xiqi Gao, Xiuming Zhu, Zixin Li, Chuan Zhang, Haiming Wang, Yongming Huang, Yunfei Chen, Harald Haas, and John S.

Thompson. On the road to 6G: Visions, requirements, key technologies and testbeds. *IEEE Communications Surveys and Tutorials*, 25(2):905–974, Feb. 2023.

Warren Weaver. Recent contributions to the mathematical theory of communication. *ETC: A Review of General Semantics*, Sep. 1949. Vol. 10, No. 4.

Zhenzi Weng and Zhijin Qin. Semantic communication systems for speech transmission. *IEEE Journal on Selected Areas in Communications*, 39(8):2434–2444, Jun. 2021.

Huiqiang Xie and Zhijin Qin. A lite distributed semantic communication system for Internet of things. *IEEE Journal on Selected Areas in Communications*, 39(1):142–153, Nov. 2020.

Huiqiang Xie, Zhijin Qin, Geoffrey Ye Li, and Biing-Hwang Juang. Deep learning enabled semantic communication systems. *IEEE Transactions on Signal Processing*, 69:2663–2675, Apr. 2021.

Huiqiang Xie, Zhijin Qin, Xiaoming Tao, and Khaled B. Letaief. Task-oriented multi-user semantic communications. *IEEE Journal on Selected Areas in Communications*, 40(9):2584–2597, Jul. 2022.

Huiqiang Xie, Zhijin Qin, and Geoffrey Ye Li. Semantic communication with memory. *IEEE Journal on Selected Areas in Communications*, 41(8):2658–2669, 2023. doi: 10.1109/JSAC.2023.3288236.

Hongyuan Ye, Feifei Gao, Jing Qian, Hao Wang, and Geoffrey Ye Li. Deep learning-based denoise network for CSI feedback in FDD massive MIMO systems. *IEEE Communications Letters*, 24(8):1742–1746, Apr. 2020.

Guangyi Zhang, Qiyu Hu, Zhijin Qin, Yunlong Cai, Guanding Yu, Xiaoming Tao, and Geoffrey Ye Li. A unified multi-task semantic communication system for multimodal data, Sep. 2022a. URL https://arxiv.org/abs/2209.07689.

Zhenguo Zhang, Qianqian Yang, Shibo He, Mingyang Sun, and Jimin Chen. Wireless transmission of images with the assistance of multi-level semantic information. In *Proceedings of International Symposium on Wireless Communication Systems (ISWCS)*, Hangzhou, China, Feb. 2022b.

Qingyang Zhou, Rongpeng Li, Zhifeng Zhao, Chenghui Peng, and Honggang Zhang. Semantic communication with adaptive universal transformer. *IEEE Wireless Communications Letters*, 11(3):453–457, Dec. 2021.

13

Imitation Learning-based Implicit Semantic-aware Communication Networks

Yiwei Liao[1], Zijian Sun[1], Yong Xiao[1], Guangming Shi[2], Yingyu Li[3], H. Vincent Poor[4], Walid Saad[5], Merouane Debbah[6], and Mehdi Bennis[7]

[1]*School of Electronic Information and Communications, Huazhong University of Science and Technology, Wuhan, Hubei, China*
[2]*School of Electronic Engineering, Xi'an University of Electronic Science and Technology, Xi'an, Shaanxi, China*
[3]*School of Mechanical Engineering and Electronic Information, China University of Geosciences, Wuhan, Hubei, China*
[4]*Department of Electrical and Computer Engineering, Princeton University, Princeton, NJ, USA*
[5]*Department of Electrical and Computer Engineering, Virginia Tech, Arlington, VA, USA*
[6]*6G Center, Khalifa University of Science and Technology, Abu Dhabi, United Arab Emirates*
[7]*Centre for Wireless Communications, University of Oulu, Oulu, Finland*

13.1 Introduction

Semantic communication is a novel communication paradigm in which the meaning of messages can be identified and utilized during communication. Compared to the traditional systems, semantic communication allows communication participants including both information source and destination to exploit commonly-shared human knowledge and experience as well as syntax, semantics, and inference rules to assist the transportation and accurate delivery of the intended meaning. Recent observations suggest that semantic communication could significantly improve the efficiency and reliability of communication, enhance users' quality-of-experience (QoE), and achieve smoother cross-protocol/domain communication [Shi et al., 2021; Strinati and Barbarossa, 2021; Liang et al., 2022; Lu et al., 2022; Xiao et al., 2022, 2023].

Previous works on semantic communication focused primarily on transporting the explicit information, e.g., signal and/or its explicit features and human-labeled objects that can be directly observed from the source signals. The intended information communicated between the source and destination users can be much

Foundations of Semantic Communication Networks, First Edition.
Edited by Walid Saad, Christina Chaccour, Christo Kurisummoottil Thomas, and Merouane Debbah.

more than just explicit information. For example, an image showing "a child is riding a bicycle" consists of explicit objects, "a child" and "a bicycle." The relationship ("ride") between these two objects, however, cannot be directly recognized from the image. To understand the exact meaning of the message, the receiver (the father), must be able to infer the implicit information, i.e., hidden information that cannot be observed from the signal itself.

From the above examples, we can observe that in addition to the explicit information, the content of communication often consists of rich implicit information that can be critical in understanding and efficiently delivering the semantic meaning of messages at the intended destinations. Compared to the traditional explicit semantic information-based communication solutions, implicit semantic-aware communication (iSAC) has multiple unique benefits. First, the implicit relationships among concepts and ideas in the communication messages can be utilized to further reduce the redundancy of the messages transmitted. It is also helpful for recovering some missing or corrupted information at the destination user. Second, the implicit information may reflect the truly important meaning of information, i.e., real intention, of (human) user's information and, thus, can be used for further enhancing communication QoE and avoiding misinterpretation/misunderstanding. Finally, implicit semantics including the context and the reasoning mechanism can be used for inferring the meaning of unknown knowledge and discovering hidden relationships with previously learned concepts. This can be very useful for developing novel self-learning and adaptive solutions for supporting cognition and alerting network abnormalities as well as autonomous decision-making under unknown scenarios.

In this chapter, we propose a novel implicit information communication framework, called iSAC, for representing, modeling, communicating, and interpreting the semantics involving both explicit and implicit meaning of messages. We first propose a novel graph-inspired representation of semantics, which includes three key components: explicit semantics (objects and relations), implicit semantics (inferred information behind entities), and reasoning mechanism (user's reasoning rule and/or inference policy). Different from existing knowledge graph-based solutions that rely on a pre-stored dictionary of terms, our proposed solution allows biased personal understanding and preference of the source user to be sequentially learned, encoded, and interpreted by the destination user during the semantic communication process. We introduce a novel projection-based semantic encoding solution to convert explicit semantics into a low-dimensional semantic constellation space in which the Euclidean distance between semantic terms (entities or relations) is proportional to their meaning dissimilarity. We also introduce a novel generative imitation-based reasoning mechanism learning (G-RML) solution for supporting automatic encoding and decoding/interpretation of implicit semantics at the destination user. One of the major differences between the existing communication solution and G-RML is that, in the latter solution, the source user does not focus on sending as much of the useful message as possible. Instead, the source user will guide the destination user in learning a reasoning mechanism that can automatically infer implicit semantics including hidden concepts, objects, and their relationships that are most relevant to the meaning of the source user.

13.1.1 Framework of Implicit Semantic Communications

13.1.1.1 Representation of Semantics

A pivotal challenge in the domain of iSAC lies in devising a universal yet detailed method for encapsulating the semantic essence of any message, capturing both its overt and covert semantics. This study adopts a broad framework, positing that the nuanced meaning intended by the sender can be articulated through a network of concepts and terms, interlinked by diverse relational types to mirror their individual message comprehension. This approach is underpinned by recent insights from cognitive neuroscience, which reveal that the human brain transcends mere transmission of isolated concepts, such as symbols and labels. Instead, it constructs intricate semantic networks by weaving together various terms and concepts through multifaceted relational dynamics.

Further inspired by another novel finding in cognitive neuroscience–that individuals intuitively navigate through concealed concepts and ideas by tracing the most directly associated knowledge terms via a personal reasoning mechanism–this paper models the transition from explicit to implicit semantics as a sequential decision-making journey. This journey is steered by a distinct reasoning mechanism, emphasizing the cognitive process's dynamic and interactive nature. Through this lens, the chapter explores the intricate dance between visible and hidden meanings, seeking to emulate the brain's adeptness at uncovering depth in communication.

13.1.1.2 Explicit Semantics

Correspond to a set of real-world entities and relations such as "child," "bicycle," "Tweety," "social media website," "canary bird" and "belong to," "is/are" etc., that can be directly detected in the source signal.

We can further define the explicit semantics as $\boldsymbol{v} = \langle e^0, r^0 \rangle$ where e^0 and r^0 are sets of entities and relations that can be directly recognized in the source signal.

13.1.1.3 Implicit Semantics $\boldsymbol{u}_v$

Correspond to the hidden information, including the hidden relations and entities that cannot be directly identified from the source signal, but are important for the interpretation of semantic meaning reflected by the explicit semantics, e.g., a child "rides" a bicycle, Tweety "is" a "canary bird," and Tweety "is" a "cartoon character" "in" a "TV show." In this paper, we consider a sequential inference process in which the relevant hidden entities and relations are sequential decided one iteration at a time, based on the observed explicit semantics. We can further define the implicit semantics as $\boldsymbol{u}_v = \langle e^i, r^i \rangle_{i=\{1,2,\ldots\}}$ where e^i and r^i are the set of hidden entities and relations inferred in the ith iteration for

13.1.1.4 Reasoning Mechanism Π

Corresponds to a user-related inference function or rule that characterizes the correlations between explicit and implicit semantics.

For example, in the previous example, the entity "Tweety" may link to several other hidden entities such as "smart phone App," "canary bird" and "cartoon character" with corresponding hidden relations. The receiver (the father) therefore needs to identify possible entities and the corresponding relations that are linked to entity "Tweety" based on the reasoning mechanism learned or updated from the communication history. For example, if, in his/her recent communications with the father, the child mentioned about reading some books about animals, the father would be able to infer the most appropriate meaning of the child's question by finding a reasoning path connecting a sequence of hidden entities and relations, i.e., "Tweety is a canary bird that is highly likely to have appeared in a book recently read by the child." It can be observed that the reasoning mechanism plays an essential role in inferring the appropriate meaning of the message. Generally speaking, the semantics of a source signal can involve at least one key explicit entity and one or multiple implicit reasoning paths associated with the explicit entity. The implicit reasoning paths generated from the explicit semantics may not be unique and can be closely related to the user's background, environment, and context of communication. We consider a sequential inference process that is dominated by a sequential decision-making policy called the reasoning policy. Let Π be the reasoning policy for inferring the most relevant hidden relations and entities based on e^j and r^j, e.g., we have $\langle e^{j+1}, r^{j+1} \rangle = \Pi \left\langle e^j, r^j \right\rangle$. We also use $\hat{\Pi}_{\pi_\theta}$ to denote the reasoning mechanism for generating possible semantic paths via sequential inference process based on Π.

13.1.2 Knowledge Base

One of the key ideas of iSAC is to incorporate a priori knowledge and personal background information including facts, concepts, and experiences (reflecting relations) to assist in the interpretation of the intended meaning. In other words, the knowledge shared among the communication participants (users) is essential in iSAC system. We refer to the collection of all the knowledge entities and relations that are accessible to each user as the *knowledge base*. The knowledge bases of different users do not have to be the same. In other words, the knowledge base of each user may consist of two parts of elements:

1) **Common knowledge:** Corresponds to knowledge entities and relations that are shared among all the communication participating users (e.g., source and destination users).
2) **Private knowledge:** Involves the entities and relations associated with personal understanding and viewpoints about some real-world concepts and relationships. Generally speaking, the private knowledge entities and relations of different users do not have to be the same.

13.1.3 Reasoning Mechanism Modeling and Learning

Deciphering implicit meanings presents a formidable challenge, renowned for its complexity. In response, we introduce a cutting-edge methodology named G-RML, designed to equip systems with the capability to autonomously deduce hidden semantic layers from explicitly presented semantic data.

In G-RML, the journey to uncover implicit semantics is framed as a series of sequential decisions, guided by a carefully formulated reasoning policy. This innovative approach enables the receiver to methodically unravel and extend potential hidden connections and entities originating from the explicit semantic groundwork laid by the sender. A notable hurdle for the receiver is the inherent difficulty in identifying a specific reward function that accurately gauges the efficacy of its inferential outcomes. To surmount this, G-RML employs an imitation learning strategy, wherein the sender plays a pivotal role during the training phase. Here, the sender mentors the receiver in adopting a policy adept at constructing reasoning pathways that align closely with the patterns observed in expertly navigated paths.

This strategy not only streamlines the inferential process but also fosters a learning environment where the receiver can progressively refine its ability to bridge explicit and implicit semantic elements, thus enhancing the fidelity and depth of communication.

13.2 System Model and Problem Formulation

13.2.1 System Model

We introduce the iSAC architecture, as illustrated in Figure 13.1a. It consists of two phrases of implementation: model training phrase and communication phrase. In the training phrase, the source and destination users will jointly train two models, an explicit semantic encoding model for efficient transmission

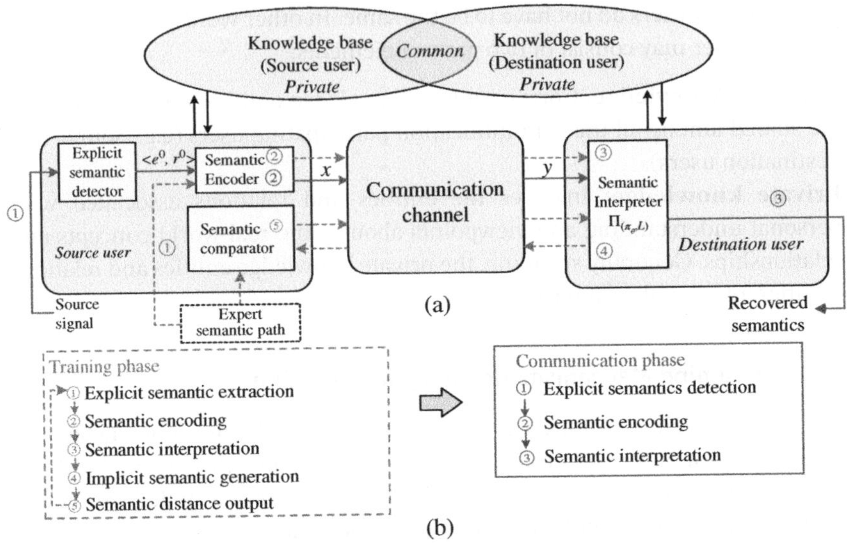

Figure 13.1 The proposed iSAC architecture: (a) A systematic review of the proposed architecture; (b) The implementations of training and communication phases.

of explicit semantics identified by the source user, and an implicit semantic inference model which allows the destination user to infer implicit semantic meaning of the source user based on the received explicit semantics.

To train the first model, we design a projection-based semantic encoder to convert the high-dimensional information of the explicit semantics into a compact low-dimensional representation for physical channel transmission. For the training process of the second model, the destination user will construct a semantic interpreter that tries to generate the possible implicit semantics to be sent to the source user for evaluating. The source user will train a semantic comparator to output the semantic distance, a metric for measuring the meaning dissimilarity between the generated implicit semantics sent from the destination user and its local set of expert semantic paths obtained from the communication history. The source user will send the semantic distance to the semantic interpreter at the destination user for the model adjustment. In the communication phrase, the source user will directly send the key explicit semantics detected from the source signal. The destination user can directly apply the trained semantic interpreter to infer the implicit semantics without consulting the source user. Our proposed architecture can be directly implemented into the existing communication systems. The only requirement is that the semantic encoding and interpretation models need to be pre-trained offline and installed into the existing communication system in which the traditional encoding solution can be replaced with the pre-trained semantic encoder and also the digital symbol-based decoding solution can be replaced with our proposed pre-trained semantic interpreter. The detailed procedures of both phrases and their implementations are illustrated in Figure 13.1b.

1) **Source User Side:**
   Explicit Semantics Detector: The encoder should be able to first identify a set of explicit semantic terms, e.g., at least one visible entity and/or relation from the observed source signal. This can be achieved by using some well-trained models such as YOLO[1] and wav2letter to identify known labels of objects from various types (e.g., image, voice, and text) of source signals. As mentioned earlier, these labels themselves cannot fully characterize the complete meaning represented by the detected objects. In other words, the meaning of semantic terms identified from the source signal can be linked to a set of hidden features and attributes including the properties as well as the relationships to other terms.
   (Explicit) Semantic Encoder: The intricate web of semantic entities, relationships, and their associated attributes often encapsulates a wealth of information that poses challenges for efficient transmission over physical channels. To address this, the source user is tasked with distilling the complex, high-dimensional landscape of explicit semantics into a more streamlined, transmission-friendly format. To facilitate this conversion, we have crafted a projection-based encoding function designed to transform explicit semantics into a condensed representation within a specially conceived low-dimensional space, termed the semantic constellation space, optimized for physical channel transmission efficiency.
   Contrary to the dual coding processes-source encoding and channel encoding-employed in conventional communication systems, which separately aim to maximize information throughput and bolster data delivery resilience against channel impairments, our approach introduces a unified projection function. This singular function is ingeniously designed to serve both purposes within the semantic encoder framework. By integrating these objectives, we not only streamline the encoding process but also enhance the overall efficacy and reliability of semantic information transmission, marking a significant leap forward in the evolution of communication systems.
   Semantic Comparator: A fundamental distinction between semantic communication and its traditional counterpart lies in the nature of message delivery outcomes. Unlike the binary success or failure metrics prevalent in conventional communication, semantic communication introduces the concept of "semantic distance." This nuanced, continuous-valued metric gauges the discrepancy between the implicit semantic meaning deciphered by the

1 YOLO stands for "You Only Look Once." It is a real-time object detection system using a single neural network to predict bounding boxes and class probabilities in one step.

receiver and the original intent of the sender, as manifested in expertly curated semantic pathways.

In light of this, our discourse advances the development of a statistically grounded distance metric designed to quantitatively assess the semantic gap. This metric meticulously compares two collections of semantic reasoning pathways: those generated by the receiver and those derived from expertly navigated semantic trajectories. By implementing this approach, we enable a more refined and precise evaluation of communication effectiveness, transcending the simplistic binary indicators to embrace the complexities inherent in semantic interpretation.

2) **Physical Communication Channel:**
We consider a discrete memoryless channel with a limited capacity. In this chapter, we assume the source user cannot know the channel fading coefficient. The destination user therefore needs to learn a semantic interpretation network based on the noisy version of received explicit semantics.

3) **Destination User Side:**
Semantic Interpreter: The destination user will be able to recover the most likely sequences of hidden relations and entities by extending various reasoning paths from its received explicit semantics.
The semantic interpreter should be able to learn a reasoning mechanism that can sequentially decide the possible relations to extend the reasoning paths to best interpret the true meaning of the source user.

13.3 iSAC Architecture

13.3.1 Source User Side

13.3.1.1 Semantic Encoding

The source user first converts the identified explicit semantics into a suitable form for channel transmission. The representation of semantics proposed in Section Semantic representation is efficient for characterizing the complex relations among different semantic entities. It is however inefficient for physical channel transmission due to its redundancy in representing rich attributes of entities and relations. Also, any error in decoding semantic entities and relations will result in failure of recovering the entire semantic message. To address the above issues, we would like to design a semantic encoder with the following ideal properties: (1) *Efficient transmission*: the encoded signal should have a much smaller dimensional size, compared to the original graphical representations of semantics, and (2) *Robustness against channel corruption*: we would like to design a projection-based encoding function to separate different semantic terms in the constellation space based on the meaning difference and usage preference of the source user. In this way, even if some semantic terms, e.g., entities or relations, have been corrupted during the channel transmission, the destination user can recover these corrupted terms using the designed projection function. More specifically, we assume the source user can observe a set of preferred entity-and-relation combinations, called the *semantic preferred set*, e.g., consisting of meaningful combinations of semantic terms that are most frequently observed from the source user's communication history. We can then design a projection function to separate different combinations of semantic terms in the constellation space according to the observed usage preference, i.e., terms that are more frequently observed to be used together should be more closely located in the constellation space, compared to the rarely observed semantic term combinations.

To achieve property (1), we propose a projection-based semantic encoding function to convert the high-dimensional representations of explicit semantic terms into the low-dimensional semantic constellation representations. More specifically, we adopt an energy-based projection function, TransE [Bordes et al., 2013], to convert high-dimensional graphical representations of semantic entities and relations into low-dimensional constellation subspaces, which are referred to as entity and relation constellation spaces, respectively.

To achieve property (2), we first define the energy, a value measuring the possibility, of a combination of an initial entity, a connected relation, and a possible tail entity in the semantic constellation space, then we can then train the semantic encoding (projection) function to separate the semantic preferred (most frequently observed) and unpreferred (rarely observed) combinations of terms by minimizing the following loss function.

To illustrate the performance of our proposed semantic encoding solution, in Figure 13.2, we consider the encoding process of a semantic message involving entities and relations associated with eight classes (city, university, drugs, musician, movie, sports, musical instrument, and football team) as well as their corresponding relations randomly sampled from the NELL-995 knowledge base dataset where each entity has 200 attributes. The graphical representation of the semantic meaning involved in the given message is illustrated in Figure 13.2. We present the projected representation of entities in the semantic constellation space in Figure 13.3. We can observe that by adopting our proposed encoding (projection) function, different symbols in the semantic constellation space can be separated and the Euclidean distance between symbols depends on their meaning difference (dissimilarity) in the associated dimensions.

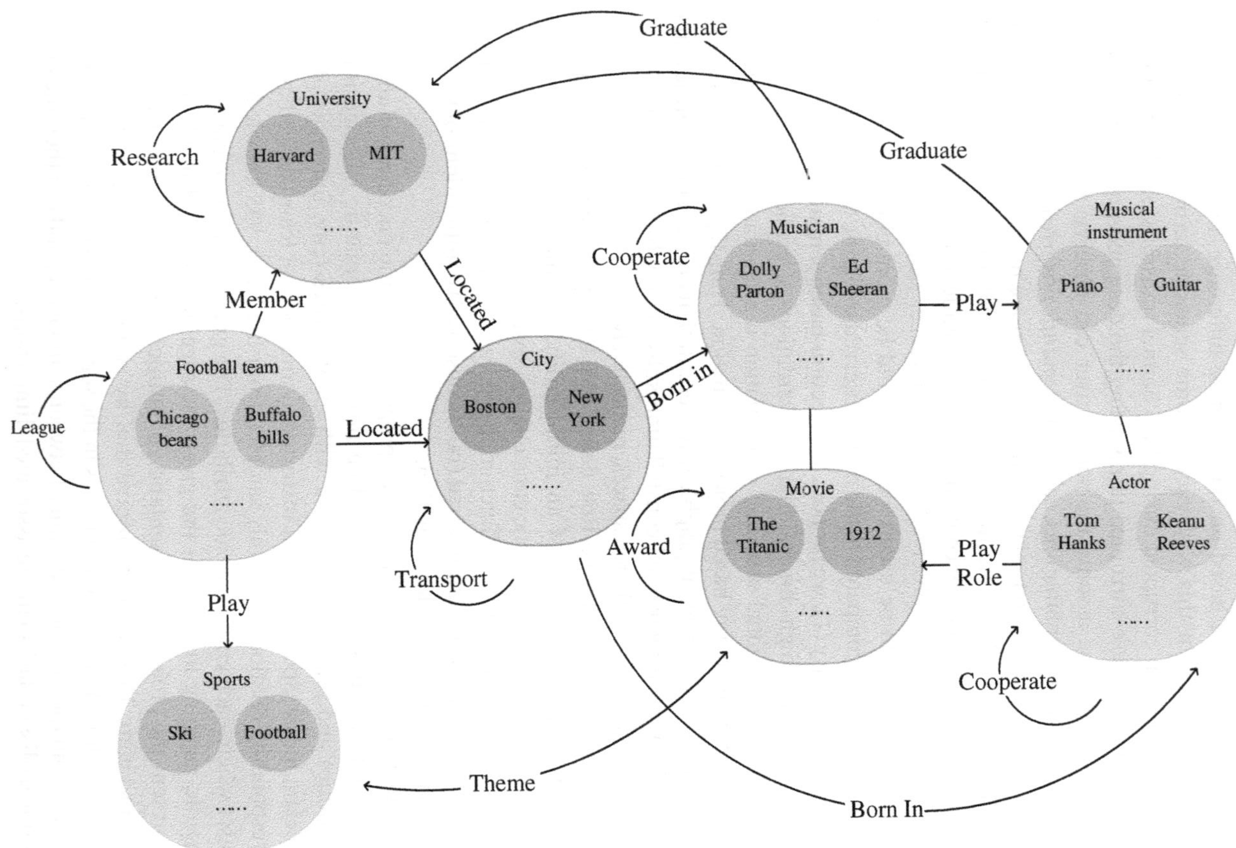

Figure 13.2 A graphical representation of the semantic meaning of 64 concepts (entities) associated with eight classes (city, sports, university, musician, movie, music instrument, actor, and football team) connected by multiple relations.

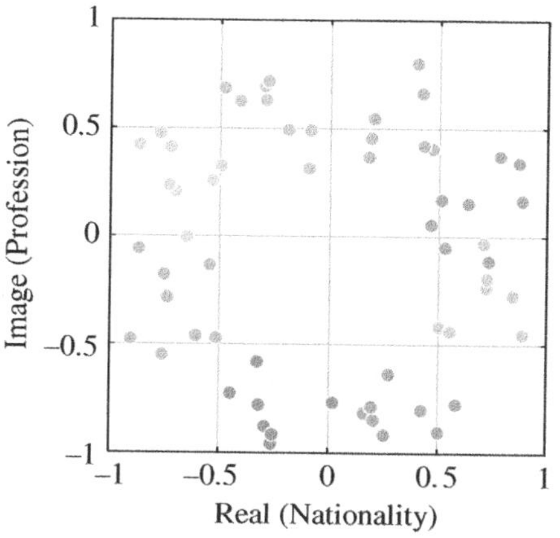

Figure 13.3 Semantic constellation of 64 symbols in two selected dimensions (Nationality and Profession).

13.3.1.2 Semantic Distance

The main objective of the destination user is to infer the implicit semantics based on the recovered explicit semantics. Different from the explicit semantics that are pre-identified and determined by the source user before encoding, the implicit semantics are generally difficult to be characterized by a deterministic path due to their randomness and polysemy. In other words, we cannot use the projection-based energy function developed for the semantic encoder to evaluate the semantic distance between implicit semantics.

Let us now develop a statistic-based distance measure to evaluate the meaning difference between the true implicit semantics and the inferred semantics interpreted by the destination user. As mentioned earlier, the implicit semantics expressed by each individual user can be considered as a set of possible semantic paths generated by a hidden expert reasoning mechanism based on the given explicit semantics. The reasoning mechanism Π is defined as a mapping function that maps a set of explicit semantic entities and relations to a set of semantic paths consisting of a sequence of possible relations. Note that neither source nor destination user can know or directly observe Π. The source user however can observe a set of expert semantic paths, e.g., sampled from the previous communication history, that can be assumed to be generated by Π. Different expert semantic paths can be initiated with different combinations of semantic terms. These initial semantic terms can be considered as the explicit semantics and the rest of the terms can be regarded as the implicit semantics generated by Π. The main objective of the destination user is then to learn an approximated reasoning mechanism $\hat{\Pi}$ that can match the implicit semantic generation process observed from the expert semantic paths.

13.3.2 Destination User Side

13.3.2.1 Semantic Interpreter

As mentioned earlier, the destination user needs to learn a reasoning mechanism that can infer the implicit semantics based on the explicit semantics received from the source user. Let us first define the implicit semantic reasoning process from the received explicit semantics of the destination user as a Markov decision process (MDP) problem in which a set of semantic reasoning paths are sequentially generated based on a reasoning mechanism, a policy, that decides a set of relations to extend the currently reasoned paths originated from explicit semantics.

- **State:** Each reasoning path is sequentially generated by choosing one relation and entity pair at a time. Suppose the maximum length of the inferred reasoning paths is given by L. The state of the reasoning process includes the current set of reasoning paths $\boldsymbol{\eta}^t$ as well as its current distance (length) to the initial (visible) entities t for $1 \leq t \leq L$, i.e.
- **Action:** Given the current state, the action of the user is to choose the next possible relations to extend the current reasoning paths.
- **Reward:** The reward function captures the main objective which is to minimize the expected semantic distance between the expert paths and the reasoning paths generated by the semantic interpreter at the destination user.
- **Policy:** We define the policy $\hat{\Pi}_{\pi_\theta}$ as a neural network, called policy network, parameterized by θ, specifying the reasoning mechanism. Our policy network maps the current state into the possible relations to extend the paths.

Note that the policy network only specifies the probability distribution of relations extended from a set of current paths. The semantic interpreter however needs to utilize the learned policy to recover a set of possible full reasoning paths, each consists of a sequence of entities and relations. We, therefore, define a reasoning mechanism to represent a probability distribution of all the possible paths generated according to the given policy π_θ with the maximum path length constraint L. The value of L can be closely related to the depth of meaning that can be expressed by the source user. For example, a shorter reasoning path represents a relatively more straightforward meaning. As the length of the path increases, the chance of disclosing a deeper meaning increases. However, this will also increase the search space and the probability of misrepresenting the true meaning of the source user due to overfitting.

13.3.3 Algorithm and Theoretical Analysis

Our proposed iSAC architecture consists of two phrases of implementation. In the communication phrase, the source user will first detect the explicit semantic terms from the source signal. It will then apply semantic encoder to convert the explicit semantics into a low-dimensional signal to be sent to the physical channel. The destination user will first recover the explicit semantics from its received signal using a pre-trained projection function. It will then generate the implicit semantics based on the explicit semantics using a pre-trained semantic interpreter.

The training phrase of iSAC architecture consists of a joint training algorithm, called G-RML algorithm, that jointly trains the semantic comparator and semantic interpreter. In particular, the semantic interpreter will try to generate a reasoning path extended from the recovered explicit semantics via a series of reasoning episodes based on the trained semantic reasoning network. Particularly, in each reasoning episode, the semantic interpreter chooses an action, a set of relations to extend the semantic paths, based on the current state. When the next relations are decided, the semantic interpreter will update the path η_t by concatenating the selected relation r_{t+1} as well as the linked hidden entity e_{t+1}. By repeating the above process, the semantic reasoning paths will be sequentially extended until the maximum length of the reasoning path L arrives. Note that the value of L can be empirically selected based on the depth of the meaning in the previous communication between the source and destination users.

The detailed procedures of semantic interpreter during communication phrase as well as the training procedures of G-RML algorithm are described in Algorithms 1 and 2, respectively (Figures 13.4 and 13.5).

We can prove that, despite the destination user cannot observe any expert semantic paths, by applying the proposed G-RML algorithm, the semantic interpreter can always generate the semantic paths that follow the same probability distribution as the paths generated by the true reasoning mechanism of the source user.

Algorithm 1 Semantic Interpreter (Communication Phrase)

Input: policy π_θ, received explicit symbols (e.g., entities and/or relations) $\boldsymbol{v}$ and maximum length L of reasoning paths
Output: Set of possible implicit semantic reasoning paths $\hat{\mathcal{K}}$
For reasoning episode $j = 0, 1, 2...$ **do**

- Initialize $\eta_0 = \boldsymbol{v}$, $s_0 = (\boldsymbol{v}, 0)$
- **For** each step t **do**
  - Get action $a_t \leftarrow r^t \sim \pi_\theta(a_t \mid s_t)$
  - **If** r^t is valid **Then**
    Choose its connected entity e^t
  - Update path $\eta_t \leftarrow$ concatenate (η_t, r^t, e^t)
  - Update state $s_{t+1} = (\eta^t, t)$
  - **If** max number of hops is reached **Then**
    Store path η_t in $\hat{\mathcal{K}}$
- **End For**

Figure 13.4 Semantic interpreter (communication phrase).

Algorithm 2 G-RML Algorithm (Training Phrase)

Input: Observed visible entities e^0, expert semantic path set $\mathcal{K}$, initial policy network $\pi_{\theta_0}(a_0|s_0)$ and comparator network $\varpi_{\phi_0}(\boldsymbol{p}^0)$ with parameters θ_0, ϕ_0, and max length of hops L
Output Learned reasoning policy π^*_θ at the destination user
For each training iteration t **do**

- Destination user generates a semantic path set $\hat{\mathcal{K}}_t$ from policy π_{θ_t} based on Algorithm 1 to be sent to the source user;
- Source user updates semantic comparator parameters ϕ_t to ϕ_{t+1} with the following gradient:

$$\Psi_{\hat{\Pi}_{\pi_\theta}}[\nabla_\phi \log \varpi_\phi(\eta)] + \Psi_{\Pi}[\nabla_\phi(1 - \log \varpi_\phi(\eta))] \tag{9}$$

- Destination user updates reasoning policy parameters θ_t to θ_{t+1} by minimizing the cost function with Monte Carlo Policy Gradient:

$$\mathbb{E}_{\boldsymbol{p}_t^D \in \hat{\mathcal{K}}}[\nabla_\theta \log(\pi_\theta(\boldsymbol{a}|\boldsymbol{s})Q(\boldsymbol{s},\boldsymbol{a}))], \text{where } Q(\boldsymbol{s},\boldsymbol{a}) = \log(\varpi_{\phi_{t+1}}(\eta)) \tag{10}$$

End For

Figure 13.5 G-RML algorithm (training phrase).

13.4 Extension to Collaborative Reasoning

13.4.1 Collaborative Reasoning Computing Network

The above mechanisms and model mainly define the one-to-one semantic communication method. In more practical scenarios, however, the encoded semantic information often needs to be inferred and decoded from the shared semantic knowledge base of multiple terminals.

When it comes to a complicated network system with several servers, we need to introduce and consider a multi-user computing network, which consists of multiple edge servers, as shown in the Figure 13.6 To support semantic communication and semantic inference between multiple users, the multi-user computing network should include the following parts:

1) **Central Coordinator:**
   A Central Coordinator functions as a centralized computing system that provides computational and storage resources accessible to users globally. It also manages a globally shared semantic knowledge base, which includes accumulated knowledge entities from all the different terminals, as well as the potential relationships between these entities. Typically, the entities and

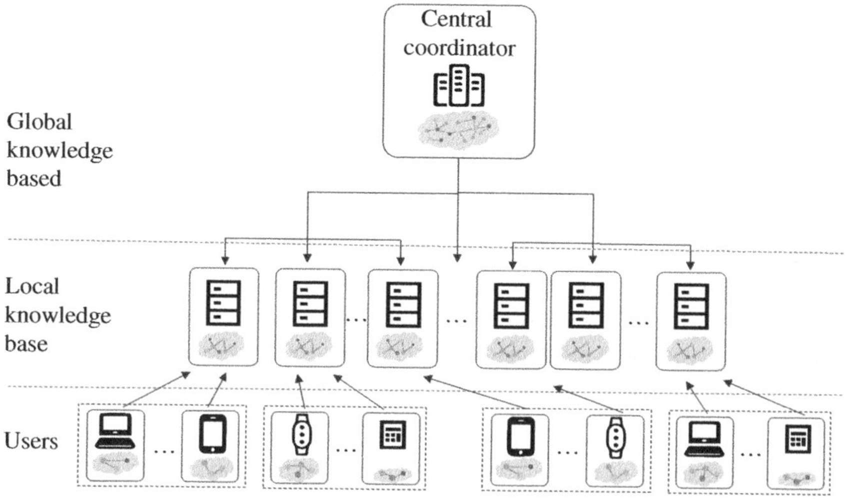

Figure 13.6 Multi-user computing network.

relations within this globally shared knowledge base represent, universally shared information that is not specific to any regional or local context.

2) **Local knowledge Base:**
   A local knowledge base is maintained by local edge servers, each of which provides computational and storage resources that are accessible to local users. Each local edge server also contains a local knowledge base that can store locally shared or personalized knowledge. This includes knowledge that is tailored to individual preferences or based on personal experiences, such as biased interpretations of knowledge concepts and relationships specific to one or a limited number of users.
3) **Users:**
   A user can function as either an information source or a destination, aiming to communicate semantic meanings with other users. We assume that each source user can access and analyze past communication history, extracting a set of semantic reasoning trajectories that include globally, regionally, locally shared, and even privately accessible semantic knowledge. The source user utilizes these extracted semantic reasoning trajectories, referred to as expert reasoning paths, to guide the Central Coordinator and Local Knowledge Base across different tiers in training models for semantic encoding, decoding, and interpretation.

13.4.2 Algorithm for Collaborative Reasoning

When it comes to the model with multi-users, we may face two primary challenges:

1) **Cross-knowledge base reasoning:** A single message received by a source user might reference semantic entities stored across multiple edge servers and/or knowledge bases;
2) **Inter-knowledge base reasoning:** Unlike traditional sample-based datasets, this approach focuses on relation-based semantic reasoning, where entities in different knowledge bases–stored on separate edge servers–may have interrelations. However, these relationships can be obscured or lost due to the decentralized nature and distribution of knowledge bases across various edge servers.

To overcome these challenges, we propose implementing an index list on each edge server that specifies the locations of all other knowledge entities associated with different edge servers. This index only contains references to knowledge entity labels and the corresponding edge servers where these entities are stored, without storing or displaying any relationships or attributes of the entities themselves. For semantic interpretation, each user submits their explicit semantic information to the nearest edge server, which then distributes the entities in the received explicit semantics to the appropriate edge servers according to the index. Each edge server subsequently generates an inference path based on the received entities, utilizing its local knowledge base and the established inference strategy.

In this scenario, the source user must also guide each edge server in training a cross-server strategy, which determines the next entity to be considered in another edge server based on the currently observed or inferred entity in the current server. The detailed process for training and coordinating models is shown in Figure 13.7.

13.5 Conclusion

In this chapter, we introduced iSAC, an innovative implicit communication framework. iSAC enables the destination user to develop a reasoning mechanism guided by the source user, automatically creating reasoning pathways that capture the message's implicit meanings. We have also introduced the G-RML framework that allows the destination user to emulate the source user's reasoning processes. By applying G-RML, the destination user can learn to generate pathways that follow the probability distribution of expertly navigated paths, effectively transmitting implicit semantics. We also provide algorithm and architecture for multi-user for future extension. Our work highlights our dedication to pushing the boundaries of semantic communication, demonstrating the potential to enhance the transmission of subtle meanings in various communication scenarios.

```
// Training Phase
```
Input: Entities and relations of previously observed expert paths
Output: The project function Θ
for *each server* $k \in [K]$ **do**
  Initialize Θ
  while $L \geq$ *Threshold* **do**
    Update Θ w.r.t
$$\sum_{\substack{\langle \tilde{e},\tilde{r},\tilde{e}' \rangle \in \mathcal{K}, \\ \langle \tilde{e},\tilde{r}',\tilde{e}'' \rangle \in \mathcal{K}^-}} \nabla\{\sigma + \|\tilde{e} + \tilde{r} - \tilde{e}'\|^2 - \|\tilde{e}_t + \tilde{r}' - \tilde{e}''\|^2\}$$
  end
end
Return Θ
```
// Encoding Phase
```
Input: Explicit semantics identified by the semantic identifier
Output: Low-dimensional representations of semantics

- Source user receives Θ
- $\hat{\boldsymbol{v}}^E \leftarrow \Theta(\boldsymbol{v}^E)$

```
// Decoding Phase
```
Input: $\hat{\omega}^E$ received by the destination user
Output: A set of reasoning paths that best interpret the implicit semantics
Destination user uploads ω^E *to edge server*
if *hard decoding* **then**
  $\omega^E \leftarrow \Theta^{-1}(\hat{\omega}^E)$
else
  $\hat{p} \leftarrow \sum_{j=1}^{J} \tilde{r}_j + \hat{\omega}^E$
end

Figure 13.7 G-RML multi-user algorithm.

Bibliography

A. Bordes, N. Usunier, A. Garcia-Duran, J. Weston, and O. Yakhnenko. Translating embeddings for modeling multi-relational data. In *International Conference on Neural Information Processing Systems*, Nevada, Dec. 2013.

J. Liang, Y. Xiao, Y. Li, G. Shi, and M. Bennis. Life-long learning for reasoning-based semantic communication. In *Proceedings of the IEEE ICC Workshop*, Seoul, South Korea, May 2022.

K. Lu, Q. Zhou, R. Li, Z. Zhao, X. Chen, J. Wu, and H. Zhang. Rethinking modern communication from semantic coding to semantic communication. *IEEE Wireless Communications*, 30(1):158–164, May 2022.

G. Shi, Y. Xiao, Y. Li, and X. Xie. From semantic communication to semantic-aware networking: Model, architecture, and open problems. *IEEE Communications Magazine*, 59(8):44–50, Aug. 2021.

E. C. Strinati and S. Barbarossa. 6G networks: Beyond Shannon towards semantic and goal-oriented communications. arXiv preprint arXiv:2011.14844, Feb. 2021.

Y. Xiao, X. Zhang, Y. Li, G. Shi, and T. Basar. Rate-distortion theory for strategic semantic communication. In *Proceedings of the IEEE Information Theory Workshop*, Mumbai, India, Nov. 2022.

Y. Xiao, Z. Sun, G. Shi, and D. Niyato. Imitation learning-based implicit semantic-aware communication networks: Multi-layer representation and collaborative reasoning. *IEEE Journal on Selected Areas in Communications*, 41(3):639–658, 2023.

14

Semantic and Goal-Oriented Communication: A Data Valuation Perspective

Shashi Raj Pandey, Van Phuc Bui, and Petar Popovski

Department of Electronic Systems, Aalborg University, Aalborg Øst, Denmark

14.1 Introduction

Semantic communication (SC) attempts to generalize the problem of communication beyond the technical problem of correct transmission of a set of bits toward accounting for the rationale behind data transmission and its utility at the receiver's end through the interpretation of content, context, and timing aspects, such as age-of-information (AoI) [Kaul et al., 2012; Kountouris and Pappas, 2021; Strinati and Barbarossa, 2021; Gündüz et al., 2022; Xu et al., 2023]. Numerous applications reflect on the significance of SemCom, for instance, when quantifying and encoding important information for transmission to obtain high-spectral satellite imagery [Bui et al., 2023a], semantic coding and recovery [Gündüz et al., 2022], value-of-information (VoI)-based metric for strategic scheduling of devices [Bui et al., 2023b; Holm et al., 2023] for application timing awareness, and semantic-aware decision making and inference [Uysal et al., 2022], to name a few. This perspective offers methodologies to limit unnecessary transmissions and prioritize data to improve goal-oriented (GO) performance; hence, SemCom is instrumental in supporting GO communications, where the communication is tailored to achieve specific *goals*.

In this chapter, we describe semantic and GO problems in communications and develop data transmission strategies based on *data valuation principles* to show how they can be addressed. Taken from an economic perspective, data valuation entails both the monetary value data holds and its usage. A parallel can be drawn in communication systems, where data value relates to the *usefulness* of computed/transmitted data at the transmitter/receiver's end, respectively. Therein, data valuation, in terms of temporal frame, can be considered over the long term or the short term. However, all these aspects have not been widely considered in the context

Foundations of Semantic Communication Networks, First Edition.
Edited by Walid Saad, Christina Chaccour, Christo Kurisummoottil Thomas, and Merouane Debbah.

of SemCom. This leads us to develop two particular regimes of communications: *strategic pull-based* and *strategic push-based*. To that end, we present two practical scenarios reflecting the aforementioned communication regimes: (i) satellite communications for high-spectral Earth imagining (details in Section 14.3) and (ii) communication-efficient edge federated learning (FL), which offers insights into the motivation behind considering data valuation approaches toward semantic and GO communications.

14.2 Data Valuation Principles

"Not all transmitted messages are of equal importance to the receiver." We highlight local and global data valuation approaches, as illustrated in Figure 14.1, to elaborate data valuation principles. In its entirety, data valuation explores the significance of data, locally or globally, and interprets its value before and after transmissions, respectively. An example, wireless sensor networks (WSNs) are often push-based, where the transmitter (TX) is oblivious to the value of transmitted data at the receiver (RX). Typically, the sensors sample observations of the environment and transmit them to a data collector periodically, often adding communication overhead with redundant or less informative data. Given knowledge of RX's utility, e.g., the objective to lower mean squared error (MSE) on a particular prediction task, maintaining the freshness of data [Kaul et al., 2012], and so on, the TX can be strategic in transmitting only the related and relevant data. In practice, the TX can exploit feedback information from the RX on the improvement of its utility through the transmitted data. Such an approach of local data valuation offers a communication regime of *strategic push*, where VoI is considered to decide the worth of data transmission by the TX based on uncertainty reduction in the observed processes at the RX. This translates into extracting the semantics of data locally and transmitting limited yet meaningful data – an approach of SemCom. However, it is nontrivial to have a shared utility between TX and RX; the utility may change over time, depending on the context of data use, and also because of the involved privacy and pricing concerns during information exchange [Pandey et al., 2023b]. Another regime is that the RX can initiate a query for particular

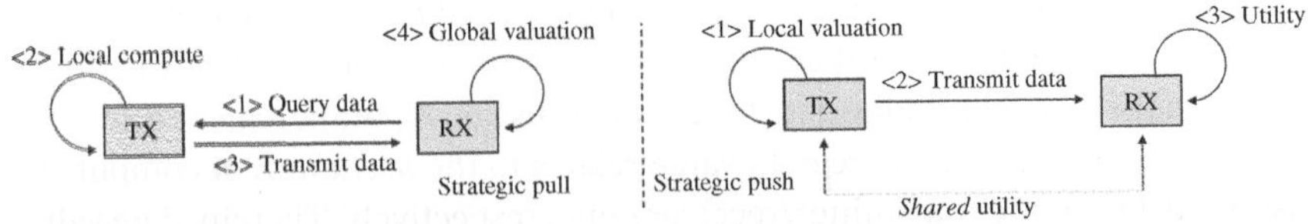

Figure 14.1 Semantic and goal-oriented transmission with strategic pull/push through data valuation.

data, which includes scheduling sensors that potentially contribute to its objective with the transmitted data. The RX can execute *strategic pull* through global valuation, i.e., assessing its utility defined as, say, MSE, then evaluating marginal contributions of data received prior from the sensors, and asking specific sensors to transmit at predefined channel uses. The RX can even employ expensive data valuation methods such as computing Shapley value (SV) [Shapley, 1953] to obtain the marginal contributions of the transmitted data and allocate channel uses to the sensor for collision-free data transmission. This limits transmission opportunities for irrelevant data and prioritizes communications to improve goals defined at the RX – an approach of GO communications.

Next, we present an overview of two semantic and GO communication problems and understand them through the lens of data valuation principles.

14.2.1 Semantic Approaches In Satellite Communications For Earth Observation

The amount of data generated by Earth observation satellites can be enormous, which poses a great challenge to the satellite-to-ground connections at a limited rate. For instance, the Sentinel-2 system acquires an extensive amount of data (2.4 terabits per day) for transmitting to the terrestrial gateway, with each surface location being captured at periodic intervals of five days [ESA, 2023]. While forwarding the complete dataset can be advantageous for the accurate detection of any alterations or anomalies, it is not regarded as efficient in terms of data storage and transmission capabilities. It is then imperative to consider the problem of efficient downlink communication of multispectral satellite images for Earth observation using change detection through learning-based data valuation techniques. We can have semantic data valuation and selective transmission through an efficient image processing pipeline that consists of the joint design of cloud removal and change encoding. This can be seen as an instance of SC, as we are encoding only important information, such as changed multispectral pixels (MPs), while aiming to minimize energy consumption. It comprises a three-stage end-to-end scoring mechanism that determines the importance (*valuation*) of each MP before deciding its transmission.

14.2.2 Goal-Oriented Problems in FL

FL McMahan et al. [2017] is an iterative model training approach done in a decentralized, privacy-preserving manner, where the global model gets updated by aggregating the local models received at each interaction round – i.e., the global communication – between the devices and the parameter server (PS). A GO communication problem in the FL scenario can be abstracted through the device selection and participation problems, which, in actuality, encompasses the goal of lowering the number of incurred relevant transmissions for a communication-efficient, high-quality model training. As a first-order consideration for such GO interactions, this means a selected device needs to decide its transmission plan on when not to participate in FL. This can be done by approximating the value of the trained global model upon participation by this device.

The improvements in the device's learning tasks through arbitrary participation strategy in the FL training can be nominal with an average model performance. This suggests the limited influence of participation of such devices; due to statistical and system-level heterogeneity, not all selected devices in each communication round contribute greatly to the trained model. Then, it is intuitive for devices to decide "when not to transmit" as well as to minimize unnecessary computing and communication costs. In addition, feedback from the PS to evaluate risks on participation enables devices to align their transmission strategies while satisfying the overall GO performances after model aggregation at the PS. Such feedback can be driven through contribution-based strategic selection [Kang et al., 2019; Nishio and Yonetani, 2019; Pandey et al., 2023a], particularly as a device selection problem, using data valuation principles. For instance, as discussed, at the PS side, game-theoretic tools such as SV [Shapley, 1953] offer insights when evaluating the contribution of the received model parameters [Singhal et al., 2023]. In principle, given N devices, SV calculates the average marginal contribution of each device $n \in \mathcal{N}$ on all possible combinations of coalition subsets as

$$\mu_{n,t}(N, U) = \frac{1}{N!} \sum_{S \subseteq \mathcal{N} \setminus \{n\}} \frac{U(S \cup \{k\}) - U(S)}{\binom{N-1}{|S|}}, \tag{14.1}$$

where the function $U(\cdot)$ gives the value for any subset of the devices, e.g., let S be a subset of $\mathcal{N}$, then $U(S)$ gives the value of that subset [Pandey et al., 2022a].

This reflects a scenario involving both strategic pull- and push-based communication regimes, where local and global data valuation is taken into account.

14.3 Semantic Communication For Earth Observation with LEO Satellites

The section discusses a learning-based semantic data valuation approach that identifies data relevance through changed pixels in a high-spectral satellite image and enables SC. Therein, an optimization problem is formulated to reflect the contributions of data valuation measures in improving the energy efficiency aspects for Earth observation with low earth orbit (LEO) satellites.

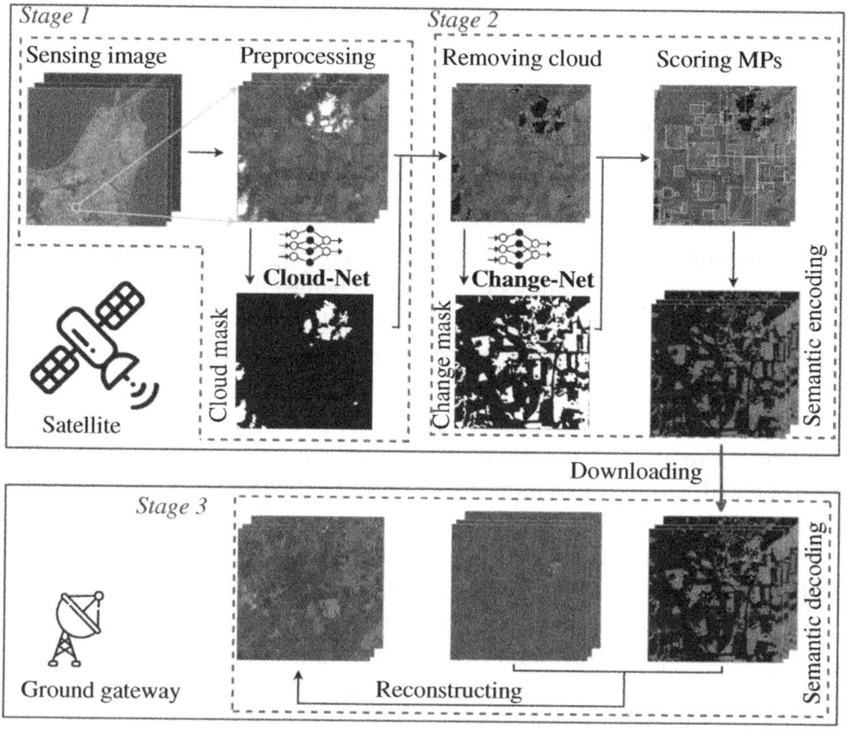

Figure 14.2 Proposed end-to-end scoring architecture for strategic push-based communication.

Consider a satellite system as illustrated in Figure 14.2, where a LEO satellite communicates with a gateway deployed at ground level, which is directly connected to a server. The satellite captures multispectral images and is equipped with an artificial intelligence (AI) module intended for image feature extraction. These features, such as potential pixel changes and their associated coordinates, must be transmitted in an energy efficient way within the limited time period when the satellite and gateway are connected. This means, for two coregistered multispectral satellite images, $\mathbf{X}^{t_0}$ and $\mathbf{X}^{t_1}$, captured at times t_0 and t_1 respectively, a learning-based image processing approach that valuates and communicates only the changed MPs from $\mathbf{X}^{t_1}$ to Earth, minimizes both irrelevant data transmission to the ground and overall energy consumption. Given $\mathcal{X}^{t_1} = \{\mathbf{x}_{ij}^{t_1}\}$ as a set of available MPs in image $\mathbf{X}^{t_1}$, this requires the selection of a subset $\mathcal{P}$ MPs among $\mathcal{X}^{t_1}$ ($|\mathcal{P}| \leq |\mathcal{X}^{t_1}|$) based on their *semantic relevance* in reconstructing the image at the gateway with high fidelity. The reference pixel set $\mathcal{X}^{t_0} = \{\mathbf{x}_{ij}^{t_0}\}$ of the reference image $\mathbf{X}^{t_0}$ is processed as analogous. We define $\mathcal{S} = \{s_{ij}\}$ as the accurate change map where $s_{ij} \in \{0, 1\}$ represents the changed/unchanged MP. In particular, $s_{ij} = 1$ if there is a pixel change at location (i,j) and vice versa. This sets the premise to build the following energy model.

14.3.1 Energy Model

Let $\alpha_{ij} \in \{0,1\}$ be a binary decision variable with $\boldsymbol{\alpha} \triangleq \{\alpha_{ij}\}$, where $\alpha_{ij} = 1$ if we select MP $\mathbf{x}_{ij}^{t_1}$ for $\mathcal{P}$. The transmission set $\mathcal{P}$ is then $\mathcal{P} \triangleq \{\alpha_{ij}\mathbf{x}_{ij}^{t_1}\}_{\forall \alpha_{ij},\mathbf{x}_{ij}^{t_1}}$. The total data transmission cost $C(\boldsymbol{\alpha})$ for $\mathcal{P}$ is given by $C(\boldsymbol{\alpha}) = \sum_{\alpha_{ij} \in \boldsymbol{\alpha}} D\alpha_{ij}$, where D is the depth of each pixel or the number of spectral bands. The model for satellite energy consumption during processing tasks, based on relevant CPU parameters [Mao et al., 2017], relates energy per clock cycle to the square of CPU clock frequency f_{CPU}. Thus, $E_{cycle}^{proc} = P^{proc}(f_{CPU})/f_{CPU}$. Considering linear power with the number of processor cores N_{CPU}, the energy to process data $E^{proc}(\boldsymbol{\alpha})$ is modeled as $E^{proc}(\boldsymbol{\alpha}) = C(\boldsymbol{\alpha})\hat{R}^{comp}E_{cycle}^{proc}$, where $\hat{R}^{comp} = e^{\kappa\rho} - e^{\kappa}$ represents compressing complexity, with κ and ρ based on the compression algorithm [Leyva-Mayorga et al., 2023]. We consider an interference-free communication channel affected by additive white Gaussian noise (AWGN) with free space path loss. Once we know the signal-to-noise ratio (SNR), a proper modulation and coding scheme is selected to achieve reliable communication following the DVB-S2 system [Morello and Mignone, 2006]. The downlink energy consumption for downloading all data $\mathcal{P}$ is therefore calculated as $E^{trans}(\boldsymbol{\alpha}) = P_{tx}C(\boldsymbol{\alpha})/R_{\mathcal{P}}$ with $R_{\mathcal{P}}$ denoting the selected rate and P_{tx} is the transmit power. Summing up, the total energy for processing and transmitting the selected data $\mathcal{P}$ is derived as $E^{tot}(\boldsymbol{\alpha}) = E^{proc}(\boldsymbol{\alpha}) + E^{trans}(\boldsymbol{\alpha})$.

14.3.2 Energy Efficient Data Downloading With Change Detection Constraint

The objective of the energy reduction in transmitting only the changes to the nearby gateway shifts the focus toward the communication aspects of the system. That is, all x_{ij} that have change map $s_{ij} = 1$ will be scheduled to the transmission set $\mathcal{P}$. This is achieved by setting the values of the decision variable $\boldsymbol{\alpha}$ as $\alpha_{ij} \geq s_{ij}$, which ensures that all the changed MPs are transmitted. Accordingly, the optimization problem is formulated as

$$\text{P1}: \min_{\boldsymbol{\alpha}} \quad E^{tot}(\boldsymbol{\alpha}) \quad \text{s.t. } \alpha_{ij} \geq s_{ij}. \tag{14.2}$$

Note that (14.2) can only be solved if the satellite knows indexes of all changed MPs beforehand, but this is not the case in reality. Instead, it is necessary to implement an AI model at the satellite to valuate, through precise estimation of s_{ij} and, hence, perform inference on the changed MPs.

Define $s_{ij}^{p} \in \{0,1\}$ as the predictor for s_{ij} given by the selected change detection algorithm, which achieves an error level ϵ. Then, (14.2) is reformulated as

$$\text{P2} : \min_{\{\alpha_{ij}\}} \sum_{i=1}^{H}\sum_{j=1}^{W} \alpha_{ij} \text{ s.t. } \alpha_{ij} \geq s_{ij}^{p} \quad \text{and} \quad \Pr(s_{ij}^{p} = 0 | s_{ij} = 1) \leq \epsilon, \tag{14.3}$$

where ϵ is the threshold that ensures a sufficient number of change MPs are selected. Accordingly, we propose an algorithm to obtain an efficient suboptimal solution to the problem (14.3).

14.3.3 Preprocessing and Semantic Encoding

This part describes steps of learning-based semantic data valuation to detect change MPs in Earth observation applications, where Algorithm 14.1 summarizes these steps.

Algorithm 14.1 Proposed algorithm for problem (14.3)

Input: Geometrical multispectral image pair $\mathbf{X}^{t_0}$ and $\mathbf{X}^{t_1}$

Output: The transmission set $\mathcal{P}$, reconstructed MP image $\hat{\mathbf{X}}^{t_1}$, and energy requirement for processing and transmission

1: *Preprocessing and Cloud Removing*

2: Select process bands (R, G, B, Nir) for cloud removal and (R, G, B) for change detection

3: Normalize $\mathbf{X}^{t_0}$ and $\mathbf{X}^{t_1}$

4: Cloud detection and removal using Cloud-Net

5: *Change Scoring and Semantic Encoding*

6: Scoring MPs using Change-Net and update $\{s_{ij}^{p}\}$

7: Perform threshold segmentation to get the transmission set $\mathcal{P} = \{\mathbf{x}_{ij}^{t_1} | s_{ij}^{p} = 1, i = 1, 2, \dots, H, j = 1, 2, \dots, W\}$

8: Update $\alpha_{ij} = s_{ij}^{p}$ $\forall i, j$ and calculate required energy $E^{tot}(\alpha)$

14.3.3.1 Preprocessing

As part of the preprocessing stage, the following steps are undertaken: (i) the selection of $N (N < D)$ bands representative of the complete set of MP images; (ii) the adjustment of illumination levels to facilitate the detection (cloud and change) in the subsequent stage; and (iii) the elimination of cloud cover from the observed image. Before removing clouds, a radiometric correction procedure is implemented to rectify radiometric disparities among multispectral images stemming from diverse imaging conditions such as sun angle, light intensity, and atmospheric circumstances. The chosen approach is relative radiometric normalization, which is founded on the z-score method, which normalizes images to zero mean and a unit standard deviation.

14.3.3.2 Cloud Removing

Cloud-Net model for identifying cloud regions in multispectral Sentinel-2 images based on U-Net using a combination of thresholding and deep learning techniques is deployed. This approach utilizes four spectral bands – red, green, blue, and near infrared (RGBNir) – for both training and prediction, making it practical for use in satellite hardware for cloud detection. The output probability map is obtained via the use of a sigmoid activation function in the final convolution layer of the network.

14.3.3.3 Change Scoring and Semantic Encoding

In this phase, we have two multispectral images that have been geometrically aligned, have consistent lighting conditions, and have had cloud cover removed. Our goal underscoring the semantic data valuation approach is to develop a computationally efficient automatic change detection method for these two images that is practical for satellite applications. Motivated by Ronneberger et al. [2015], the Change-Net based on U-Net model is developed to perform semantic segmentation of images. We use a negative log-likelihood loss function to classify every MP in the observed images as either changed or not. In our case, the system output $\bar{s}_{ij} \in [0, 1]$ is defined as the change score of the corresponding cloud-removed MP $\tilde{\mathbf{x}}_{ij}^{t_1}$, which is the probability of $p(s_{ij} = 1|\tilde{\mathbf{x}}_{ij}^{t_1}, \boldsymbol{\theta}_{model})$ under the parameter set $\boldsymbol{\theta}_{model}$. As $\bar{s}_{ij}$ is close to 1, it is likely that there is a change in location (i, j). Thus, to make a good prediction, we would like to learn a scoring system such that $s_{ij}^{p} = \mathbf{1}(\bar{s}_{ij} \geq \tau)$, where τ is the predefined threshold, derived experimentally to satisfy the last constraint in (14.3).

14.3.4 Numerical Results

We evaluate our proposed algorithm in the Sentinel-2 system when the transmission process initiates upon the satellite's entry into the gateway's coverage area. Other parameters used follow Bui et al. [2023a]. We use the 38-Cloud dataset [Mohajerani and Saeedi, 2019] for Cloud-Net, and the Onera Satellite Change Detection dataset [Daudt et al., 2019] for Change-Net. We also apply the trained algorithm to real-world data collected from a Sentinel-2 image to identify changes at the pixel level relative to reference images.

Figure 14.3a plots the receiver operating characteristics (ROC) of 3 representative images in the Onera Satellite Change Detection (OSCD) dataset with the highest, median, and lowest area under the ROC curve (AUC) levels, respectively. The efficacy of our trained Change-Net model can be readily observed, as it achieves a high level of performance with a maximum AUC of 0.98 and a minimum AUC of 0.80. Figure 14.3b depicts the percentage of downloads accumulated and the probability of the image that has the median AUC score, respectively, while also presenting a threshold score of 0.6 to guarantee the selection of 95% of

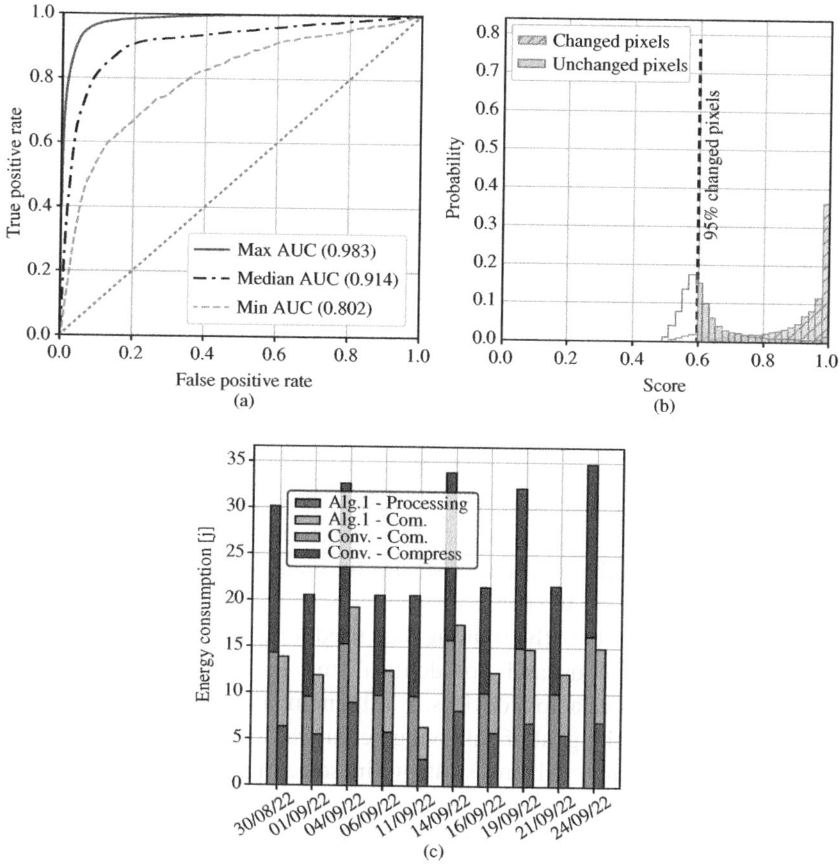

Figure 14.3 (a) ROC curve over change dataset (the AUC scores are shown in the legends). (b) Probability of changed and unchanged MPs versus score of multispectral image with median AUC. (c) Energy requirement for processing and transmission.

modified MPs. Remarkably, to guarantee the selection of a sufficient number of altered MPs (at a rate of 95%), we have labeled approximately 50% of unchanged MPs with a value of 1, leading to the requirement of encoding and transmitting approximately 50% of the entire sensing data. Figure 14.3c shows the required energy to compress and transmit real-world data from the Sentinel-2. Compared to the conventional scheme (transmit the whole sensing image), our architecture saves at least 40% in required energy, thereby directly saving both energy for data processing and transmission from satellites, as well as storage space and processing time at Earth gateways.

14.4 Goal-Oriented Communications In FL

The section discusses the data valuation approach to stir the transmission strategies of distributed devices toward a common goal of communication-efficient training for a better-generalized, high-quality FL model; hence, improving the overall GO performance. In FL, the collaborative training with N devices aims at finding an optimal global model parameter $w \in \mathbb{R}^d$ that minimizes the empirical risk on all of the distributed training data samples D, i.e., finding the solution to the following optimization problem: $\min_{w \in \mathbb{R}^d} F(w)$, where $F(w) := \sum_{n=1}^{N} \frac{D_n}{D} f_n(w)$, and $f_n(w)$ captures the cost of prediction with the model parameter w; In practice, the training examples (x_i, y_i) are drawn from $\mathcal{D}_n$, implying $f_n(w) = \mathbb{E}_{(x_i,y_i)\sim\mathcal{D}_n}[l(w; (x_i, y_i))]$ as the expected loss. Here, $l(w; (x_i, y_i))$ is the measuring loss function for the prediction [Konečný et al., 2016; McMahan et al., 2017] over data samples $(x_i, y_i), \forall i \in \mathcal{D}_n$. The general FL problem can be solved with standard methods, such as federated averaging (FedAvg) [McMahan et al., 2017], where a periodic aggregation of local updates leveraging the variants of gradient descent methods, such as stochastic gradient descent (SGD) algorithms, is made. However, in such a distributed setting, implicit challenges are imposed due to the quality of the training data used for generating local updates, for instance, heterogeneous and nonidentical and independently distributed (non-i.i.d.) data properties across devices, – influencing the quality of the trained model. Therein, recall the goal of achieving better training and communication efficiency translates into PS being strategic in selecting quality updates from devices.

One such instance, an efficient incentive design is a fundamental stimulus for quality participation in a collaborative operation in FL over wireless [Khan et al., 2020; Pandey et al., 2020]. On that pretext, another natural question on incentive design to tailor performance toward GO communication in FL is: *How to ensure quality participation with contribution-based incentives that meet the objective of training and communication efficiency simultaneously?* This leads to (i) investigating ways to quantify the expected contribution of clients' shared model parameters in improving the model performance during aggregation, given system-level constraints; and (ii) transparent distribution of incentives as per the contribution, both, in terms of participation and the contribution made in the FL training [Pandey et al., 2020; Nguyen et al., 2021].

As the contributions of devices in enhancing the global model performance are often unknown a priori, mechanism designs for data valuation methods act as a fundamental tool to offer an estimation of the device's local updates contribution in FL, which is followed by the contribution-based model aggregation to reduce communication overhead.

14.4.1 Contribution-Based Aggregation

A cost-efficient Shapley-based scheme [Jia et al., 2019; Pandey et al., 2022b] allows approximation of the contribution of local models during model aggregation to lower the overall communication overhead and design incentives based on that. Once getting the local updates from N clients at iteration t, a plain model aggregation scheme is FedAvg [McMahan et al., 2017], i.e., we compute the average of received local updates $\Delta\phi^t_{[n]}, \forall n \in [N]$ as $\phi^{t+1} := \phi^t + \frac{1}{N}\sum_{n=1}^{N} \Delta\phi^t_{[n]}$, and pass ϕ^{t+1} back to the $[M] \subseteq [N]$ random clients to compute local updates through SGD, and iteratively as such, thereafter. A cost-efficient Shapley-based scheme [Jia et al., 2019; Pandey et al., 2022b] allows approximation of the contribution of local models during model aggregation to lower the overall communication overhead and design incentives based on that. The PS performs δ random permutations over the received updates associated with the global model parameters, where the average marginal contribution is evaluated as

$$u_n = \frac{1}{|M|!}\sum_{p\in\mathcal{P}} \left[V(S_{n,p} \cup \{n\}) - V(S_{n,p})\right]. \tag{14.4}$$

Here, $p \in \mathcal{P}$ is the permutation of device updates, i.e., $\Delta\phi^t_{[n]}, \forall n \in [M]$, with a probability of $\frac{1}{|M|!}$; $V(\cdot)$ is monotone quantifying performance score evaluated on the test dataset with the permuted subset of device updates, which translates to the improvement in global model accuracy and is measured by ϵ; $S_{n,p}$ is a set of preceding devices before n for a permutation p selected for model aggregation – then, all possible orders of devices contribute to the average contribution (14.4). Then, formally, for a test dataset $(x_i, y_i) \sim \mathcal{D}$, $V(\mathcal{D}) = V_{ref} - \frac{1}{|\mathcal{D}|}\sum_{i=1}^{|\mathcal{D}|} l(w(\alpha); (x_i, y_i))$, such that $\mathcal{F}(\alpha^*) \leq l(;) \leq V_{ref}$ with $\mathbb{E}\left[\mathcal{F}(\alpha) - \mathcal{F}(\alpha^*)\right] < \epsilon$ for a small ϵ and the variable α related as a dual mapping with global model parameter [Pandey et al., 2022b].

14.4.2 Contribution-Based Participation

Likewise, being strategic in contribution-based aggregation, device selection, and participation is fundamental in addressing problems of GO communications in FL. *"Can feedback on the instantaneous valuation of local updates from the PS enable risk-averse participation of devices to lower the number of irrelevant transmissions?"* *Strategic selection* of devices can be done by reformulating the learning objective as multiarm bandits (MAB) problem ([Pandey et al., 2023a; Singhal et al., 2023]), in which choosing each device n corresponds to an arm pulled for an obtained valuation, i.e., the instantaneous reward $\mu_{n,t}$, plus a penalty factor that captures the number of times that arm has been selected before, as $I_{n,t}$. Upper confidence bound (UCB) algorithm assigns the following value to each arm $f_{n,t} = \mu_{n,t} + \sqrt{\frac{\ln t}{I_{n,t}}}$. Then, selecting an arm is equivalently solving: $\mathcal{M}_t = \underset{n \in \{1,2,\dots,N(t)\}}{\arg\max} f_{n,t}$, where $N(t)$ is the number of possible devices available at time instance t, and $\mathcal{M}_t$ is the set of M_t devices selected among $N(t)$ choices, i.e., $\mathcal{M}_t \subseteq \{1, 2, \dots, N(t)\}$. SV calculates the average marginal contributions $\mu_{n,t}$ as discussed before. Similarly, following strategic selection, *strategic participation* can be defined with conditional value-at-risk (CVaR) [Rockafellar and Uryasev, 2000] as a data valuation measure for deciding transmission of local updates to the PS. In principle, the relevance of local updates can be cast as *semantic information* for developing local transmission strategies, i.e., making a decision on when to "not transmit." The devices use feedback about the state of the PS and evaluate their contributions in training the learning model in each aggregation period, which eventually lowers the number of occupied connections. With the action profile on device participation upon random selection $\mathcal{A}_t$ with probability ϵ at time t, i.e., selecting a subset of devices from a pool of $N(t)$ devices, we use CVaR-regret at global aggregation round τ as $\mathcal{R}_{\tau,\theta} = \tau \max_n \left[CVaR_\theta(F_{\mu_n}) \right] - \mathbb{E}\left\{ \sum_{t=1}^{\tau} CVaR_\theta(F_{A_t}) \right\}$, to capture the contribution profile of selected devices. Here, devices employ a local regret threshold defined as participation risk factor γ, and the average θ risk on the global learning model is updated in each global communication round with the resulting CVaR-based risk-averse device selection problem as $\mathcal{M}_{t \in \{1,2,\dots,\tau\}} = \underset{n \in \{1,2,\dots,N(t)\}}{\arg\max} f_{n,t} - \mathbb{E}[\mathcal{R}_{t,\theta}]$. The derivation follows [Dvoretzky et al. [1956], Theorem 1. Tamkin et al. [2019]].

This approach applies to any closed loop system and is shown to be effective in limiting unnecessary transmissions and improving GO performance.

14.4.3 Performance Evaluations

Figure 14.4a demonstrates the impact of random permutation in evaluating the value of local updates in model aggregation. We observe a fundamental limit on imposing larger local computations versus the number of permutations at the aggregator for improving the convolution neural network (CNN) model's performance trained for well-known CIFAR-10 datasets [Krizhevsky and Hinton, 2009]. This means the PS can compensate the costs of local computation with added δ. We used MNIST dataset [LeCun et al., 1998] and compare with (i) *FedSGD* [McMahan et al., 2017] (with *full* participation); (ii) *FedSGD* (with *partial* participation), where 60% of the devices are selected, and (iii) *UCB* [Yoshida et al., 2020], where devices are selected purely on the UCB metric without considering risk-averse characteristics of the devices, ***strategic pull-push*** approach based on data valuation principles offers: (i) a reduction in the number

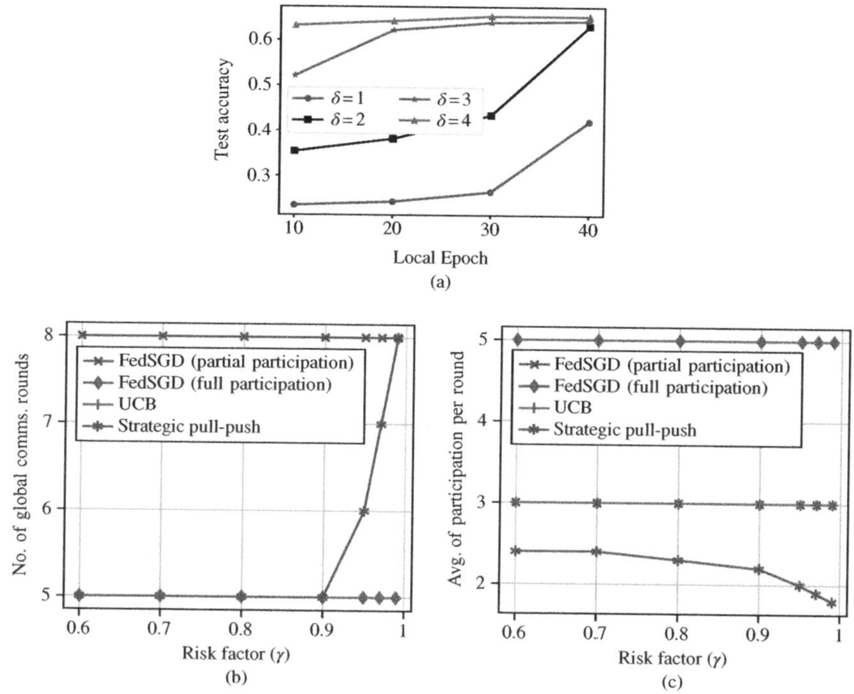

Figure 14.4 Impact of: (a) $\delta = \{1, 2, 3, 4\}$ on training performance, (b) risk factor (γ) on the number of communication rounds, and (c) communication links per communication round.

of global communication rounds (Figure 14.4b) and (ii) a gain in the average communication links per iteration for a target model accuracy of 80% (Figure 14.4c), when the risk factor $\gamma \geq 0.9$. This suggests a significant reduction in the number of transmissions to satisfy the KPIs with GO communications in FL training.

14.5 Conclusion

In this book chapter, we present the data valuation perspective on semantic and GO communication. In doing so, we highlight and discuss two specific classes of problems that encompass two regimes of communications we argue for, respectively, strategic *pull-based* and *push-based*. Our results reflect the potential of adopting data valuation principles in redesigning data transmissions and query strategies that address several problems in semantic and GO communication.

Bibliography

Van-Phuc Bui, Thinh Q. Dinh, Israel Leyva-Mayorga, Shashi Raj Pandey, Eva Lagunas, and Petar Popovski. On-board change detection for resource-efficient earth observation with leo satellites. In *Proceedings of the IEEE GLOBECOM*, 2023a.

Van-Phuc Bui, Shashi Raj Pandey, Federico Chiariotti, and Petar Popovski. Scheduling policy for value-of-information (VoI) in trajectory estimation for digital twins. *IEEE Communications Letters*, 27(6):1654–1658, 2023b.

Rodrigo Caye Daudt, Bertrand Le Saux, Alexandre Boulch, and Yann Gousseau. OSCD - Onera satellite change detection, 2019. URL https://dx.doi.org/10.21227/asqe-7s69.

Aryeh Dvoretzky, Jack Kiefer, and Jacob Wolfowitz. Asymptotic minimax character of the sample distribution function and of the classical multinomial estimator. *The Annals of Mathematical Statistics*, 27(3):642–669, 1956.

ESA. Sentinel-2 operations. Accessed: 17 Feb. 2023. URL https://www.esa.int/Enabling_Support/Operations/Sentinel-2_operations.

Deniz Gündüz, Zhijin Qin, Inaki Estella Aguerri, Harpreet S. Dhillon, Zhaohui Yang, Aylin Yener, Kai Kit Wong, and Chan-Byoung Chae. Beyond transmitting bits: Context, semantics, and task-oriented communications. *IEEE Journal on Selected Areas in Communications*, 41(1):5–41, 2022.

Josefine Holm, Federico Chiariotti, Anders E. Kalτr, Beatriz Soret, Torben Bach Pedersen, and Petar Popovski. Goal-oriented scheduling in sensor networks with application timing awareness. *IEEE Transactions on Communications*, 71(8):4513–4527, 2023.

Ruoxi Jia, David Dao, Boxin Wang, Frances Ann Hubis, Nick Hynes, Nezihe Merve Gürel, Bo Li, Ce Zhang, Dawn Song, and Costas J. Spanos. Towards efficient data valuation based on the Shapley value. In *The 22nd International Conference on Artificial Intelligence and Statistics*, pages 1167–1176. PMLR, 2019.

Jiawen Kang, Zehui Xiong, Dusit Niyato, Shengli Xie, and Junshan Zhang. Incentive mechanism for reliable federated learning: A joint optimization approach to combining reputation and contract theory. *IEEE Internet of Things Journal*, 6(6):10700–10714, 2019.

Sanjit Kaul, Roy Yates, and Marco Gruteser. Real-time status: How often should one update? In *2012 Proceedings IEEE INFOCOM*, pages 2731–2735. IEEE, 2012.

Latif U. Khan, Shashi Raj Pandey, Nguyen H. Tran, Walid Saad, Zhu Han, Minh N. H. Nguyen, and Choong Seon Hong. Federated learning for edge networks: Resource optimization and incentive mechanism. *IEEE Communications Magazine*, 58(10):88–93, 2020. doi: 10.1109/MCOM.001.1900649.

Jakub Konečný, H. Brendan McMahan, Daniel Ramage, and Peter Richtárik. Federated optimization: Distributed machine learning for on-device intelligence. arXiv preprint arXiv:1610.02527, 2016.

Marios Kountouris and Nikolaos Pappas. Semantics-empowered communication for networked intelligent systems. *IEEE Communications Magazine*, 59(6):96–102, 2021.

Alex Krizhevsky and Geoffrey Hinton. Learning multiple layers of features from tiny images, 2009.

Yann LeCun, Léon Bottou, Yoshua Bengio, and Patrick Haffner. Gradient-based learning applied to document recognition. *Proceedings of the IEEE*, 86(11):2278–2324, 1998.

Israel Leyva-Mayorga, Marc Martinez-Gost, Marco Moretti, Ana Pérez-Neira, Miguel Ángel Vázquez, Petar Popovski, and Beatriz Soret. Satellite edge computing for real-time and very-high resolution Earth observation. *IEEE Transactions on Communications*, 71(10):6180–6194, 2023. doi: 10.1109/TCOMM.2023.3296584.

Yuyi Mao, Changsheng You, Jun Zhang, Kaibin Huang, and Khaled B. Letaief. A survey on mobile edge computing: The communication perspective. *IEEE Communications Surveys and Tutorials*, 19(4):2322–2358, 2017.

Brendan McMahan, Eider Moore, Daniel Ramage, Seth Hampson, and Blaise Aguera y Arcas Communication-efficient learning of deep networks from decentralized data. In *International Conference on Artificial Intelligence and Statistics (AISTATS)*, Fort Lauderdale, FL, USA, Apr. 2017.

Sorour Mohajerani and Parvaneh Saeedi. Cloud-Net: An end-to-end cloud detection algorithm for Landsat 8 imagery. In *IEEE Proceedings of IGARSS*, pages 1029–1032. IEEE, 2019.

Alberto Morello and Vittoria Mignone. DVB-S2: The second generation standard for satellite broad-band services. *Proceedings of the IEEE*, 94(1):210–227, 2006.

Lam Duc Nguyen, Shashi Raj Pandey, Soret Beatriz, Arne Broering, and Petar Popovski. A marketplace for trading AI models based on blockchain and incentives for IoT data. arXiv preprint arXiv:2112.02870, 2021.

Takayuki Nishio and Ryo Yonetani. Client selection for federated learning with heterogeneous resources in mobile edge. In *IEEE International Conference on Communications (ICC)*, Shanghai, China, May 2019.

Shashi Raj Pandey, Nguyen H. Tran, Mehdi Bennis, Yan Kyaw Tun, Aunas Manzoor, and Choong Seon Hong. A crowdsourcing framework for on-device federated learning. *IEEE Transactions on Wireless Communications*, 19(5):3241–3256, 2020.

Shashi Raj Pandey, Lam Nguyen, and Petar Popovski. FedToken: Tokenized incentives for data contribution in federated learning. In *Workshop on Federated Learning: Recent Advances and New Challenges (in Conjunction with NeurIPS 2022)*, 2022a.

Shashi Raj Pandey, Lam D. Nguyen, and Petar Popovski. A contribution-based device selection scheme in federated learning. *IEEE Communications Letters*, 26(9):2057–2061, 2022b.

Shashi Raj Pandey, Van Phuc Bui, and Petar Popovski. Goal-oriented communications in federated learning via feedback on risk-averse participation. In

2023 IEEE 34th Annual International Symposium on Personal, Indoor and Mobile Radio Communications (PIMRC), pages 1–6, 2023a. doi: 10.1109/PIMRC56721.2023.10293926.

Shashi Raj Pandey, Pierre Pinson, and Petar Popovski. Strategic coalition for data pricing in IoT data markets. *IEEE Internet of Things Journal*, 11(4):6454–6468, 2023b. doi: 10.1109/JIOT.2023.3310660.

R. Tyrrell Rockafellarand Stanislav Uryasev. Optimization of conditional value-at-risk. *Journal of Risk*, 2:21–42, 2000.

Olaf Ronneberger, Philipp Fischer, and Thomas Brox. U-Net: Convolutional networks for biomedical image segmentation. In Medical *Image Computing* and *Computer-Assisted Intervention*—MICCAI 2015: 18th *International Conference*, Munich, Germany, October 5—9, 2015, proceedings, Part III 18, pages 234–241. Springer, 2015.

Lloyd S. Shapley. A value for n-person games, 1953.

Pranava Singhal, Shashi Raj Pandey, and Petar Popovski. Greedy Shapley client selection for communication-efficient federated learning. arXiv preprint arXiv:2312.09108, 2023.

Emilio Calvanese Strinati and Sergio Barbarossa. 6G networks: Beyond Shannon towards semantic and goal-oriented communications. *Computer Networks*, 190:107930, 2021.

Alex Tamkin, Ramtin Keramati, Christoph Dann, and Emma Brunskill. Distributionally-aware exploration for CVaR bandits. In *NeurIPS 2019 Workshop on Safety and Robustness on Decision Making*, 2019.

Elif Uysal, Onur Kaya, Anthony Ephremides, James Gross, Marian Codreanu, Petar Popovski, Mohamad Assaad, Gianluigi Liva, Andrea Munari, Beatriz Soret, Touraj Soleymani, and Karl Henrik Johansson. Semantic communications in networked systems: A data significance perspective. *IEEE Network*, 36(4):233–240, 2022.

Wei Xu, Zhaohui Yang, Derrick Wing Kwan Ng, Marco Levorato, Yonina C. Eldar, and Muane Debbah. Edge learning for B5G networks with distributed signal processing: Semantic communication, edge computing, and wireless sensing. *IEEE Journal of Selected Topics in Signal Processing*, 17(1):9–39, 2023. doi: 10.1109/JSTSP.2023.3239189.

Naoya Yoshida, Takayuki Nishio, Masahiro Morikura, and Koji Yamamoto. MAB-based client selection for federated learning with uncertain resources in mobile networks. In *IEEE Globecom Workshops (GC Wkshps)*, Taipei, Taiwan, Dec. 2020.

Part IV

Security of Semantic Networks

In the fourth section of the book, the exploration of security challenges presented by semantic communications unfolds. Part IV embarks on a comprehensive examination, commencing with an incisive analysis of the multi-domain security risks inherent in deploying deep neural networks for semantic communications.

The intricate landscape of adversarial machine learning is dissected, revealing the nuances of test-time adversarial attacks, both targeted and nontargeted, and their profound impacts on the integrity of deep neural networks. Through meticulous analysis, vulnerabilities exposed by manipulating inputs at various stages of semantic communication are laid bare, unveiling potential avenues for exploitation.

Emerging from this scrutiny is the introduction of the encrypted, privacy-preserving semantic communication system (EnSCS), a groundbreaking framework seamlessly integrating universality and confidentiality. At its core lies an innovative adversarial encryption training scheme, meticulously crafted to ensure the unwavering accuracy of semantic communication in both encrypted and unencrypted modalities.

Through the deployment of robust adversarial training methodologies, the EnSCS stands as a bastion of integrity and accuracy in conveying semantics, erecting a secure and private sanctuary for information exchange. Within this fortified fortress, the preservation of data integrity and the safeguarding of privacy reign supreme herald a new era of trust and reliability in semantic communications.

Foundations of Semantic Communication Networks, First Edition.
Edited by Walid Saad, Christina Chaccour, Christo Kurisummoottil Thomas, and Merouane Debbah.

15

Securing Semantic Communications Against Adversarial Attacks

Yalin Evren Sagduyu[1], *Aylin Yener*[2], *and Sennur Ulukus*[3]

[1] *Nexcepta, Gaithersburg, MD, USA*
[2] *Department of Electrical and Computer Engineering, INSPIRE@OhioState Research Center, The Ohio State University, Columbus, OH, USA*
[3] *Department of Electrical and Computer Engineering, University of Maryland, College Park, MD, USA*

15.1 Introduction

We study the susceptibility of semantic communications to adversarial attacks compromising the preservation of semantic information. The objective of traditional communications systems is to recover information transmitted over a channel by minimizing a reconstruction loss (e.g., symbol error probability) at the receiver. For that purpose, the transmitter executes source coding, channel coding, and modulation operations to generate the transmitted signals, whereas the receiver executes demodulation, channel coding, and source decoding operations to reconstruct source inputs at the receiver. These operations can be performed individually or jointly, and deep learning has been shown to be highly effective in optimizing the design of transmitter–receiver functionalities [O'Shea and Hoydis, 2017; Bourtsoulatze et al., 2019].

In addition to the traditional objective of data recovery, semantic communications aims to preserve the meaning during the information transfer [Guler and Yener, 2014; Güler et al., 2018; Kountouris and Pappas, 2021; Qin et al., 2021; Strinati and Barbarossa, 2021; Xie et al., 2021; Chaccour et al., 2022; Uysal et al., 2022; Farshbafan et al., 2023; Gündüz et al., 2023]. Semantic communications can be set up by employing an encoder at the transmitter and a decoder at the receiver that are jointly trained. The encoder encapsulates joint operations of source coding, channel coding, and modulation, while the decoder performs two tasks, namely (i) a deep learning task with the underlying labels that capture whether semantic information is preserved or not, and (ii) a task of reconstructing data samples from the transmitter. By pursuing a multitask learning approach, the goal is to achieve

Foundations of Semantic Communication Networks, First Edition.
Edited by Walid Saad, Christina Chaccour, Christo Kurisummoottil Thomas, and Merouane Debbah.

high accuracy in task completion as well as low reconstruction loss while keeping the number of channel uses (determined by the encoder output size) at a low level for a delay and energy-efficient operation.

While semantic communications offers major benefits, we highlight that semantic communications is highly vulnerable to various attacks. In the context of adversarial machine learning, the semantic information can be falsified in test time by an adversarial (evasion) attack that manipulates inputs to the deep neural networks (DNNs) and fools them into making errors in recovering semantic information, namely identifying correct labels. Adversarial attacks aim to change the label that captures semantic information to any other label (different from the intended one) during information transfer. We consider both untargeted and targeted attacks to change the label that captures semantic information to any other label or a particular target label (different from the intended one), respectively.

We consider different methods for generating adversarial attacks, namely fast gradient sign method (FGSM), projected gradient descent (PGD), basic iterative method (BIM), momentum iterative method (MIM), and DeepFool. We use CIFAR-10 image data as the input at the transmitter and consider additive white Gaussian noise (AWGN) and Rayleigh channels to model communications from the transmitter to the receiver. In this setting, we demonstrate that adversarial attacks can effectively disrupt the delivery of semantic information from the transmitter to the receiver, i.e., reduce the accuracy of the deep learning task that is used to verify whether the semantic information is preserved at the receiver. We show that small adversarial perturbations (relative to input data) are sufficient for that purpose.

Subsequently, we delve into defense, featuring adversarial training, as we endeavor to fortify semantic communication systems against the evolving landscape of adversarial exploits. Adversarial training equips the DNN models with the ability to not only perform well under normal conditions but also to exhibit resilience in the face of potential adversarial attacks [Goodfellow et al., 2014; Madry et al., 2017]. This proactive approach enhances the overall security and reliability of DNN models. We show that it is feasible to reduce the attack performance by making the semantic communication robust to adversarial manipulations via adversarial training. Overall, our results provide insights into potential threats to semantic communications and highlight the need to bolster the robustness of semantic communications against subtle but effective threats of adversarial attacks.

The remainder of the chapter is organized as follows. Section 15.2 discusses semantic communications as a deep learning paradigm. Section 15.3 presents the multitask learning framework for semantic communications. Section 15.4 describes adversarial machine learning as an emerging attack vector and discusses methods to launch adversarial attacks. Section 15.5 evaluates vulnerabilities of semantic communications to adversarial attacks. Section 15.6 describes methods to protect DNNs against adversarial attacks. Section 15.7 evaluates the defense performance of adversarial training to protect semantic communications against adversarial attacks. Section 15.8 discusses future research directions. Section 15.9 concludes the chapter.

15.2 Semantic Communications

The emergence of 6G communication networks marks the onset of a transformative era in connectivity, offering unparalleled speeds, extremely low latency, and increased capacity to support an ever-growing array of applications, from augmented reality/virtual reality (AR/VR) and vehicle-to-everything (V2X) networks to Internet of Things (IoT) deployments. As the demand for seamless and efficient communication intensifies, the need for innovative solutions becomes paramount. Semantic communications emerge as a promising paradigm, leveraging the capabilities of DNNs and multitask learning to redefine the way information is encoded, transmitted, and decoded.

Semantic information in the context of 6G communications embodies a deeper layer of meaning and understanding within the transmitted data. Unlike traditional communication paradigms that focus primarily on encoding and decoding raw information, semantic communications strive to capture the inherent context, relationships, and significance of the data being transmitted. In the context of 6G, the emphasis on semantic information becomes crucial, as it enables a more intelligent, adaptive, and context-aware communication environment.

- VR/AR applications within 6G networks benefit immensely from semantic information. In immersive experiences, semantic understanding allows for more realistic interactions and seamless integration of virtual elements with the physical world. For example, in a healthcare scenario, a surgeon using AR-assisted procedures can leverage semantic information to accurately overlay medical data onto a patient's anatomy, providing real-time insights and enhancing surgical precision.
- The significance of semantic information amplifies in the context of V2X communications. In autonomous driving scenarios, vehicles communicating with each other need to exchange more than just location and speed; they must share a semantic understanding of the road conditions, traffic rules, and potential hazards. This semantic layer ensures safer and more reliable communication, contributing to the overall success of autonomous transportation systems.
- 6G's embrace of semantic communications translates into a more nuanced interaction between devices in IoT. For instance, consider a smart city where a

network of IoT devices collaborates to manage traffic flow. Semantic information here goes beyond simple data exchange, incorporating the understanding of traffic patterns, real-time conditions, and even anticipating potential congestion points. This enriched semantic layer enables more efficient traffic management, leading to reduced congestion, lower emissions, and overall improved urban mobility.

Semantic communications marks a significant departure from conventional communications by not only ensuring accurate information reconstruction but also prioritizing the preservation of intrinsic meaning throughout the communication process. We consider a deep learning framework, where the transmitter and receiver functions are modeled as DNNs, as shown in Figure 15.1. To that end, the training loss encompasses traditional reconstruction loss alongside a novel semantic loss, specifically designed to gauge the fidelity of conveyed meaning. This semantic loss is evaluated based on the success of completing mission-critical tasks with the transmitted information, ensuring its intended significance is retained.

We build semantic communications as a training process of DNNs at the transmitter and the receiver. An encoder DNN at the transmitter unifies source coding, channel coding, and modulation into a cohesive operation. This integrated approach allows the encoder to adapt and optimize the encoded data for transmission across diverse and dynamic channels. In the meantime. the receiver employs two decoder DNNs that undertake a dual responsibility – evaluating the preservation of semantic information through a deep learning task and reconstructing the original data samples. This twofold objective ensures that the transmitted information not only reaches its destination intact but also retains its intrinsic semantic

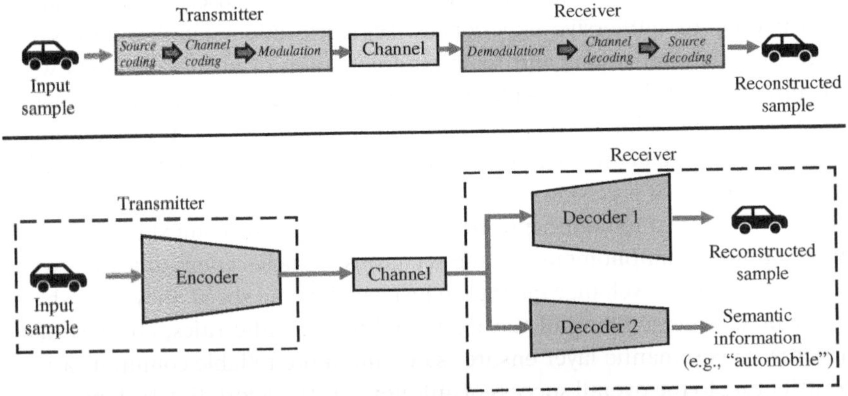

Figure 15.1 (a) Conventional communications (bottom) versus (b) semantic communications (top). Source: Choi et al. (2024)/IEEE/CC BY 4.0.

meaning. One decoder focuses on the reconstruction process, incorporating joint demodulation, channel decoding, and source decoding. In the meantime, a second decoder executes a deep learning task, namely image classification, to assess semantic preservation.

Through joint training of three DNNs (one encoder and two decoders), this approach effectively captures latent feature spaces and transfers compressed feature vectors reliable with minimal channel uses. The outcome is a system that balances low reconstruction loss with minimal semantic loss, demonstrating proficiency in preserving both fidelity and meaning during information transfer.

The encoder and the two decoders are jointly trained through multitask learning. This collaborative learning paradigm optimizes the entire system to simultaneously maximize task accuracy and minimize reconstruction loss, while also managing channel resources efficiently. The goal is to strike a delicate balance, where the number of channel uses (determined by the encoder output size) needs to be kept low, ensuring delay and energy-efficient operations without compromising the integrity of the transmitted semantic information. This holistic approach signifies a paradigm shift from conventional communication methods, where semantic content is often neglected during the encoding and decoding processes.

15.3 Multitask Learning For Semantic Communications

Multitask learning represents a departure from the conventional single-task approach. Multitask learning, in essence, involves training a model to excel at multiple related tasks concurrently. In semantic communications, where the preservation of semantic information and the accurate reconstruction of data coexist as primary objectives, multitask learning emerges as a powerful strategy. It enables the encoder and decoder to collaboratively navigate the intricacies of both tasks, promoting a holistic understanding of the transmitted information. Multitask learning operates on the premise that certain aspects of knowledge and representation are shared among different tasks in semantic communications:

Shared feature extraction: The initial layers of the neural network, governed by shared parameters W_{shared}, extract features that encapsulate commonalities between semantic evaluation and data reconstruction tasks. This shared understanding lays the foundation for a comprehensive learning process.

Task-specific refinement: Subsequent layers, guided by task-specific parameters $W_{task}^{(i)}$, delve into intricacies unique to each task. This task-specific refinement allows the model to adapt and specialize, ensuring a nuanced approach to semantic preservation and data reconstruction.

Loss function framework: The learning process is steered by task-specific loss functions denoted as L_{task_i}, one for the semantic evaluation task gauging the accuracy of semantic information preservation and another one for the data reconstruction task measuring the dissimilarity between the original and reconstructed data.

Let L_{total} represent the combined loss function for both semantic evaluation and data reconstruction tasks. In a scenario with two tasks, the objective function L_{total} is expressed as:

$$L_{total} = \lambda_{SE} \cdot L_{SE}(W_{shared}, W_{task}^{(SE)}) + \lambda_{DR} \cdot L_{DR}(W_{shared}, W_{task}^{(DR)}). \tag{15.1}$$

Here, L_{SE} and L_{DR} represent the loss functions for the semantic evaluation and data reconstruction tasks, respectively. The parameters W_{shared}, $W_{task}^{(SE)}$, and $W_{task}^{(DR)}$ denote shared and task-specific parameters. The hyperparameters λ_{SE} and λ_{DR} control the influence of each task on the overall learning process.

This formulation encapsulates the essence of multitask learning in semantic communications, where the collaborative optimization of shared and task-specific parameters strives to achieve a delicate balance between semantic accuracy and data fidelity. Multitask learning has been applied to both semantic and task-oriented communications along with extensions to integrated sensing and communications [Sagduyu et al., 2023c,2023d]. In the context of 6G communication systems, this integrated approach contributes to the efficiency and adaptability required for next-generation connectivity.

We assume that the transmitter has image samples from the CIFAR-10 dataset and needs to deliver them to a receiver over a wireless channel. The semantic information conveyed by the received data samples is captured by their labels. CIFAR-10 is a widely employed benchmark in computer vision, consisting of 60,000 32×32 color images distributed across 10 classes, each containing 6000 images. These classes encompass a diverse range of objects and animals, including airplanes, automobiles, birds, cats, deer, dogs, frogs, horses, ships, and trucks as 10 classes. The dataset is divided into a training set, containing 50,000 images, and a test set with 10,000 images, facilitating the training and evaluation of machine learning models. The training and test inputs are normalized to [0,1] before feeding them to the DNNs.

We consider binary cross entropy (BCE) as the loss function of Decoder 1 that reconstructs data samples, and categorical cross entropy (CCE) as the loss function of Decoder 2 that checks the semantic information. Adam optimizer is used with learning rate 10^{-3} and batch size is selected as 64. In multitask learning, the weights for reconstruction and semantic task are set 0.95 and 0.05, respectively.

The encoder–decoder architectures for semantic communications are shown in Table 15.1. The selection of DNN architectures begins with a small number of

Table 15.1 Encoder–decoder architectures for semantic communications.

Network	Layer	Properties
Encoder	Input	Size: 32 × 32 × 3
	Conv2D	Filter size: 32, kernel size: (3,3), strides: (1, 1), activation: ReLU
	Conv2D	Filter size: 32, kernel size: (3,3), strides: (1, 1), activation: ReLU
	MaxPooling2D	Pool size: (2,2)
	Dropout	Dropout rate: 0.25
	Conv2D	Filter size: 64, kernel size: (3,3), strides: (1, 1), activation: ReLU
	Conv2D	Filter size: 64, kernel size: (3,3), strides: (1, 1), activation: ReLU
	MaxPooling2D	Pool size: (2,2)
	Dropout	Dropout rate: 0.25
	Conv2D	Filter size: 128, kernel size: (3,3), strides: (1, 1), activation: ReLU
	Conv2D	Filter size: 128, kernel size: (3,3), strides: (1, 1), activation: ReLU
	MaxPooling2D	Pool size: (2,2)
	Dropout	Dropout rate: 0.25
	Flatten	—
	Dense	Size: 512, activation: ReLU
	Dropout	Dropout rate: 0.25
	Dense	Size: n_c, activation: linear
Decoder 1	Input	Size: n_c
	Dense	Size: 8192, activation: ReLU
	Reshape	Size: (8,8,128)
	Conv2DTranspose	Filter size: 32, kernel size: (3,3), strides: (2, 2), activation: ReLU
	Conv2DTranspose	Filter size: 16, kernel size: (3,3), strides: (2, 2), activation: ReLU
	Conv2DTranspose	Filter size: 3, kernel size: (3,3), strides: (2, 2), activation: Sigmoid
Decoder 2	Input	Size: n_c
	Dense	Size: n_c, activation: ReLU
	Dense	Size: n_c, activation: ReLU
	Dense	Size: 10, activation: Softmax

layers and compact layer sizes, progressively expanding them until optimal performance is attained. The overall DNN is parameterized by the size of encoder output (number of channel uses), n_c. Total number of trainable parameters is 863,131 (565,054 for Encoder, 295,907 for Decoder 1, and 2170 for Decoder 2) for $n_c = 20$, and it grows to 3,394,921 for $n_c = 300$, demonstrating the need for small n_c also from the memory requirement point of view in addition to latency, energy, and resource-efficiency benefits.

Figures 15.2 and 15.3 show the semantic task accuracy, namely, classifier accuracy of Decoder 2, and the mean squared error (MSE) of image reconstruction by Decoder 1, as a function of the signal-to-noise ratio (SNR) of communication channel. Figures 15.4 and 15.5 show the semantic task accuracy and the MSE as a function of the encoder output size, namely the number of channel uses n_c that reflects the compression ratio. Both semantic task accuracy and MSE improve (i.e., the semantic information is better preserved and the input samples are better reconstructed at the receiver) with the SNR and n_c. As expected, the performance decreases when we switch the channel model from the AWGN channel to the Rayleigh channel. Note that the performance quickly improves and converges with increasing SNR or n_c under the AWGN channel. Due to the more complex channel conditions, we need relatively better channels and higher compression ratios under the Rayleigh channel (compared to Gaussian channel) to achieve the performance improvements.

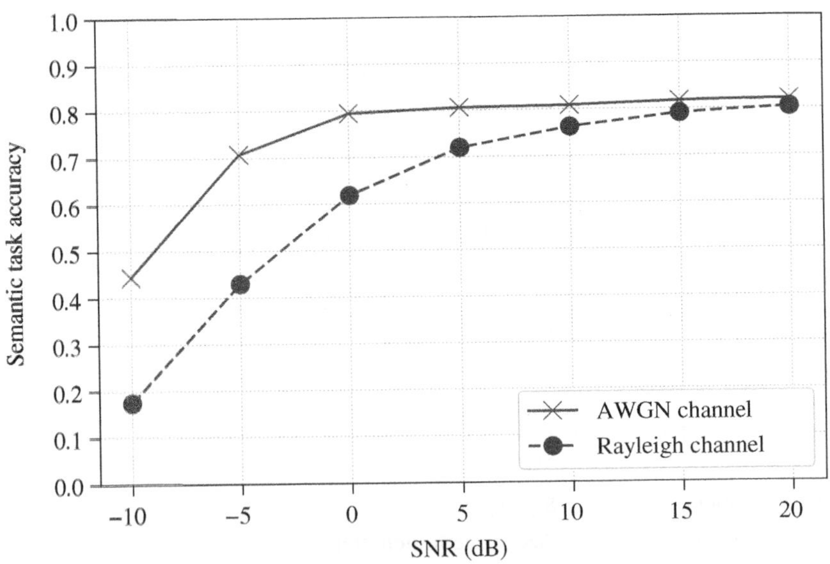

Figure 15.2 Semantic task accuracy as a function of the SNR.

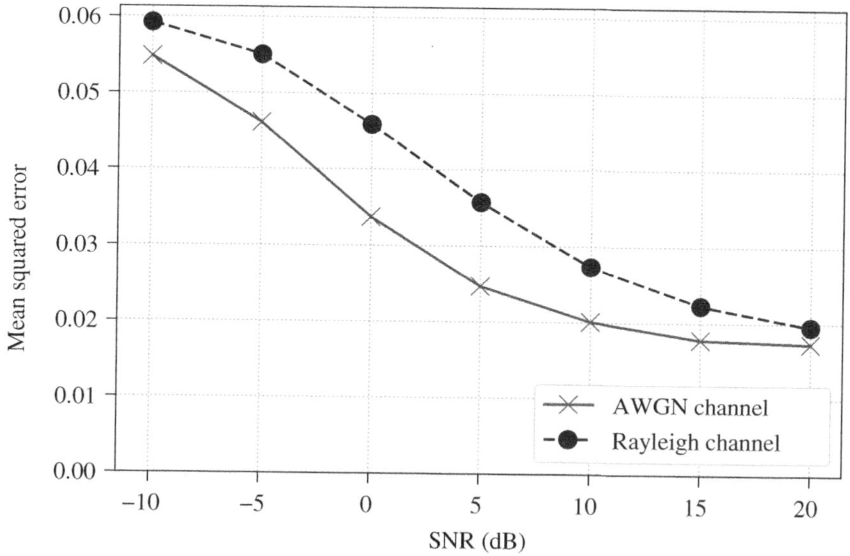

Figure 15.3 MSE of reconstruction as a function of the SNR.

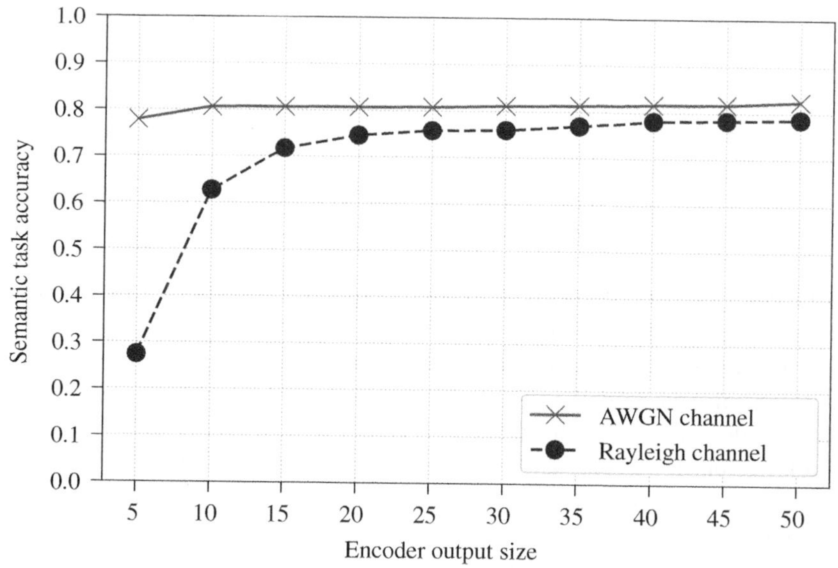

Figure 15.4 Semantic task accuracy as a function of encoder output size.

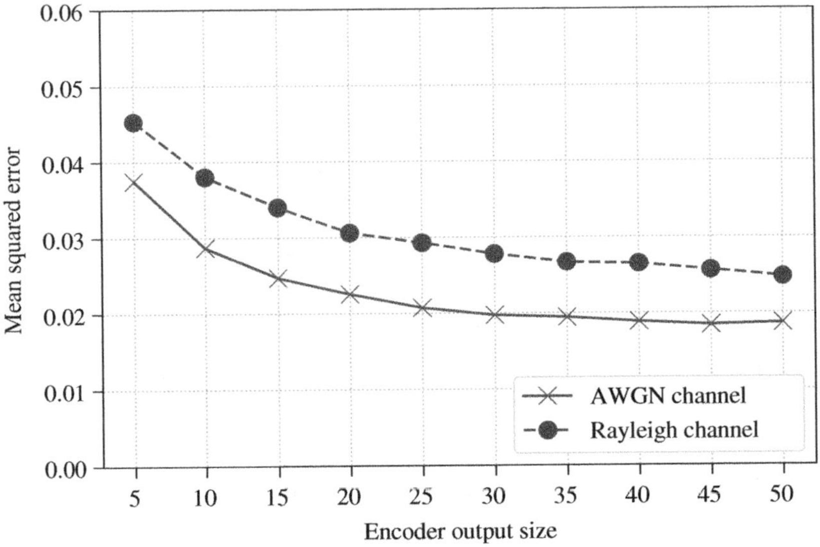

Figure 15.5 MSE of reconstruction as a function of the encoder output size.

15.4 Adversarial Attacks

Adversarial machine learning studies vulnerabilities in machine learning models and corresponding defense mechanisms. Malicious actors can manipulate models by exploiting their weaknesses during both training and testing phases. These attacks aim to compromise the model's functionality, leading to incorrect predictions, misclassifications, or unauthorized access to sensitive information.

Adversarial (evasion) attacks: Manipulate input data to induce misclassifications during model evaluation.

Poisoning (causative) attacks: Inject malicious data during the training phase to compromise the model's performance such that model's predictions are influenced by the injected poisoned data in test time.

Backdoor (trojan) attacks: Embed hidden triggers or patterns in the model during training and activate them to cause model behavior to deviate from normal functionality.

Membership inference attacks: Determine if a specific data sample is part of the model's training dataset.

Model inversion attacks: Recover sensitive information about training data from the model's outputs by exploiting model predictions to infer characteristics of the training data and reverse-engineer sensitive attributes from model outputs.

Model extraction attacks: Seek to duplicate or clone the architecture and functionality of the target model.

Attribute inference attacks: Infer sensitive attributes of input data through the analysis of model predictions.

Vulnerabilities of wireless systems to adversarial machine learning have been extensively studied with a particular focus on DNN-driven signal classification and spectrum monitoring applications [Sagduyu et al., 2020, 2023e; Adesina et al., 2023]. While wireless security has been investigated under poisoning attacks [Sagduyu et al., 2019], backdoor attacks [Davaslioglu and Sagduyu, 2019], and membership inference attacks [Shi and Sagduyu, 2023], adversarial attacks have been the primary focus of security studies for a variety of wireless systems, including modulation classification [Sadeghi and Larsson, 2018; Kim et al., 2021a], power control [Kim et al., 2021b; Manoj et al., 2022], auto-encoder communications [Sadeghi and Larsson, 2019], and 5G spectrum sensing and authentication [Sagduyu et al., 2021]. Recently, there have been efforts to study semantic and task-oriented communications under adversarial machine learning attacks, including adversarial attacks [Sagduyu et al., 2023a,2023f–2024] and backdoor attacks [Sagduyu et al., 2023b].

In this chapter, we focus on adversarial attacks as strategic manipulations of input data designed to subvert the decision-making process of the machine learning model in test (inference) time. The attacker introduces carefully crafted perturbations to input samples with the aim of inducing misclassifications or altering the model's outputs. These perturbations are often small and can significantly impact the model's predictions, leading to potentially erroneous results.

15.4.1 Untargeted Adversarial Attack

In an untargeted attack, the adversary seeks to cause misclassification without specifying a particular target class. The perturbation δ is introduced to the input x with the objective of maximizing the chosen loss function L under a constraint, typically bounded by a small perturbation budget ϵ. Mathematically, it can be expressed as:

$$\text{argmax}_{\delta} L(f_{\theta}(x+\delta), y) \text{ subject to } ||\delta|| \leq \epsilon. \tag{15.2}$$

Here, $f_{\theta}(x)$ represents the model's output for input x, θ is the set of DNN model parameters, y is the true label, and L is the selected loss function. The perturbation δ is constrained by the perturbation budget ϵ to ensure that the alterations remain imperceptible.

15.4.2 Targeted Adversarial Attack

In a targeted attack, the adversary aims to guide the model toward misclassification into a specific, predefined target class y_{target}. The perturbation δ is optimized to minimize the loss function L with respect to the target class. The formulation is as follows:

$$\text{argmin}_{\delta}\, L(f_{\theta}(x+\delta), y_{target}) \text{ subject to } \|\delta\| \leq \epsilon. \tag{15.3}$$

Similar to the untargeted case, $f_{\theta}(x)$ is the model's output for input x, and L is the chosen loss function. The perturbation δ remains constrained by the perturbation budget ϵ, ensuring that the manipulation remains stealthy while achieving the desired misclassification into the specified target class.

15.4.3 Fast Gradient Sign Method (FGSM) Attack

In an untargeted FGSM attack, the perturbation δ is crafted to amplify the loss for the correct class, steering the model toward an incorrect prediction [Goodfellow et al., 2014]. For that purpose, the FGSM attack takes a single step in the direction of the gradient of the loss with respect to the input, multiplied by a small perturbation magnitude.

Given an original input data point x, a neural network model f, a loss function L, and a perturbation magnitude ϵ:

1. **Compute the gradient**: Compute the gradient of the loss with respect to a given input sample by back-propagating the loss from the output layer to the input layer:

$$\nabla J(x) = \frac{\partial L}{\partial x}(f_{\theta}(x), y). \tag{15.4}$$

2. **Sign the gradient**:

$$\delta = \epsilon \cdot \text{sign}(\nabla J(x)), \tag{15.5}$$

where ϵ is a small positive constant.

3. **Add perturbation to the original sample**:

$$x_{adv} = x_{original} + \delta. \tag{15.6}$$

4. **Clipping**: Clip the values of the adversarial sample x_{adv} to ensure they stay within a valid input range:

$$x_{adv} = \text{clip}(x+\delta, x_{min}, x_{max}), \tag{15.7}$$

where $\text{clip}(\cdot, x_{min}, x_{max})$ clips the sample to x_{min} or x_{max} or keeps it as is depending on its value.

Note that if the attack is targeted, the perturbation δ is tailored to reduce the loss for the target class, steering the model toward a misclassification in that specific direction for that target class. For that purpose, the gradient is taken with respect to a particular target label y_{target} and the perturbation is generated in the opposite direction of the gradient such that steps 1 and 2 are changed as the following, while steps 1 and 2 remain the same.

1. **Compute the gradient**: Compute the gradient of the loss with respect to a given input sample and a target label y_{target} by back-propagating the loss from the output layer to the input layer:

$$\nabla J(x) = \frac{\partial L(f_\theta(x), y_{target})}{\partial x}. \tag{15.8}$$

2. **Sign the gradient**:

$$\delta = -\epsilon \cdot \text{sign}(\nabla J(x)), \tag{15.9}$$

where ϵ is a small positive constant.

15.4.4 Projected Gradient Descent (PGD) Attack

In a PGD attack, an adversarial example is generated through iterative optimization, taking into account both the gradient information and the constraint set to ensure that the perturbed data remains within acceptable bounds [Madry et al., 2017]. The projection step helps maintain the adversarial examples within the specified constraint set, contributing to the model robustness.

Given an original input data point x, a neural network model f, a loss function L, a perturbation magnitude ϵ, a step size or learning rate α, and a perturbation constraint set S:

1. **Initialization**:

$$x^{(0)} = x. \tag{15.10}$$

2. **Iteration**: For $t = 1, 2, \ldots, T$ (where T is the number of iterations or until convergence):

   (a) **Compute gradient**:

$$\nabla J(x^{(t-1)}) = \frac{\partial L}{\partial x}(f_\theta(x^{(t-1)}), y). \tag{15.11}$$

   (b) **Perturbation**:

$$\delta^{(t)} = \text{clip}(\epsilon, \alpha, \nabla J(x^{(t-1)})). \tag{15.12}$$

(c) **Projection**:

$$x^{(t)} = \text{Proj}_S(x^{(t-1)} + \delta^{(t)}), \tag{15.13}$$

where Proj_S is the projection operator that enforces the perturbed data to lie within the constraint set S.

3. **Output adversarial example**: The final adversarial example is given by $x_{adv} = x^{(T)}$.

15.4.5 Basic Iterative Method (BIM) Attack

BIM is an iterative variant of FGSM [Kurakin et al., 2018]. Similar to PGD, BIM performs multiple iterations of perturbation, but unlike PGD, it does not involve projection onto a constraint set. BIM iteratively applies small perturbations to the input to maximize the loss and generate adversarial examples.

Given an original input data point x, a neural network model f, a loss function L, a perturbation magnitude ϵ, a step size or learning rate α, and a number of iterations T:

1. **Initialization**:

$$x^{(0)} = x. \tag{15.14}$$

2. **Iteration**: For $t = 1, 2, \ldots, T$:
(a) **Compute gradient**:

$$\nabla J(x^{(t-1)}) = \frac{\partial L}{\partial x}(f_\theta(x^{(t-1)}), y). \tag{15.15}$$

(b) **Perturbation**:

$$\delta^{(t)} = \epsilon \cdot \text{sign}(\nabla J(x^{(t-1)})). \tag{15.16}$$

(c) **Clipping**:

$$x^{(t)} = \text{clip}(x^{(t-1)} + \alpha \cdot \text{sign}(\nabla J(x^{(t-1)})), x - \epsilon, x + \epsilon). \tag{15.17}$$

3. **Output adversarial example**: The final adversarial example is given by $x_{adv} = x^{(T)}$.

15.4.6 Momentum Iterative Method (MIM) Attack

The MIM attack is an extension of the PGD attack that incorporates momentum into the optimization process [Dong et al., 2018]. The use of momentum helps to accelerate the optimization process and can be particularly effective in escaping local minima.

Given an original input data point x, a neural network model f, a loss function L, a perturbation magnitude ϵ, a step size or learning rate α, a momentum parameter μ, and a perturbation constraint set S:

1. **Initialization**:

$$x^{(0)} = x, \quad v^{(0)} = 0. \tag{15.18}$$

2. **Iteration**: For $t = 1, 2, \ldots, T$ (where T is the number of iterations or until convergence):
   (a) **Compute gradient**:

$$\nabla J(x^{(t-1)}) = \frac{\partial L}{\partial x}(f_\theta(x^{(t-1)}), y). \tag{15.19}$$

   (b) **Update momentum**:

$$v^{(t)} = \mu \cdot v^{(t-1)} + \frac{\nabla J(x^{(t-1)})}{\|\nabla J(x^{(t-1)})\|_2}. \tag{15.20}$$

   (c) **Perturbation**:

$$\delta^{(t)} = \text{clip}(\epsilon, \alpha, v^{(t)}). \tag{15.21}$$

   (d) **Projection**:

$$x^{(t)} = \text{Proj}_S(x^{(t-1)} + \delta^{(t)}), \tag{15.22}$$

   where Proj_S is the projection operator that enforces the perturbed data to lie within the constraint set S.
3. **Output adversarial example**: The final adversarial example is given by $x_{adv} = x^{(T)}$.

15.4.7 DeepFool Attack

DeepFool is an iterative attack method that aims to find the minimum perturbation needed to misclassify an input sample [Moosavi-Dezfooli et al., 2016]. It iteratively computes the linear approximation of the decision boundaries and generates perturbations in the direction of the decision boundary until the sample is misclassified.

Given an original input data point x, a neural network model f, a loss function L, a maximum perturbation magnitude ϵ, and maximum number of iterations T:

1. **Initialization**:

$$x^{(0)} = x. \tag{15.23}$$

2. **Iteration**: For $t = 1, 2, \ldots, T$:
   (a) **Compute gradient**:

$$\nabla J(x^{(t-1)}) = \nabla f_{true}(x^{(t-1)}) - \nabla f_{target}(x^{(t-1)}). \tag{15.24}$$

(b) **Compute adversarial perturbation**:

$$\delta^{(t)} = -\frac{f_{true}(x^{(t-1)}) - f_{target}(x^{(t-1)})}{||\nabla J(x^{(t-1)})||_2^2} \cdot \nabla J(x^{(t-1)}). \tag{15.25}$$

(c) **Update adversarial example**:

$$x^{(t)} = \text{clip}(x^{(t-1)} + \delta^{(t)}, x - \epsilon, x + \epsilon). \tag{15.26}$$

3. **Output adversarial example**: The final adversarial example is given by $x_{adv} = x^{(T)}$.

15.5 Adversarial Attacks On Semantic Communications

For numerical results, we use the following parameters. Number of iterations, T, is 20, for the PGD, BIM, MIM, and DeepFool attacks. Step size α is ϵ/T for the PGD and BIM attacks. Step size α is 0.01 and momentum parameter (decay factor) μ is 0.9 for the MIM attack. The attack perturbation is upper bounded with respect to the input signal power by the perturbation-to-signal ratio (PSR) that is measured in dB.

In the iterative methods, the perturbation is updated over multiple iterations. Therefore, the resulting PSR is not directly given by ϵ but it is determined as a function of ϵ. For all these attacks, we vary the perturbation magnitude ϵ and measure the resulting PSR and the attack performance.

We start with the untargeted FGSM attack and show its effectiveness in fooling the conveyed semantic information compared to using Gaussian noise as perturbation (that we refer to as Gaussian attack). The attack performance is measured by the attack success probability that is defined as the probability that the attack fools the classifier into making classification errors (toward any unintended label). We assume that the SNR of communication channel is 10 dB. Figures 15.6 and 15.7 show the attack success probability and the MSE of reconstruction, respectively, as a function of PSR. The attack performance is similar under both AWGN and Rayleigh channels. Note that the adversarial attack is highly effective in reducing the semantic task accuracy even with small PSRs, whereas the Gaussian attack remains ineffective unless PSR is high. In the meantime, the side effect of adversarial perturbation on the MSE of reconstruction is minor for both attack types unless the PSR is high, indicating that the adversarial attack can manipulate the semantic information stealthily (e.g., with −10 dB PSR) without changing the reconstruction performance at any significant level.

Figures 15.8 and 15.9 compare the attack success probability of untargeted attacks based on different adversarial attack methods (FGSM, BIM, MIM, and DeepFool) under AWGN and Rayleigh channels, respectively. Once the parameters of iterative methods are properly selected, they can all improve

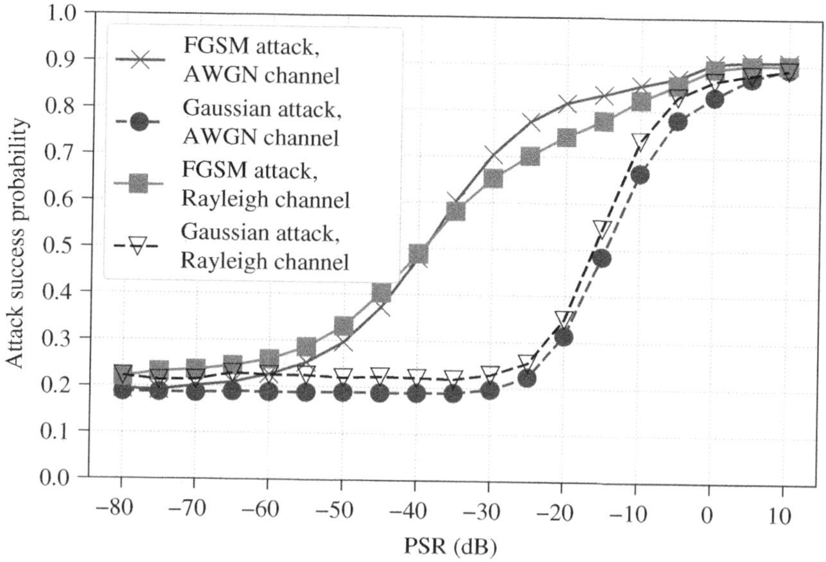

Figure 15.6 Attack success probability for untargeted FGSM attack versus Gaussian perturbation under AWGN and Rayleigh channels.

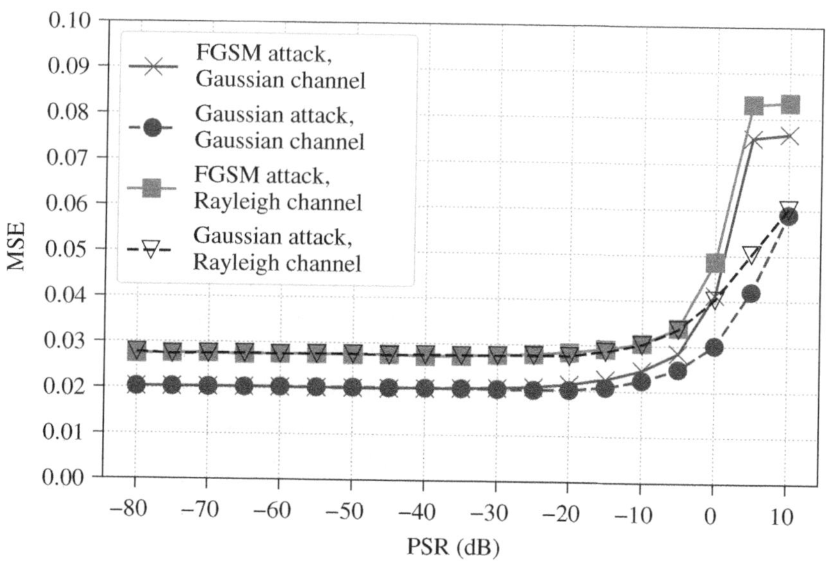

Figure 15.7 MSE for untargeted FGSM attack versus Gaussian perturbation under AWGN and Rayleigh channels.

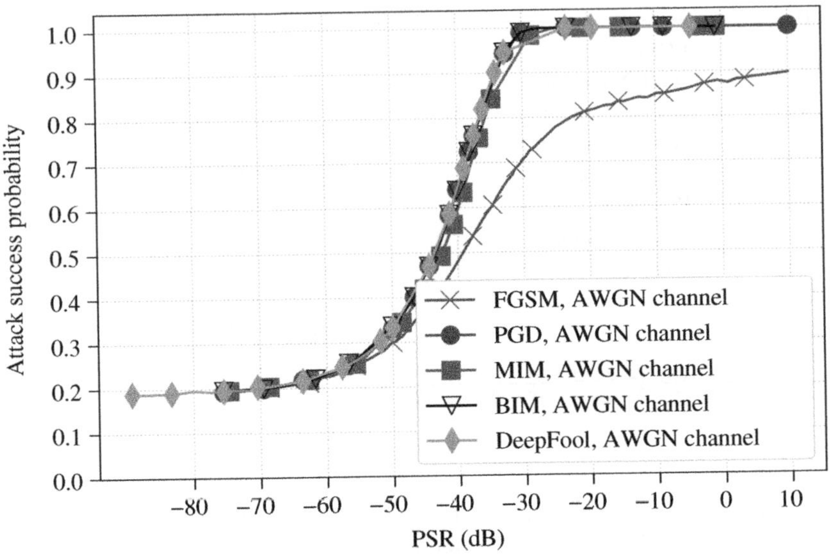

Figure 15.8 Attack success probability for different untargeted attack methods under AWGN channel.

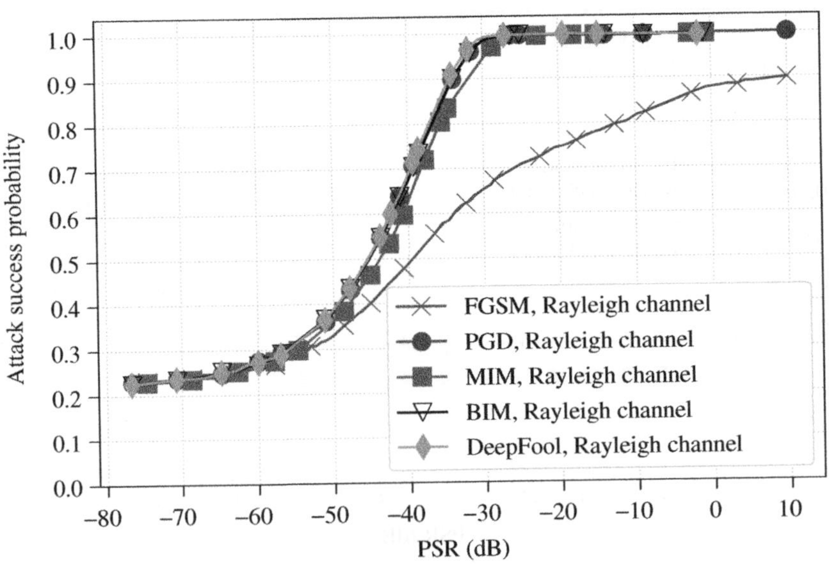

Figure 15.9 Attack success probability for different untargeted attack methods under Rayleigh channel.

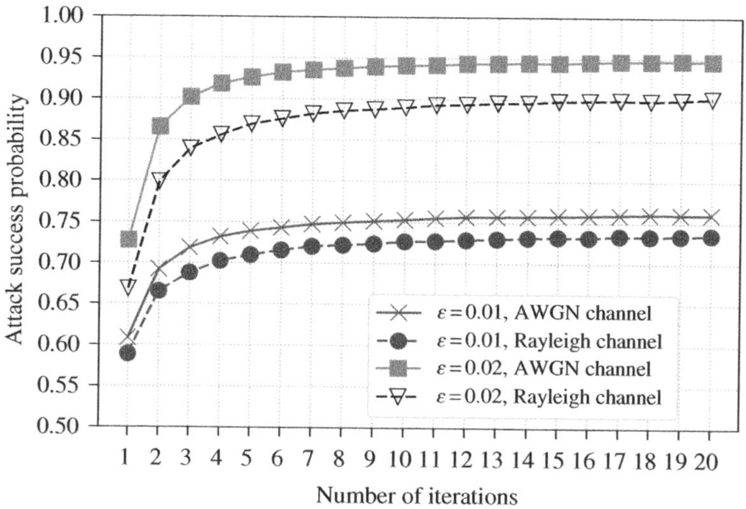

Figure 15.10 Attack success probability for untargeted PGD attack as a function of the number of attack iterations.

the attack success probability significantly over the FGSM attack such that the classifier accuracy drops to zero as the PSR increases.

The number of iterations T plays a crucial role in attack success. As an example, Figure 15.10 shows how the attack success probability increases as T increases under the untargeted PGD attack. While a large T means high complexity and long convergence time, a small T renders the attack ineffective such that T emerges as an important design parameter.

Next, we consider the targeted attack, where the objective is to flip the label from a given original label $y_{original}$ to a given target label y_{target} (different from the original label). For targeted attacks, the attack success probability is defined as the probability that the classifier is fooled into classifying the input samples of a given $y_{original}$ to a given y_{target}. For numerical results, we assume that the $y_{original}$ is "automobile" and y_{target} is "truck," and the SNR of communication channel is 10 dB. To compare the targeted FGSM attack with Gaussian attack, Figures 15.11 and 15.12 show the attack success probability and the MSE of reconstruction, respectively, as a function of PSR. While the targeted FGSM attack is less effective than the untargeted attack, it still outperforms the Gaussian attack significantly across all PSRs. Different from the untargeted FGSM attack, the targeted FGSM attack is noticeably better under the AWGN channel compared to the Rayleigh channel. Similar to the untargeted FGSM attack, the targeted FGSM attack starts affecting the MSE of reconstruction only when the PSR is high.

Figures 15.13 and 15.14 compare the attack success probability of targeted attacks based on different adversarial attack methods (FGSM, BIM, MIM, and

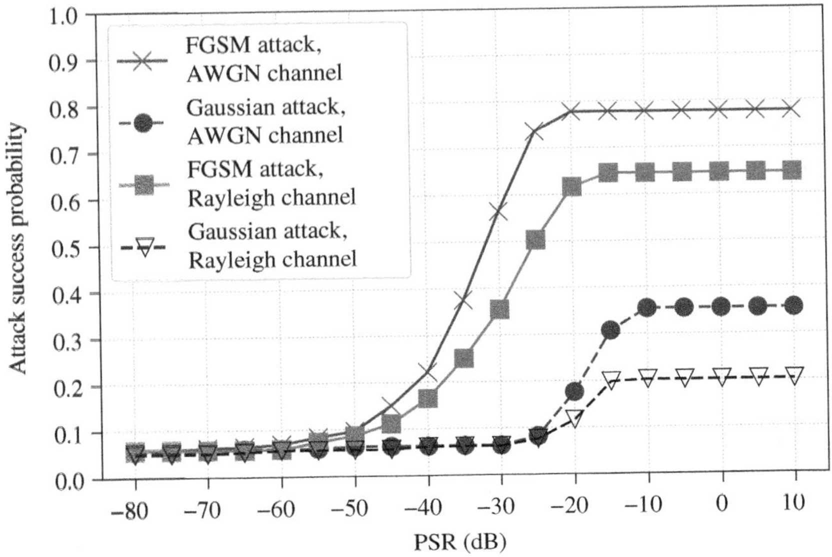

Figure 15.11 Attack success probability for targeted FGSM attack versus Gaussian perturbation under AWGN and Rayleigh channels.

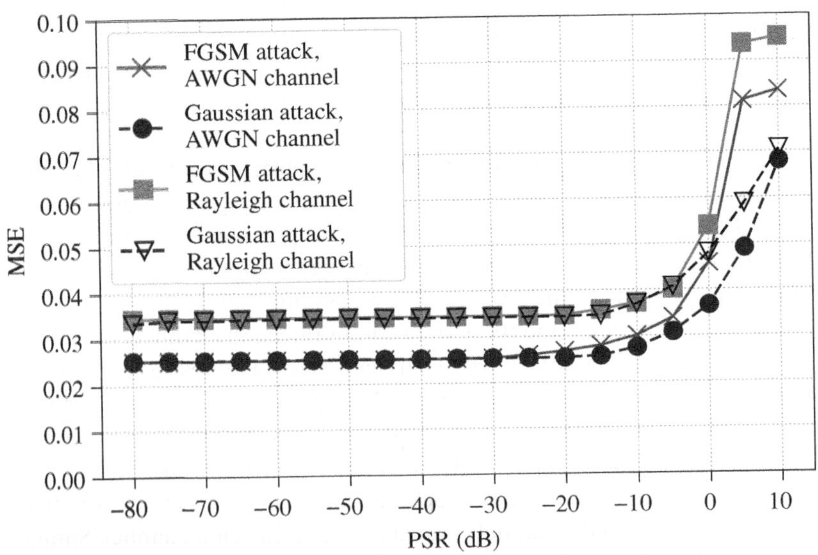

Figure 15.12 MSE for targeted FGSM attack versus Gaussian perturbation under AWGN and Rayleigh channels.

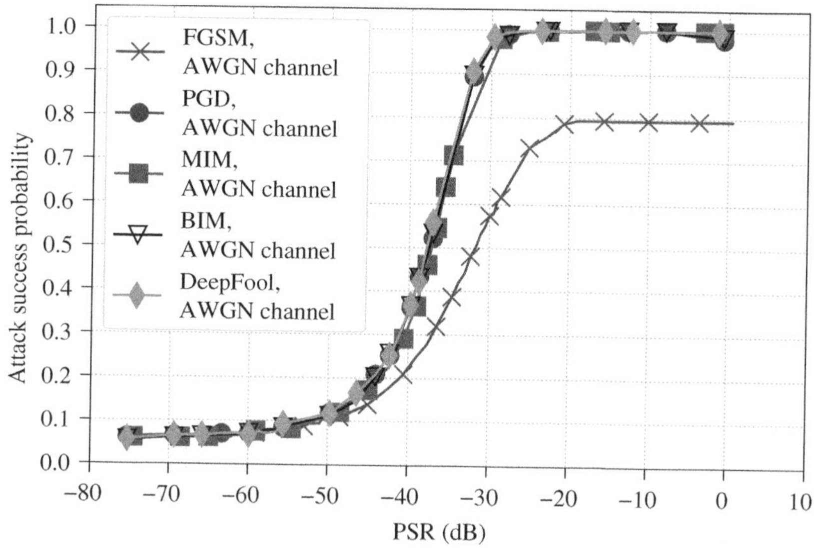

Figure 15.13 Attack success probability for different targeted attack methods under AWGN channel.

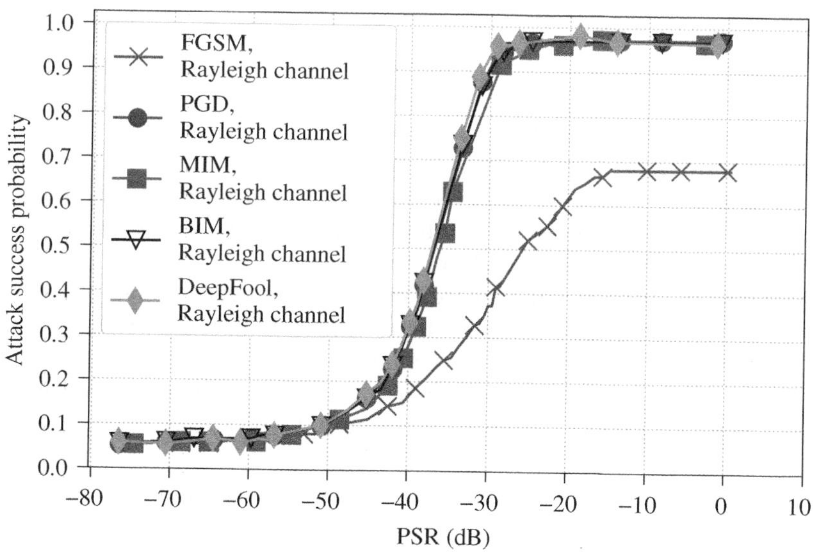

Figure 15.14 Attack success probability for different targeted attack methods under Rayleigh channel.

DeepFool) under AWGN and Rayleigh channels, respectively. Similar to the untargeted attacks, iterative methods are highly effective for targeted attacks. In particular, they can flip the predicted labels from the given $y_{original}$ to the given y_{target} as the PSR increases.

15.6 Defense Against Adversarial Attacks

Defense against adversarial attacks is a critical aspect of ensuring the robustness and reliability of machine learning models. As machine learning systems become increasingly integral to various applications such as in semantic communications, their vulnerability to adversarial perturbations poses a significant challenge. Adversarial attacks involve manipulating input data with small changes to deceive the model and induce misclassifications. In response to this threat, the field of defense against adversarial attacks has emerged, aiming to develop strategies and techniques that enhance a model's resilience to such manipulations. These defenses encompass a diverse range of methodologies.

Adversarial training: Adversarial training aims to enhance model robustness by incorporating adversarial examples during the training process. The goal is to expose the model to a mixture of clean and adversarial examples, allowing it to learn to withstand perturbations. During training, the dataset is augmented with adversarial examples generated using methods like FGSM, providing the model with a more comprehensive understanding of potential adversarial inputs.

Gradient masking: The goal of gradient masking is to limit access to the model's gradients during the training process, making it harder for attackers to estimate the gradient information necessary for crafting effective adversarial examples. Various techniques can be applied to obscure the gradient information, adding an extra layer of defense against adversarial attacks.

Randomized smoothing: Randomized smoothing introduces controlled random noise to input features during both training and inference. The goal is to create uncertainty in the model's predictions, making it less sensitive to small perturbations. By incorporating random noise, the model becomes more robust against adversarial attacks, as it is less likely to be misled by small, crafted perturbations.

Ensemble methods: Ensemble methods incorporate multiple models, providing a diversified defense strategy. By training and deploying an ensemble of models with different architectures or initializations, the likelihood of adversarial examples simultaneously fooling all models is reduced. This collective decision-making process contributes to a more robust defense against adversarial attacks.

Input preprocessing: Input preprocessing involves modifying input data to remove or reduce adversarial perturbations. Techniques such as input normalization, feature squeezing, or image preprocessing are applied to remove high-frequency components that are often manipulated by adversarial attacks. This proactive approach helps mitigate the impact of potential adversarial perturbations.

Certified defenses: Certified defenses aim to provide guarantees on model robustness within a certain region. By employing mathematical methods, these defenses establish certified robustness bounds for the model. If the perturbation is within this bound, the model guarantees correct predictions, enhancing its reliability in the face of adversarial attacks.

Defense by detection: Defense by detection focuses on identifying and rejecting adversarial examples. Anomaly detection techniques are employed to flag samples likely to be adversarial. Once detected, these samples are treated with caution, contributing to a defense strategy that prioritizes accurate predictions and rejects potential adversarial inputs.

Transfer learning from pretrained models: Transfer learning leverages pretrained models as a starting point to enhance robustness. Models pretrained on large datasets capture more robust features, providing a solid foundation against adversarial attacks. Fine-tuning these pretrained models on specific tasks leverages the learned features to improve the model's resilience to adversarial perturbations.

Robust training objectives: Robust training objectives involve optimizing models explicitly for worst-case scenarios. By designing training objectives that account for adversarial scenarios, such as minimizing worst-case loss functions, models can be trained to be more resilient to adversarial perturbations. This approach aims to prepare models for a broad range of potential adversarial inputs.

Regularization techniques: Regularization techniques aim to penalize complex models to reduce overfitting to noise present in adversarial examples. Incorporating regularization terms, such as L1 or L2 regularization, in the loss function helps prevent models from fitting to noise, thereby improving their generalization and resilience to adversarial attacks.

15.7 Adversarial Training as Defense Against Adversarial Attacks on Semantic Communications

We consider adversarial training as the defense mechanism to protect semantic communications against adversarial attacks. Adversarial training stands out among defense strategies due to its direct integration of adversarial examples during the training process, exposing the model to potential manipulations it may face in real-world scenarios. This approach fosters improved robustness by optimizing model parameters with respect to both clean and adversarial losses, resulting in models that are more resilient to diverse adversarial attacks. Adversarial training is adaptable across various model architectures, dynamically adjusting to evolving attack strategies and offering flexibility in combination with other defense methods. Its effectiveness lies in its proactive nature, actively engaging with adversarial scenarios during training, thereby enhancing the overall security and trustworthiness of machine learning models. Below, we describe how adversarial training is applied.

Given a training dataset $\{(x_i, y_i)\}_{i=1}^{N}$ with input–output pairs, where x_i is the input data, y_i is the corresponding true label, and $f_\theta(x)$ represents the model with parameters θ. The goal of adversarial training is to find optimal model parameters θ^* by minimizing a combined loss function:

$$\theta^* = \arg\min_{\theta} \sum_{i=1}^{N} \left(L_{cls}(f_\theta(x_i), y_i) + \lambda \cdot L_{adv}(f_\theta(x_i + \delta_i), y_i) \right), \tag{15.27}$$

where L_{cls} is the classification loss on clean examples, L_{adv} is the adversarial loss on perturbed examples, δ_i is the adversarial perturbation for the ith example, and λ is a hyperparameter controlling the influence of the adversarial loss. The optimization process seeks to find model parameters that minimize the combined loss, encouraging the model to be robust against adversarial perturbations while maintaining good performance on clean examples.

Adversarial training is run over multiple iterations as a proactive defense mechanism to enhance the robustness of semantic communications against adversarial attacks. Adversarial training involves augmenting the training dataset with adversarially perturbed examples, forcing the model to learn from these deceptive inputs and thereby improving its resilience. The iterative nature of this process implies that, in each training cycle, the model encounters new (iteratively adjusted) adversarial samples, gradually adapting to a more diverse set of potential attacks.

The benefits of employing multiple iterations of adversarial training are multifaceted. First and foremost, it fortifies the model against a broader spectrum of adversarial manipulations by continuously exposing it to evolving attack strategies. This adaptive learning approach enables the model to generalize better and enhances its ability to discern genuine inputs from adversarial ones. Additionally, multiple iterations contribute to the model's increased robustness over time, improving its defensive capabilities against sophisticated attacks. Finally, it is feasible to augment the training data not necessarily with a large set of synthetic samples (such as one synthetic sample per real sample) but with a smaller set at

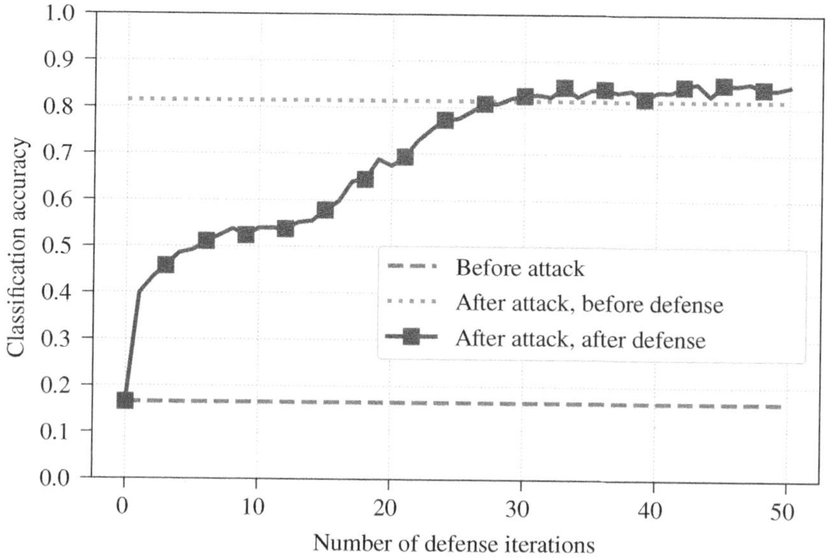

Figure 15.15 Classifier accuracy under the AWGN channel as a function of the defense iterations against the FGSM attack.

each iteration, reducing the computational memory requirements. However, it is essential to strike a balance, considering the computational cost associated with an escalating number of training iterations. Fine-tuning the number of iterations becomes crucial to optimize the trade-off between enhanced robustness and computational efficiency for practical deployment scenarios.

For numerical results, we assume for each iteration that 10,000 input samples are randomly selected, adversarially perturbed samples (adversarial examples) are generated for those input samples, and adversarial examples are added to the training data set. Figures 15.15 and 15.16 show the classifier accuracy subject to the FGSM attack ($\epsilon = 0.1$) as a function of the number of defense iterations under the AWGN and Rayleigh channels, respectively. The attack reduces the classifier accuracy significantly. While the defense is mostly ineffective at the beginning, it starts increasing the classifier accuracy as the number of defense iterations increases. More iterations are needed under the Rayleigh channel to build a robust classifier for semantic task due to its more challenging channel environment compared to the AWGN channel. With sufficient number of defense iterations, the defense brings the classification accuracy for perturbed input samples to the level before the attack is launched. We note that the defense does not reduce the classification accuracy for clean input samples (those without any perturbation added).

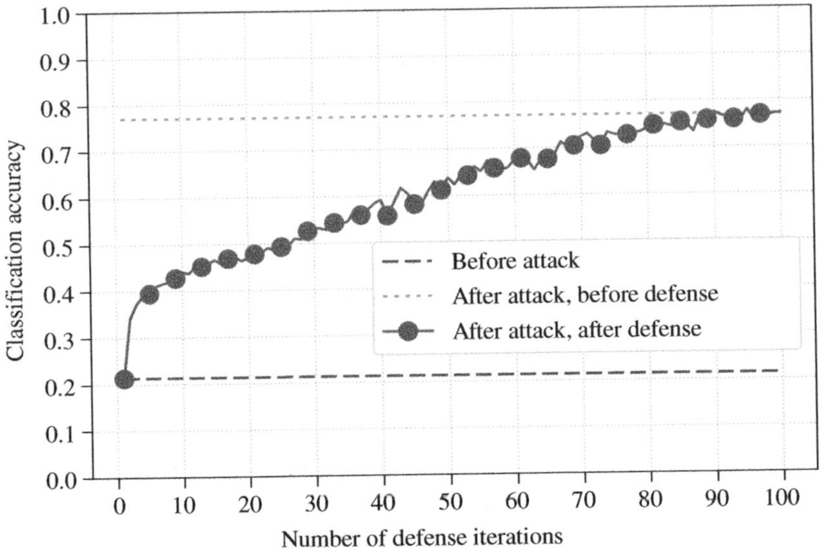

Figure 15.16 Classifier accuracy under the Rayleigh channel as a function of the defense iterations against the FGSM attack.

15.8 Future Research Directions

There are various avenues to explore for a better understanding of security vulnerabilities and defense mechanisms for semantic communications.

- Future work can delve deeper into addressing additional security challenges, especially in the context of multidomain attacks that involve the manipulation of diverse inputs, such as images, as well as wireless signals transmitted over the air. To that end, it is necessary to explore advanced strategies to fortify the system against evolving adversarial threats.
- Investigating more sophisticated defense mechanisms beyond adversarial training, such as randomized smoothing or incorporating cryptographic techniques, could enhance the robustness of semantic communication systems.
- Future research directions include exploring the transferability of adversarial attacks between different communication modalities and domains. Understanding how attacks devised for one type of semantic communication system may generalize to others is crucial for devising comprehensive defense strategies.
- Additionally, it is essential to study the effect of adversarial attacks on other performance metrics and develop strategies to mitigate these effects. These metrics are not limited to physical layer considerations, and can extend to higher layers

to better understand the attack implications on network performance, e.g., the Age-of-Information (AoI).

- Current studies on semantic communications security have primarily focused on point-to-point communications, warranting exploration into the unique challenges posed by relay and multireceiver communications. Bridging this gap is essential for advancing the security of semantic communication systems in complex, real-world network configurations.

15.9 Conclusion

In conclusion, this chapter has investigated the susceptibility of semantic communications to adversarial attacks, emphasizing the preservation of semantic information during information transfer. While semantic communications aim for both accurate task completion and minimal reconstruction loss, our exploration has unveiled a high vulnerability to adversarial manipulations. Employing diverse attack methods on CIFAR-10 image data and modeling communication with AWGN and Rayleigh channels, we demonstrated the efficacy of adversarial attacks in disrupting the delivery of semantic information. To counteract these threats, we explored adversarial training as a defense mechanism, showcasing its potential to reduce the impact of attacks and fortify the robustness of semantic communications. This proactive security approach not only safeguards semantic information but also contributes to the overall reliability of DNN models in real-world scenarios. The findings underscore the imperative need to enhance the security posture of semantic communications in the face of subtle yet potent adversarial threats.

Bibliography

Damilola Adesina, Chung-Chu Hsieh, Yalin E. Sagduyu, and Lijun Qian. Adversarial machine learning in wireless communications using RF data: A review. *IEEE Communications Surveys and Tutorials*, 25(1):77–100, 2023.

Eirina Bourtsoulatze, David Burth Kurka, and Deniz Gündüz. Deep joint source-channel coding for wireless image transmission. *IEEE Transactions on Cognitive Communications and Networking*, 5(3):567–579, 2019.

Christina Chaccour, Walid Saad, Merouane Debbah, Zhu Han, and H. Vincent Poor. Less data, more knowledge: Building next generation semantic communication networks. arXiv preprint arXiv:2211.14343, 2022.

Kemal Davaslioglu and Yalin E. Sagduyu. Trojan attacks on wireless signal classification with adversarial machine learning. In *IEEE International Symposium on Dynamic Spectrum Access Networks (DySPAN)*, 2019.

Yinpeng Dong, Fangzhou Liao, Tianyu Pang, Hang Su, Jun Zhu, Xiaolin Hu, and Jianguo Li. Boosting adversarial attacks with momentum. In *Proceedings of the IEEE Conference on Computer Vision and Pattern Recognition*, pages 9185–9193, 2018.

Mohammad Karimzadeh Farshbafan, Walid Saad, and Merouane Debbah. Curriculum learning for goal-oriented semantic communications with a common language. *IEEE Transactions on Communications*, 71(3):1430–1446, 2023.

Ian J. Goodfellow, Jonathon Shlens, and Christian Szegedy. Explaining and harnessing adversarial examples. arXiv preprint arXiv:1412.6572, 2014.

Basak Guler and Aylin Yener. Semantic index assignment. In *IEEE International Conference on Pervasive Computing and Communication (PERCOM) Workshops*, 2014.

Başak Güler, Aylin Yener, and Ananthram Swami. The semantic communication game. *IEEE Transactions on Cognitive Communications and Networking*, 4(4):787–802, 2018.

Deniz Gündüz, Zhijin Qin, Inaki Estella Aguerri, Harpreet S. Dhillon, Zhaohui Yang, Aylin Yener, Kai Kit Wong, and Chan-Byoung Chae. Beyond transmitting bits: Context, semantics, and task-oriented communications. *IEEE Journal on Selected Areas in Communications*, 41(1):5–41, 2023.

Brian Kim, Yalin E. Sagduyu, Kemal Davaslioglu, Tugba Erpek, and Sennur Ulukus. Channel-aware adversarial attacks against deep learning-based wireless signal classifiers. *IEEE Transactions on Wireless Communications*, 21(6):3868–3880, 2021a.

Brian Kim, Yi Shi, Yalin E. Sagduyu, Tugba Erpek, and Sennur Ulukus. Adversarial attacks against deep learning based power control in wireless communications. In *2021 IEEE Globecom Workshops (GC Wkshps)*, pages 1–6. IEEE, 2021b.

Marios Kountouris and Nikolaos Pappas. Semantics-empowered communication for networked intelligent systems. *IEEE Communications Magazine*, 59(6):96–102, 2021.

Alexey Kurakin, Ian J. Goodfellow, and Samy Bengio. Adversarial examples in the physical world. In *Artificial Intelligence Safety and Security*, pages 99–112. Chapman and Hall/CRC, 2018.

Aleksander Madry, Aleksandar Makelov, Ludwig Schmidt, Dimitris Tsipras, and Adrian Vladu. Towards deep learning models resistant to adversarial attacks. arXiv preprint arXiv:1706.06083, 2017.

B. R. Manoj, Meysam Sadeghi, and Erik G. Larsson. Downlink power allocation in massive MIMO via deep learning: Adversarial attacks and training. *IEEE Transactions on Cognitive Communications and Networking*, 8(2):707–719, 2022.

Seyed-Mohsen Moosavi-Dezfooli, Alhussein Fawzi, and Pascal Frossard. DeepFool: A simple and accurate method to fool deep neural networks. In *Proceedings of the*

IEEE Conference on Computer Vision and Pattern Recognition, pages 2574–2582, 2016.

T. J. O'Shea and J. Hoydis. An introduction to deep learning for the physical layer. *IEEE Transactions on Cognitive Communications and Networking*, 3(4):563–575, 2017.

Zhijin Qin, Xiaoming Tao, Jianhua Lu, and Geoffrey Ye Li. Semantic communications: Principles and challenges. arXiv preprint arXiv:2201.01389, 2021.

Meysam Sadeghi and Erik G. Larsson. Adversarial attacks on deep-learning based radio signal classification. *IEEE Wireless Communications Letters*, 8(1):213–216, 2018.

Meysam Sadeghi and Erik G. Larsson. Physical adversarial attacks against end-to-end autoencoder communication systems. *IEEE Communications Letters*, 23(5):847–850, 2019.

Yalin E. Sagduyu, Yi Shi, and Tugba Erpek. Adversarial deep learning for over-the-air spectrum poisoning attacks. *IEEE Transactions on Mobile Computing*, 20(2):306–319, 2019.

Yalin E. Sagduyu, Yi Shi, Tugba Erpek, William Headley, Bryse Flowers, George Stantchev, and Zhuo Lu. When wireless security meets machine learning: Motivation, challenges, and research directions. arXiv preprint arXiv:2001.08883, 2020.

Yalin E. Sagduyu, Tugba Erpek, and Yi Shi. Adversarial machine learning for 5G communications security. In *Game Theory and Machine Learning for Cyber Security*, pages 270–288, 2021.

Yalin E. Sagduyu, Tugba Erpek, Sennur Ulukus, and Aylin Yener. Is semantic communication secure? A tale of multi-domain adversarial attacks. *IEEE Communications Magazine*, 61(11):50–55, 2023a.

Yalin E. Sagduyu, Tugba Erpek, Sennur Ulukus, and Aylin Yener. Vulnerabilities of deep learning-driven semantic communications to backdoor (Trojan) attacks. In *Conference on Information Sciences and Systems (CISS)*, 2023b.

Yalin E. Sagduyu, Tugba Erpek, Aylin Yener, and Sennur Ulukus. Joint sensing and semantic communications with multi-task deep learning. arXiv preprint arXiv:2311.05017, 2023c.

Yalin E. Sagduyu, Tugba Erpek, Aylin Yener, and Sennur Ulukus. Multi-receiver task-oriented communications via multi-task deep learning. In *IEEE Future Networks World Forum*, 2023d.

Yalin E. Sagduyu, Yi Shi, Tugba Erpek, William Headley, Bryse Flowers, George Stantchev, Zhuo Lu, and Brian Jalaian. Adversarial machine learning: A new threat paradigm for next-generation wireless communications. In *AI, Machine Learning and Deep Learning: A Security Perspective*, pages 35–52. CRC Press, 2023e.

Yalin E. Sagduyu, Sennur Ulukus, and Aylin Yener. Task-oriented communications for nextG: End-to-end deep learning and AI security aspects. *IEEE Wireless Communications*, 30(3):52–60, 2023f.

Yalin E. Sagduyu, Tugba Erpek, Aylin Yener, and Sennur Ulukus. Will 6G be semantic communications? Opportunities and challenges from task oriented and secure communications to integrated sensing. arXiv preprint arXiv:2401.01531, 2024.

Yi Shi and Yalin Sagduyu. Membership inference attack and defense for wireless signal classifiers with deep learning. *IEEE Transactions on Mobile Computing*, 22(7):4032–4043, 2023.

Emilio Calvanese Strinati and Sergio Barbarossa. 6G networks: Beyond Shannon towards semantic and goal-oriented communications. *Computer Networks*, 190:107930, 2021.

Elif Uysal, Onur Kaya, Anthony Ephremides, James Gross, Marian Codreanu, Petar Popovski, Mohamad Assaad, Gianluigi Liva, Andrea Munari, Touraj Soleymani, Beatriz Soret Soret, and Henrik Johansson. Semantic communications in networked systems. *IEEE Network*, 36(4):233–240, 2022.

Huiqiang Xie, Zhijin Qin, Geoffrey Ye Li, and Biing-Hwang Juang. Deep learning enabled semantic communication systems. *IEEE Transactions on Signal Processing*, 69:2663–2675, 2021.

16

Encrypted Semantic Communications for Privacy Preserving

Zhiyong Chen[1], Meixia Tao[2], Zhongwei Wang[1], and Xinlai Luo[1]

[1]*Cooperative Medianet Innovation Center, Shanghai Jiao Tong University, Shanghai, China*
[2]*Department of Electronic Engineering and the Cooperative Medianet Innovation Center, Shanghai Jiao Tong University, Shanghai, China*

16.1 Introduction

By directly transmitting the semantics of information, semantic communications (SCs) can significantly reduce the amount of data required for transmission while preserving the meaning of the information, thereby improving communication efficiency. Consequently, semantic communications are envisioned to have transformative potential in 6G [Zhang et al., 2022]. Semantic communications can facilitate a wide range of intelligent and efficient communication services, such as remote image sensing and fusion, multiagent interaction and collaboration, intelligent transportation, and intelligent healthcare. It can effectively compress information redundancy, reduce transmission delay, and improve transmission rate, and ultimately meet a series of requirements for future communication [Qin et al., 2021; Dai et al., 2022; Dong et al., 2022; Gndz et al., 2023; Zhang et al., 2023; Chen et al., 2021b].

At the same time, the rapid advancement of wireless communication technology has led to numerous security challenges and privacy breaches [Yang et al., 2023b; Luo et al., 2022b]. Unlike traditional methods that focus on the mere transmission of data, semantic communications aim to transmit the underlying meaning or semantic information. This requires not only safeguarding the confidentiality and integrity of the transmitted data but also ensuring that the semantics of the information are preserved and correctly interpreted by the receiver. In this context, conventional security measures face several challenges. First, while traditional security measures primarily focus on the encryption of data, semantic communications

Foundations of Semantic Communication Networks, First Edition.
Edited by Walid Saad, Christina Chaccour, Christo Kurisummoottil Thomas, and Merouane Debbah.

pose the additional challenge of ensuring the correct interpretation and protection of semantics, even when the data itself is encrypted. Second, semantic communications often depend on artificial intelligence (AI) models to generate and interpret semantic information [Lan et al., 2021]. These AI models can be vulnerable to adversarial attacks, where attackers manipulate the semantic information. Classical security solutions, not designed to counter such AI-targeted attacks, fall short in this new context. Thirdly, the reliance on knowledge bases for the accurate interpretation of information in semantic communications necessitates security measures that can protect the integrity and confidentiality of these knowledge bases to prevent misleading interpretations [Choi and Park, 2022]. Therefore, to design a secure semantic communication system, it is crucial to address these unique challenges and hence ensure the security and reliability of semantic communications.

This chapter aims to provide an overview of the security issues in semantic communications and introduce an encrypted semantic communication (EnSC) system for privacy preserving [Luo et al., 2023]. Specifically, we first introduce the basics of SC systems. We then present the security challenges in SC. After that, we propose EnSC system where a key is integrated into the original information and then followed by the fusion of semantic features based on machine learning (ML) algorithm in the physical layer, to prioritize privacy preservation while seamlessly combining universality and confidentiality. By using adversarial training, we design an adversarial encryption training scheme to ensure the accuracy of semantic communications. Experimental results demonstrate that the proposed EnSC, coupled with the adversarial encryption training scheme, enables the receiver to successfully decode the precise semantic message, regardless of whether the transmitted semantic information is encrypted at the transmitter. Moreover, it proves challenging for potential eavesdropper to reconstruct the original semantic information from intercepted messages.

16.2 Basics Of Semantic Communication Systems

We first present the principle and system architecture of semantic communication, which lays a foundational underpinning for the subsequent investigation into the security issues of semantic communication.

Unlike traditional communication systems which primarily focus on the exact recovery of the source data at the destination, semantic communication is concerned with whether the destination can correctly understand the information and make the right decisions. The basic design idea of semantic communication is to obtain the "meaning" of the transmitted information by extracting the "semantic" information from the source [Yang et al., 2022]. The receiver relies on the same knowledge background as the sender to "interpret" the transmitted semantic information instead of recovering the original bits [Dai et al., 2022].

Semantic communication often utilizes ML techniques to autonomously identify and extract semantic features based on the given scene. A typical system diagram of learning-based semantic communication is illustrated in Figure 16.1. It differs from the traditional digital communication systems in the following aspects. First, it employs neural networks (NNs) to replace the traditional source encoder and source decoder [Xie et al., 2021] (Note that NNs are also often used for joint source-channel coding). The semantic encoder analyzes the source information and extracts its semantic content through NNs, while the semantic decoder recovers semantics from the received symbols and reconstructs the source data or executes tasks. This approach overcomes the barrier of lack of an explicit mathematical expression for semantic extraction, thereby making the system more flexible. NN-based semantic encoder and decoder can be utilized for the transmission of various types of data and the execution of corresponding tasks, e.g., images, audio, text, and video. Different ML are employed for different types of data. For example, transformer can be utilized for text or image data, while convolutional neural network (CNN) is used for processing image and video data. Second, the performance metrics of systems are no longer solely evaluated based on bit error rate (BER), instead, they are associated with the type of data and pragmatic tasks, such as bilingual evaluation understudy (BLEU), peak signal-to-noise ratio (PSNR), intersection over union (IoU), and video multimethod assessment fusion (VMAF).

Another feature different from the traditional communication system is that the core of the semantic communication system lies in the background knowledge base, which the semantic encoder and decoder are trained based upon [Shi et al., 2021]. The transceiver can independently learn to establish an independent and unique knowledge base. The knowledge base can take various forms, and those based on video, text, audio, image, and even multimodal fusion differ, which makes the learning process for the background knowledge base complex and time-consuming. In addition, a mismatched background knowledge base can lead to semantic ambiguity, which directly affects the efficiency of semantic communication. In order to ensure that all source data can be well understood and reasoned, a common practice is for the source and destination in a semantic communication system to share the knowledge base in real-time. This approach aims to achieve semantic matching as much as possible, while reducing the time required to construct the background knowledge base.

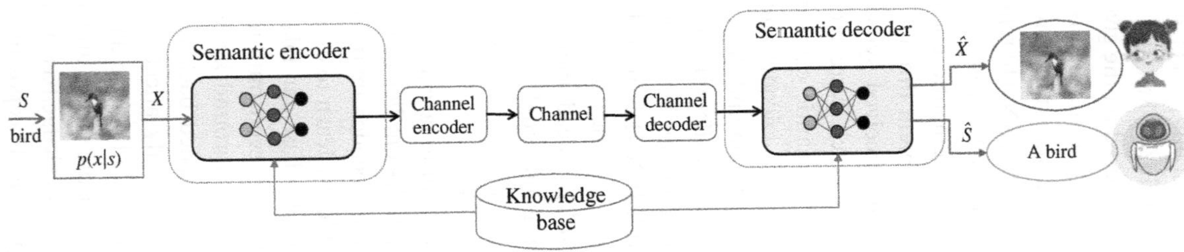

Figure 16.1 A typical diagram of learning-based semantic communication systems.

16.3 Security Issues Of Semantic Communication

The objectives of secure semantic communication encompass two aspects: effective transmission and privacy protection.

16.3.1 Security Risk Of Effective Transmission

Effective transmission refers to the ability of legitimate recipients to accurately recover and interpret the transmitted semantic information. The achievement of effective transmission mainly relies on two aspects:

(1) **Robustness of semantic encoder and decoder**: The process of extracting and recovering semantic information should be able to overcome the influence of physical noise and semantic noise.
(2) **Security of the knowledge base**: The data within the background knowledge base should be reliable and authentic, ensuring that it has not been tampered with or forged.

Attacks on either of these aspects can potentially result in the failure of semantic communication.

The first type of security risk for effective transmission is adversarial attack [Zhu et al., 2022]. Adversarial attacks can be employed by malicious attackers to introduce additional semantic noise into the communication system, as shown in Figure 16.2. These attacks can be categorized based on their location of occurrence. Adversarial attacks can take place at the physical channel [Kim et al., 2021] by adding specific interference signals to alter the input of the semantic decoder. They can also occur at the transmitter's end, such as before semantic encoding, by injecting adversarial perturbations into the original data, thereby influencing the output of the semantic encoder [Wang et al., 2023].

Depending on the attacker's knowledge of the model's parameters [Hu et al., 2023], adversarial attacks can be categorized into gradient-based white-box attacks and observation-based black-box attacks. In white-box attacks, where the attacker has access to the structure and parameters of the semantic encoding and decoding model, they can compute the gradient of the target function with respect to the input and modify the input of the semantic encoder. On the other hand, in black-box attacks, the attacker can only observe the input and output of the semantic encoder. In this case, the attacker needs to infer model-related information and create an adversarial model to generate adversarial examples. Black-box attacks are considered to be closer to real-world scenarios as they assume limited knowledge about the model's internals and parameters.

The second type of security risk is knowledge base poisoning attack. The shared knowledge base between the source and receiver plays a crucial role in the feature

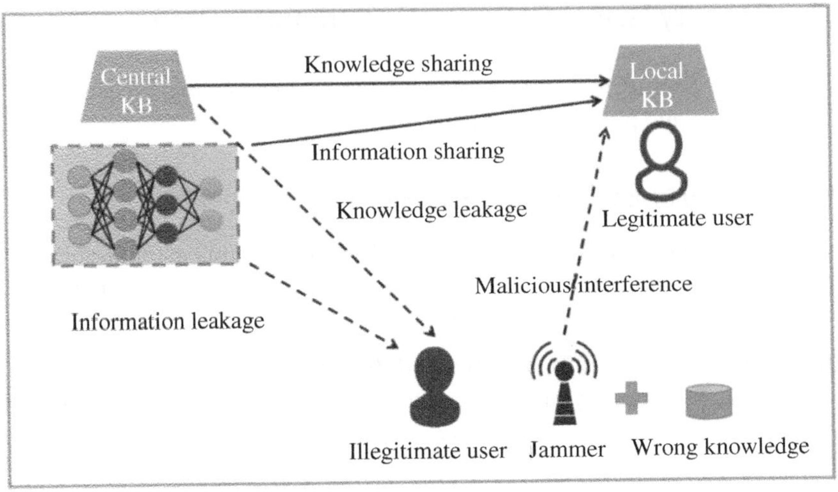

Figure 16.2 Semantic communication system under malicious attacks.

extraction and recovery process of semantic information. If this knowledge base is maliciously attacked, it can have significant consequences on the accuracy and reliability of the recovered information.

Poisoning attacks encompass malicious data injection and backdoor attacks. Malicious data injection refers to the deliberate injection of manipulated data into the knowledge base, which influences the decision boundaries of the model and leads to erroneous predictions by the semantic decoder [Du et al., 2023]. For example, in a semantic communication with knowledge base focused on images of dogs, an attacker could collect unrelated images, modify their pixel distributions to resemble the semantic information of dogs, and iteratively train the model using a loss function that measures the semantic similarity between unrelated images and dog images. This process generates a series of poisoned data, contaminating the knowledge base.

On the other hand, backdoor attacks refer to the injection of specific backdoors into ML models used for semantic communication, making them sensitive to specific content and enabling malicious manipulation of the entire semantic communication process.

To defend against poisoning attacks, several methods can be employed [Papernot et al., 2016; Mahmood et al., 2021], including:

(1) **Data sanitization**: In Chen et al. 2021a, a detector is trained to differentiate between clean data and poisoned data. They also train a mimic model to emulate the behavior of clean data on the target model. By comparing

the prediction disparities between the mimic and target models, they can effectively detect and eliminate instances of poisoning data.

(2) **Activation smoothing**: Activation smoothing aims to mitigate the impact of poisoned data by reducing the influence of individual instances on the model's decision-making process [Rosenfeld et al., 2020]. It involves incorporating regularization techniques or modifying the loss function to dampen the effect of extreme or anomalous activations caused by poisoned data.

(3) **Trigger reversal and backdoor elimination**: If a backdoor attack is suspected or identified, trigger reversal and backdoor elimination techniques can be employed [Tao et al., 2022].

16.3.2 Security Risk Of Privacy Protection

Privacy protection primarily focuses on preventing unauthorized entities such as malicious attackers or potential eavesdroppers from accessing sensitive information. This includes private knowledge bases, ML models of semantic encoders and decoders, transmitted semantic features, training gradients, and datasets, among others.

The security risk of privacy protection is eavesdropping and inference attack. In both traditional communication and semantic communication, eavesdroppers on the physical channel may have the ability to intercept the encoded information, including semantic information. Meanwhile, if the knowledge base is eavesdropped upon, there is a risk of leaking semantic features at the source [Yang et al., 2023b].

Inference attacks refer to the process where eavesdroppers, based on the intercepted information, combine it with publicly available knowledge to infer the sender's privacy, particularly sensitive information such as medical conditions, financial status, and personal preferences. For instance, by utilizing intermediate gradients of the model and perturbed data, partial training data for the semantic encoder and decoder can be recovered. Similarly, knowledge graph-based techniques can be employed for deanonymization and privacy inference in anonymous social networks [Qian et al., 2017].

To prevent eavesdropping, traditional security methods such as encryption techniques can be employed in semantic communication. Traditional encryption techniques can operate at a higher layer to encrypt the data, ensuring that even if the transmitted semantic information is compromised, eavesdroppers would struggle to recover the original content. In Section 16.4, we will introduce an encryption mechanism at physical layer to protect privacy.

To defend against inference attacks, differential privacy and data anonymization are commonly employed as an effective privacy-preserving technique. The utilization of differential privacy in semantic communication involves the addition of controlled noise to the information after semantic encoding, making it difficult for eavesdroppers to infer sensitive information. Zhao et al. [2023] propose a joint online inference and offline training framework for semantic communication, which covers privacy-protected codec training and personalized codec deployment. Data anonymization involves modifying or removing identifying information from datasets to reduce the risk of sensitive information disclosure and minimize the potential for attackers to infer personal attributes. The authors in Moqurrab et al. [2022] put forth a lightweight semantic anonymization framework designed specifically for Internet of Things (IoT) scenarios. Their proposed approach aims to alleviate the negative effects of noise on the original data.

16.4 Encrypted Semantic Communications

In this section, we introduce an EnSC system to protect user privacy, while considering both the system's versatility and privacy protection aspects. Versatility is evident as all network module structures are publicly accessible and trained using a shared background database, making it suitable for widespread deployment in real-world scenarios. Privacy protection for communication is achieved through a symmetric encryption mechanism. The confidential semantic information is encrypted by the transmitter and the receiver can recover the semantic message, thereby thwarting any attempts by potential eavesdroppers to reconstruct the original semantic information from intercepted messages.

16.4.1 Overall System Architecture

We consider a semantic communication system, which involves one transmitter (Alice) and two receivers (Bob and Eve). Bob has confidential message to receive, while Eve has no confidential messages and does not care about security issues. Both Bob and Eve are legitimate users and have their own data transmission with Alice, but Eve may potentially act as an eavesdropper, attempting to intercept communication information between Alice and Bob. Therefore, Alice and Bob aim to establish a secure semantic communication in the presence of the potential eavesdropper Eve. In this chapter, the security property mainly prevents eavesdropping because Eve is limited to intercepting information and cannot inject or modify transmitted messages.

As depicted in Figure 16.3, the objective of Alice is to transmit a confidential semantic source to Bob and an open semantic message to Eve by processing messages through the same semantic encoding and channel encoding. We consider that the messages intended for Bob and Eve are transmitted on different resource

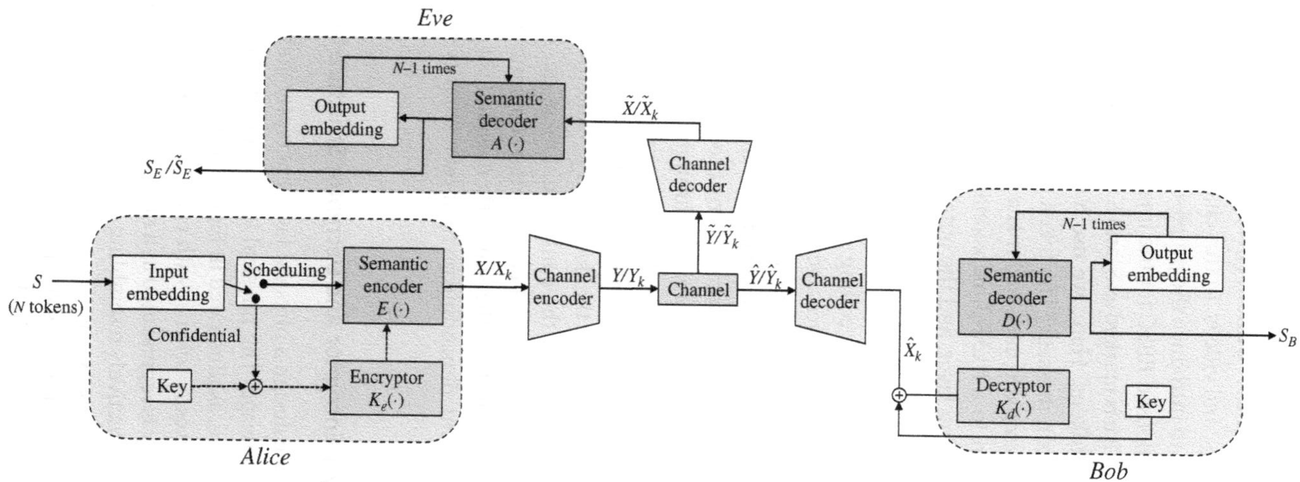

Figure 16.3 The structure of encrypted semantic communication system.

blocks. For the confidential message intended for Bob, Alice not only semantically encodes the message but also encrypts it. For the open message intended for Eve, Alice only semantically encodes it without encryption. The message can be multi-modal, such as text, images, and videos. In this chapter, without loss of generality, we use text as an example to introduce the proposed EnSC, while we also provide the performance of the proposed EnSC for the image modality in the simulation.

The input sentence at Alice is marked as S, and S is decomposed into a sequential combination of multiple tokens. Here, we use $S = S_B$ for the confidential message and $S = S_E$ for the open message. The token refers to the storage unit in the background knowledge base, and each token is a one-hot vector. The input sentence can map each token to a fixed-dimensional floating-point number vector through the word embedding layer, and the output is recorded as $\tilde{S}$. Then, when the message is confidential, the semantic message $\tilde{S}$ and the key are input into the encryptor $K_e(\cdot)$ for encryption, and then the encrypted message is input into the semantic encoder $E(\cdot)$ for semantic coding. X_k denotes the semantic vector for the encrypted message. When the message is open, it is not encrypted and the semantic vectors X are output. Finally, channel coding is performed to obtain the output Y_k or Y. The semantic codec is designed with transformer [Wu et al., 2023; Yang et al., 2023a], and the channel codec uses an autoencoder [Luo et al., 2022a]. After the encoded information is transmitted through the additive white Gaussian noise (AWGN) channel, Eve decodes the unencrypted message directly using its own semantic decoder $A(\cdot)$ to obtain the open message. Meanwhile, Bob first decrypts the encrypted message with the decoder $K_d(\cdot)$ and then decodes it using the semantic decoder $D(\cdot)$ to obtain the confidential message. At the same time, Eve attempts to reconstruct the original confidential semantic message through a semantic decoder $A(\cdot)$. Note that the semantic decoders $A(\cdot)$ and $D(\cdot)$ at the two receivers can be different, depending on their own security requirements and channel conditions.

16.4.2 Physical-Layer Encryptor And Decryptor Structure In EnSC

In the proposed EnSC, we design a physical layer encryptor and decryptor for the confidential semantic message. The physical layer encryptor and decryptor have the same network structure, as illustrated in Figure 16.4. The original confidential semantic message comprises N tokens, with each token being embedded into M dimensions. The secret key is represented as a random vector with M-dimensional floating-point numbers, effectively resembling one of the tokens within the semantic message. These elements serve as inputs for both the encryptor and decryptor. The first layer of the encryptor is the reshape layer, which concatenates the original confidential message and the key, followed by a dimensional transformation. Subsequently, the second layer, known as the hidden layer, reduces the

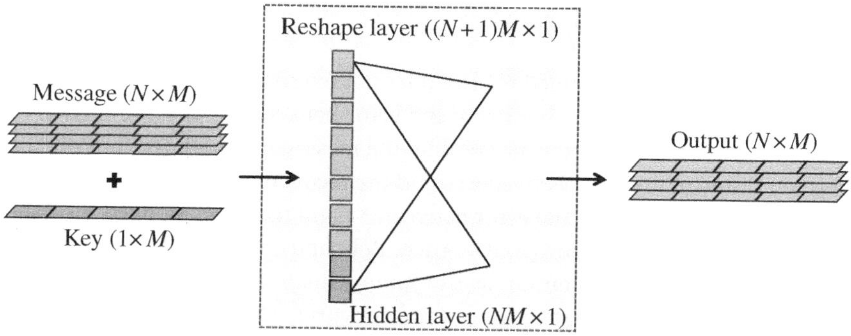

Figure 16.4 Physical layer encryptor and decryptor network structure.

dimensionality to match the length of the original confidential message. Finally, the output reshapes the vector to align with the original message's format. It is crucial to emphasize that the encryption and decryption mechanisms are learned from data rather than being generated by a predefined algorithm. The choice of a token length for the key is not indicative of the encryption's strength. The encryption's efficacy is determined by the method learned by the NN, not the data volume within the key. This assertion has been substantiated through experimentation.

16.4.3 Semantic Encoder And Decoder Structure In EnSC

We adopt the symmetric structure, as shown in Figure 16.3, to establish a stronger connection between encryption (decryption) and semantic encoding (decoding). Within the semantic communication system, Bob and Eve share the same semantic decoder structure. However, Eve may assume roles as potential attackers. In this case, the semantic attacker at Eve and the semantic decoder at Bob also identically adopt the same structures. The semantic decoder processes each word sequentially, utilizing the output of the first $N-1$ steps as additional inputs for the subsequent decoding steps. In essence, the decoder performs a number of decoding operations equal to the number of tokens in the initial input. Nevertheless, there is one single semantic encoder at Alice, to encode the semantic messages for both Bob and Eve.

16.5 Adversarial Encryption Training

16.5.1 Loss Functions

The semantic communication network utilizing key encryption requires complex training procedures and the meticulous crafting of loss functions. In order to guarantee both the feasibility and security of encryption, we implement an adversarial training approach.

The open message S_E intended for Eve does not require confidentiality for Eve. Therefore, the objective is to minimize the error between S_E and the recovered message at Eve. Meanwhile, for the confidential message S_B intended for Bob, the goal is to minimize the error between S_B and the recovered message at Bob while simultaneously maximizing the error between S_B and the eavesdropped message at Eve. Consequently, we train jointly Alice, Bob, and Eve given that Eve might also want to maximize its potential eavesdropping capability during the joint training process. Similar to generative adversarial networks (GANs), our objective is to ensure that the transmitter and receiver outperform the most skilled attackers, rather than a static adversary.

For the distance function, we use cross-entropy D_{CE}, which can be formulated as:

$$\mathscr{D}_{CE} = -\frac{1}{m}\sum_{i=1}^{m}\sum_{j=1}^{l} p\left(x_{ij}\right)\log\left(q\left(x_{ij}\right)\right), \tag{16.1}$$

where $p(x_{ij})$ and $q(x_{ij})$ are the real probability and the predicted probability, respectively, of the jth word in the ith sample, l represents the number of tokens in the sentence, and m represents the number of samples in one batch.

The loss function of the semantic communication link without encryption for the open message is given by the following formula:

$$\mathscr{L}_{ED}\left(\theta_E,\theta_D\right) = \mathscr{D}_{CE}\left(S, D\left(E\left(S;\theta_E\right);\theta_D\right)\right), \tag{16.2}$$

where $E\left(\cdot;\theta_E\right)$ and $D\left(\cdot;\theta_D\right)$ represent the outputs of the semantic encoder parameterized by θ_E at Alice and the semantic decoder at Bob parameterized by θ_D, respectively, and the optimal semantic encoder and decoder are obtained by minimizing this loss. Note that $E\left(S;\theta_E\right)$ is not directly the input of $D\left(\cdot;\theta_D\right)$ during training, it needs to pass through the channel, but in order to simplify the expression, it is not expressed in the formula. Therefore, we have the optimal semantic encoder and decoder by minimizing this loss as following

$$\left(O_E, O_D\right) = \text{argmin}_{\left(\theta_E,\theta_D\right)}\left(\mathscr{L}_{ED}\left(\theta_E,\theta_D\right)\right). \tag{16.3}$$

Similarly, the loss function of the EnSC link is defined as:

$$\mathscr{L}_{K_d}\left(\theta_{K_e},\theta_E,\theta_{K_d},\theta_D\right) = \mathscr{D}_{CE}\left(S, D\left(K_d\left(E\left(K_e(S;\theta_{K_e}\right);\theta_E\right);\theta_{K_d}\right);\theta_D\right)\right), \tag{16.4}$$

where $K_e(\theta_{K_e};\cdot)$ and $K_d(\theta_{K_d};\cdot)$ are the outputs of the encryptor and decryptor, respectively. We obtain the optimal decryptor for receiver by minimizing the loss as following

$$O_{K_d}\left(\theta_{K_d}\right) = \text{argmin}_{\theta_{K_d}}\left(\mathscr{L}_{K_d}\left(\theta_{K_e}, \theta_E, \theta_{K_d}, \theta_D\right)\right). \tag{16.5}$$

The potential eavesdropper Eve can use its own semantic decoder to directly reconstruct the confidential information, and the loss function is:

$$\mathscr{L}_A\left(\theta_{K_e}, \theta_E, \theta_A\right) = \mathscr{D}_{CE}\left(S, A\left(E\left(K_e\left(S; \theta_{K_e}\right); \theta_E\right); \theta_A\right)\right). \tag{16.6}$$

The optimal attacker can be obtained by minimizing the loss as following

$$O_A\left(\theta_A\right) = \text{argmin}_{\theta_A}\left(\mathscr{L}_A\left(\theta_{K_e}, \theta_E, \theta_A\right)\right). \tag{16.7}$$

Combining $\mathscr{L}_{Kd}$ and $\mathscr{L}_A$, give the loss function of the encryptor, and decryptor can be obtained as the following formula

$$\mathscr{L}_{K_e}\left(\theta_{K_e}\right) = \mathscr{L}_{K_d}\left(\theta_{K_e}, \theta_E, \theta_{K_d}, \theta_D\right) - \lambda\mathscr{L}_A\left(\theta_{K_e}, \theta_E, O_A\left(\theta_A\right)\right). \tag{16.8}$$

Here, the hyper-parameter λ balances the utility and confidentiality. We obtain the optimal encryptor by minimizing this loss

$$O_{K_e}\left(\theta_{K_e}\right) = \text{argmin}_{\theta_{K_e}}\left(\mathscr{L}_{K_e}\left(\theta_{K_e}\right)\right). \tag{16.9}$$

16.5.2 Training

The training of the system consists of two main steps. In the initial step, we train a channel encoder and a channel decoder with a symmetric structure. We employ randomly generated vectors to train the network module, akin to the encoded semantic vector. In the training process, the channel signal-to-noise ratio (SNR) parameter is dynamically adjusted within a certain range, improving the robustness of the network. Mean square error (MSE) serves as the loss function to reduce the distortion of the vector.

In the second step, similar to GANs, we alternately train the attacker with Alice and Eve. With a few steps of training, the semantic encoder $E(\cdot)$ and decoder $A(\cdot)$ find a way to satisfy the normal semantic communication requirements. Furthermore, the decryptor $K_d(\cdot)$ continues to learn a stable decryption method, but the semantic decoder $A(\cdot)$ gradually learns a way to decode the encrypted semantic message directly. In this process, Bob and Eve attempt to minimize reconstruction error at the same time, so we alternately update the encryptor $K_e(\cdot)$ to reduce Bob's reconstruction error but increase Eve's reconstruction error. Intuitively, learning makes the encryption method more receiver friendly. Note that, if nonadversarial training is used, only $\mathscr{L}_{ED}$, $\mathscr{L}_{K_d}$, and $\mathscr{L}_A$ are involved in the training updates, while $\mathscr{L}_{K_e}$ no longer participates in the updates.

16.5.3 Performance Evaluation

We evaluate the performance of the EnSC in this subsection. For the text input, we utilize the standard European Parliament proceedings dataset [Koehn, 2005], comprising approximately 2.0 million sentences. The learning rate is configured to be 10^{-4}, hyper-parameter λ is set to 0.2, and the SNR of the wireless channel for both Bob and Eve is fixed at 10 dB. We employ BLEU score as the evaluation metrics [Qin et al., 2023a]. The weights of one-gram and two-gram in BLEU are set to 0.6 and 0.4, respectively.

Figures 16.5–16.7 show the change of $\mathscr{L}_{ED}$, $\mathscr{L}_{K_d}$, and $\mathscr{L}_A$, respectively, with the increase of training steps using adversarial encryption training and nonadversarial encryption training. We can see that with the adversarial encryption training scheme, $\mathscr{L}_{ED}$ and $\mathscr{L}_{K_d}$ can converge to 0. However, $\mathscr{L}_A$ cannot converge to 0 and eventually fluctuate around 0.5. Because the cross-entropy is used as the distance function, if the loss $\mathscr{L}_A$ cannot converge to 0, Eve will not be able to reconstruct the original confidentiality message correctly. On the other hand, using the non-adversarial training scheme, that is, when $\mathscr{L}_{K_e}$ does not participate in the training update, $\mathscr{L}_{ED}$, $\mathscr{L}_{K_d}$, and $\mathscr{L}_A$ can all converge to 0. It means that although the system has high utility, confidentiality is inferior. Eve can easily reconstruct the original semantic message from the encrypted messages.

We can also observe that in the adversarial encryption training scheme, the rate of convergence for each loss value is notably slower compared to that of the non-adversarial encryption training scheme. In order to outsmart the most skilled Eve, the encryptor will periodically adjust its network parameters after a certain number of training steps. While this adjustment benefits Bob, it poses a challenge for

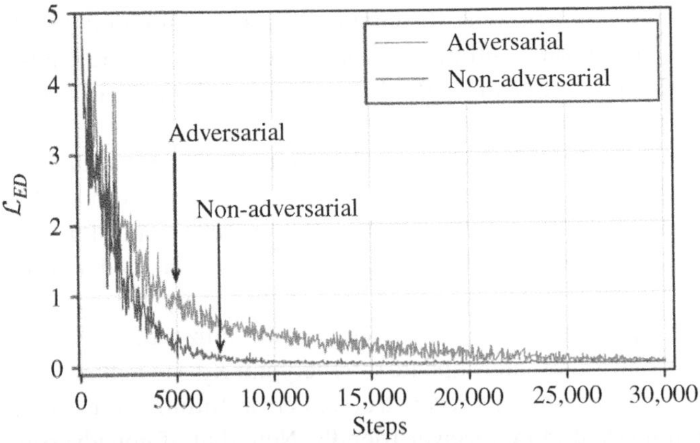

Figure 16.5 The values of $\mathscr{L}_{ED}$ under adversarial encryption training and nonadversarial encryption training.

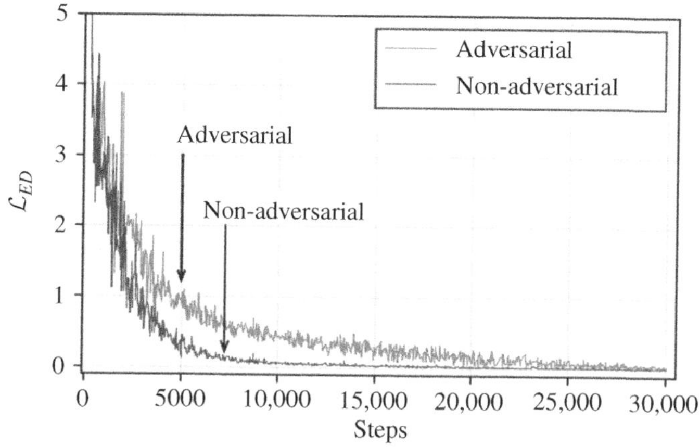

Figure 16.6 The values of $\mathscr{L}_{Kd}$ under adversarial encryption training and nonadversarial encryption training.

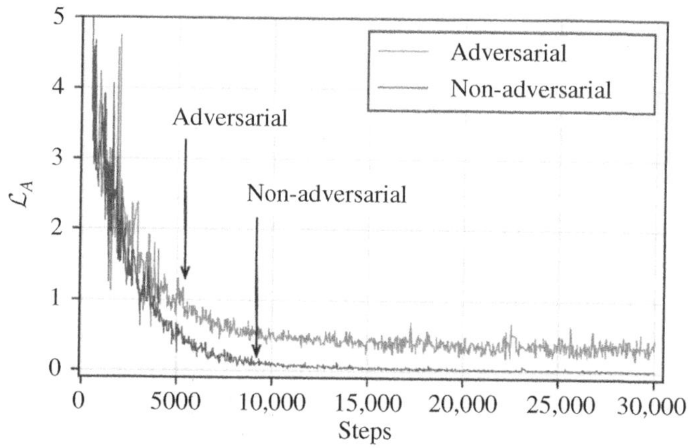

Figure 16.7 The values of $\mathscr{L}_A$ under adversarial encryption training and nonadversarial encryption training.

Eve, leading to erratic fluctuations in her loss value. What is of utmost significance is that this adversarial training process results in a more gradual reduction in all loss values, ultimately thwarting Bob's loss from converging to zero.

In Figure 16.8, we present the comparison of BLEU scores versus SNR with the two different training schemes. Notably, when employing the adversarial encryption training scheme, Bob consistently achieves significantly higher BLEU scores compared to Eve across all SNR values. Particularly, in high SNR channel conditions, Bob's BLEU score approaches nearly 1, whereas Eve's score remains

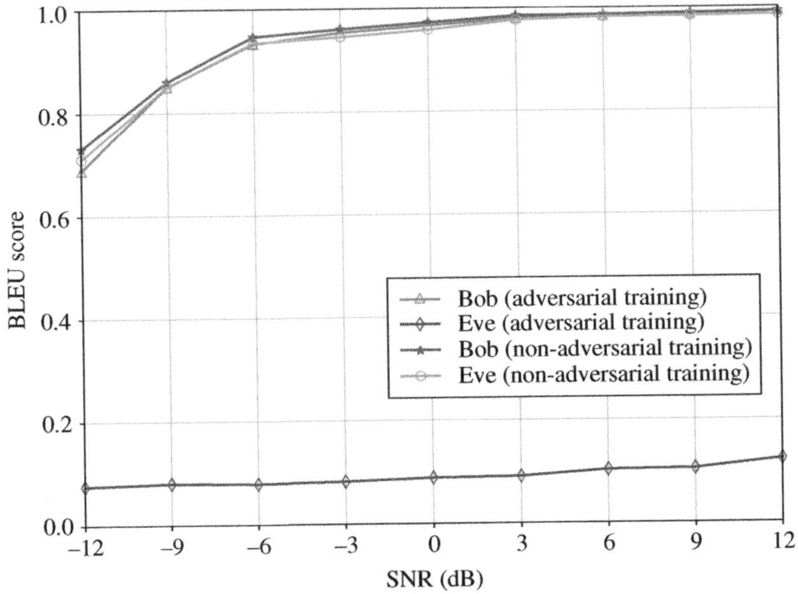

Figure 16.8 BLEU score versus SNR for Bob and Eve in EnSC with different training schemes.

below 0.2. To provide a comprehensive perspective, we also assessed Eve's performance in intercepting semantic information when trained without adversarial encryption. In this scenario, Bob's BLEU score closely aligns with Eve's, indicating the practical vulnerability of semantic attacks and emphasizing the effectiveness of adversarial encryption training in preserving privacy. Additionally, as depicted in Figure 16.8, it is evident that the adoption of adversarial encryption training does not compromise Bob's performance.

For image type inputs, the DIV2K dataset is used in the experiment, which contains 2000 high-resolution images with different scenes, such as cities, buildings, landscapes, nature, and so on. The learning rate is set to 10^{-4}, hyper-parameter λ is set to 0.02.

In Figure 16.9, we illustrate the impact of the adversarial encryption training scheme for image communication. By incorporating the agreed key as an additional input during the semantic decoding process, the legitimate receiver's utilization of the key becomes apparent, leading to a notable enhancement in the quality of the reconstructed image. Conversely, when the attacker endeavors to perform direct semantic decoding, the resulting reconstructed image exhibits severe distortion and lacks crucial information from the original image.

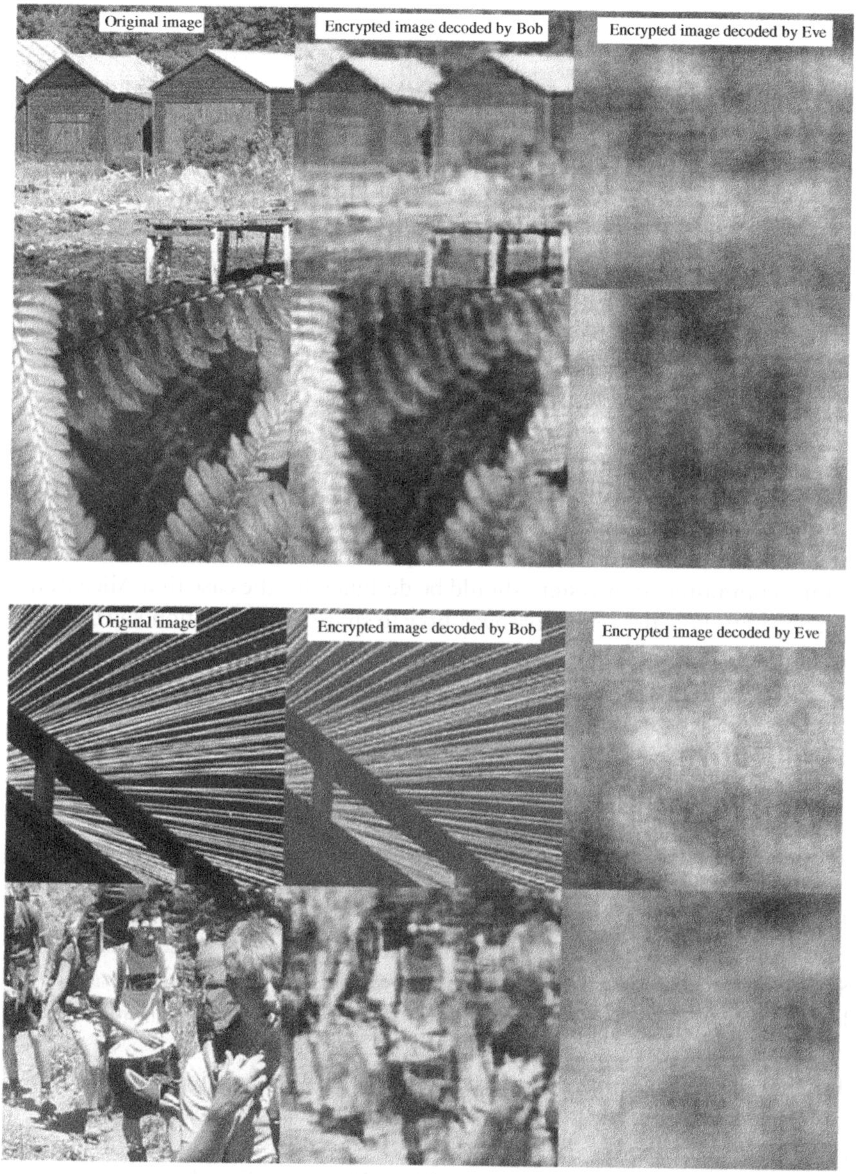

Figure 16.9 From left to right, the original confidential image, the encrypted image decoded by Bob, and the encrypted image decoded by Eve.

16.6 Conclusion

In this chapter, we describe the basics of semantic communications and discuss the security issues associated with semantic communication systems. Accordingly, we have designed an EnSC system. The proposed EnSC supports physical layer encryption for the confidential semantic information transmitted to Bob, while it does not apply physical layer encryption for the open information sent to Eve. Consequently, this approach prevents potential eavesdroppers, such as Eve, from obtaining the confidential information of Bob. We introduce an adversarial encryption training scheme that ensures the accuracy of semantic communication for both Bob and Eve, while also preventing Bob from eavesdropping on confidential semantic information. Simulation results have demonstrated that EnSC, utilizing the adversarial training scheme, significantly enhances the privacy protection of the semantic communication system. Throughout this chapter we assume that Alice, Bob, and Eve use the centralized adversarial training, yet implementing such an assumption in practical systems is challenging. Therefore, further research is required to explore how a physical layer encryption semantic communication system should be designed for the case that Alice trains separately with Bob and Eve. Moreover, how to share the encryption key between Alice and Bob also necessitates further investigation.

Bibliography

Jian Chen, Xuxin Zhang, Rui Zhang, Chen Wang, and Ling Liu. De-Pois: An attack-agnostic defense against data poisoning attacks. *IEEE Transactions on Information Forensics and Security*, 16:3412—3425, 2021a. doi: 10.1109/TIFS.2021.3080522.

Mingzhe Chen, Zhaohui Yang, Walid Saad, Changchuan Yin, H. Vincent Poor, and Shuguang Cui. A joint learning and communications framework for federated learning over wireless networks. *IEEE Transactions on Wireless Communications*, 20(1):269—283, 2021b. doi: 10.1109/TWC.2020.3024629.

Jinho Choi and Jihong Park. Semantic communication as a signaling game with correlated knowledge bases. In *2022 IEEE 96th Vehicular Technology Conference (VTC2022-Fall)*, pages 1—5. IEEE, 2022.

Jincheng Dai, Ping Zhang, Kai Niu, Sixian Wang, Zhongwei Si, and Xiaoqi Qin. Communication beyond transmitting bits: Semantics-guided source and channel coding. *IEEE Wireless Communications*, 30(4):170—177, 2022.

Peihao Dong, Qihui Wu, Xiaofei Zhang, and Guoru Ding. Edge semantic cognitive intelligence for 6G networks: Novel theoretical models, enabling framework, and typical applications. *China Communications*, 19(8):1—14, 2022.

Hongyang Du, Jiacheng Wang, Dusit Niyato, Jiawen Kang, Zehui Xiong, Mohsen Guizani, and Dong In Kim. Rethinking wireless communication security in semantic Internet of Things. *IEEE Wireless Communications*, 30(3):36—43, 2023.

Deniz Gndz, Zhijin Qin, Inaki Estella Aguerri, Harpreet S. Dhillon, Zhaohui Yang, Aylin Yener, Kai Kit Wong, and Chan-Byoung Chae. Beyond transmitting bits: Context, semantics, and task-oriented communications. *IEEE Journal on Selected Areas in Communications*, 41(1):5—41, 2023. doi: 10.1109/JSAC.2022.3223408.

Qiyu Hu, Guangyi Zhang, Zhijin Qin, Yunlong Cai, Guanding Yu, and Geoffrey Ye Li. Robust semantic communications with masked VQ-VAE enabled codebook. *IEEE Transactions on Wireless Communications*, 22(12):8707—8722, 2023.

Brian Kim, Yalin E. Sagduyu, Kemal Davaslioglu, Tugba Erpek, and Sennur Ulukus. Channel-aware adversarial attacks against deep learning-based wireless signal classifiers. *IEEE Transactions on Wireless Communications*, 21(6):3868—3880, 2021.

Philipp Koehn. Europarl: A parallel corpus for statistical machine translation. In *Proceedings of Machine Translation Summit X: Papers*, pages 79—86, 2005.

Qiao Lan, Dingzhu Wen, Zezhong Zhang, Qunsong Zeng, Xu Chen, Petar Popovski, and Kaibin Huang. What is semantic communication? A view on conveying meaning in the era of machine intelligence. *Journal of Communications and Information Networks*, 6(4):336—371, 2021.

Xinlai Luo, Benshun Yin, Zhiyong Chen, Bin Xia, and Jiangzhou Wang. Autoencoder-based semantic communication systems with relay channels. In *2022 IEEE International Conference on Communications Workshops (ICC Workshops)*, pages 711—716. IEEE, 2022a.

Xuewen Luo, Hsiao-Hwa Chen, and Qing Guo. Semantic communications: Overview, open issues, and future research directions. *IEEE Wireless Communications*, 29(1):210—219, 2022b.

Xinlai Luo, Zhiyong Chen, Meixia Tao, and Feng Yang. Encrypted semantic communication using adversarial training for privacy preserving. *IEEE Communications Letters*, 27(6):1486—1490, 2023.

Kaleel Mahmood, Rigel Mahmood, and Marten van Dijk. On the robustness of vision transformers to adversarial examples. In *Proceedings of the IEEE/CVF International Conference on Computer Vision (ICCV)*, pages 7838—7847, Oct. 2021.

Syed Atif Moqurrab, Adeel Anjum, Noshina Tariq, and Gautam Srivastava. Instantanonymity: A lightweight semantic privacy guarantee for 5G-enabled IIoT. *IEEE Transactions on Industrial Informatics*, 19(1):951—959, 2022.

Nicolas Papernot, Patrick McDaniel, Xi Wu, Somesh Jha, and Ananthram Swami. Distillation as a defense to adversarial perturbations against deep neural networks. In *2016 IEEE Symposium on Security and Privacy (SP)*, pages 582—597, 2016. doi: 10.1109/SP.2016.41.

Jianwei Qian, Xiang-Yang Li, Chunhong Zhang, Linlin Chen, Taeho Jung, and Junze Han. Social network de-anonymization and privacy inference with knowledge graph model. *IEEE Transactions on Dependable and Secure Computing*, 16(4):679—692, 2017.

Zhijin Qin, Xiaoming Tao, Jianhua Lu, Wen Tong, and Geoffrey Ye Li. Semantic communications: Principles and challenges. https://arxiv.org/abs/2201.01389, 2021.

Qi Qin, Yankai Rong, Guoshun Nan, Shaokang Wu, Xuefei Zhang, Qimei Cui, and Xiaofeng Tao. Securing semantic communications with physical-layer semantic encryption and obfuscation. In *ICC 2023 - IEEE International Conference on Communications*, pages 5608—5613, 2023a. doi: 10.1109/ICC45041.2023.10279807.

Zhijin Qin, Tantan Zhao, Fan Li, and Xiaoming Tao. Survey of research on multimodal semantic communication. *Journal on Communications*, 44(5):28, 2023b.

Elan Rosenfeld, Ezra Winston, Pradeep Ravikumar, and Zico Kolter. Certified robustness to label-flipping attacks via randomized smoothing. In *International Conference on Machine Learning*, pages 8230—8241. PMLR, 2020.

Guangming Shi, Yong Xiao, Yingyu Li, and Xuemei Xie. From semantic communication to semantic-aware networking: Model, architecture, and open problems. *IEEE Communications Magazine*, 59(8):44—50, 2021.

Guanhong Tao, Guangyu Shen, Yingqi Liu, Shengwei An, Qiuling Xu, Shiqing Ma, Pan Li, and Xiangyu Zhang. Better trigger inversion optimization in backdoor scanning. In *Proceedings of the IEEE/CVF Conference on Computer Vision and Pattern Recognition*, pages 13368—13378, 2022.

Yuntao Wang, Zhou Su, Shaolong Guo, Minghui Dai, Tom H. Luan, and Yiliang Liu. A survey on digital twins: Architecture, enabling technologies, security and privacy, and future prospects. *IEEE Internet of Things Journal*, 10(17):14965—14987, 2023.

Tong Wu, Zhiyong Chen, Dazhi He, Liang Qian, Yin Xu, Meixia Tao, and Wenjun Zhang. CDDM: Channel denoising diffusion models for wireless communications. arXiv preprint arXiv:2305.09161, 2023.

Huiqiang Xie, Zhijin Qin, Geoffrey Ye Li, and Biing-Hwang Juang. Deep learning enabled semantic communication systems. *IEEE Transactions on Signal Processing*, 69:2663—2675, 2021.

Wanting Yang, Hongyang Du, Zi Qin Liew, Wei Yang Bryan Lim, Zehui Xiong, Dusit Niyato, Xuefen Chi, Xuemin Sherman Shen, and Chunyan Miao. Semantic communications for future internet: Fundamentals, applications, and challenges. *IEEE Communications Surveys and Tutorials*, 25(1):213—250, 2022.

Ke Yang, Sixian Wang, Jincheng Dai, Kailin Tan, Kai Niu, and Ping Zhang. WITT: A wireless image transmission transformer for semantic communications. In *ICASSP 2023-2023 IEEE International Conference on Acoustics, Speech and Signal Processing (ICASSP)*, pages 1—5. IEEE, 2023a.

Zhaohui Yang, Mingzhe Chen, Gaolei Li, Yang Yang, and Zhaoyang Zhang. Secure semantic communications: Fundamentals and challenges. arXiv preprint arXiv:2301.01421, 2023b.

Ping Zhang, Wenjun Xu, Hui Gao, Kai Niu, Xiaodong Xu, Xiaoqi Qin, Caixia Yuan, Zhijin Qin, Haitao Zhao, Jibo Wei, and Fangwei Zhang. Toward wisdom-evolutionary and primitive-concise 6G: A new paradigm of semantic communication networks. *Engineering*, 8:60—73, 2022.

Hongwei Zhang, Shuo Shao, Meixia Tao, Xiaoyan Bi, and Khaled B. Letaief. Deep learning-enabled semantic communication systems with task-unaware transmitter and dynamic data. *IEEE Journal on Selected Areas in Communications*, 41(1):170—185, 2023. doi: 10.1109/JSAC.2022.3221991.

Lindong Zhao, Dan Wu, and Liang Zhou. Data utilization versus privacy protection in semantic communications. *IEEE Wireless Communications*, 30(3):44—50, 2023.

Tianqing Zhu, Dayong Ye, Wei Wang, Wanlei Zhou, and Philip S. Yu. More than privacy: Applying differential privacy in key areas of artificial intelligence. *IEEE Transactions on Knowledge and Data Engineering*, 34(6):2824—2843, 2022. doi: 10.1109/TKDE.2020.3014246.

Appendix A

A.1 Proof of Lemma 11.3

It is straightforward to show that $\mathcal{L}$ is a monoidal preorder, where the order is imposed by $\boldsymbol{x} \leq \boldsymbol{y}$, whenever there is a morphism from $\boldsymbol{x}$ to $\boldsymbol{y}$. For each object $\boldsymbol{x}$ in $\mathcal{L}$, the representable copresheaf $h^{\boldsymbol{x}} = hom(\boldsymbol{x}, -)$ is given by the following conditional probability:

$$h^{\boldsymbol{x}}(\boldsymbol{y}) = \begin{cases} \pi(\boldsymbol{y} \mid x, c), & \text{if } \boldsymbol{x} \leq \boldsymbol{y}, \\ 0, & \text{otherwise.} \end{cases} \tag{A.1}$$

Assignment $\boldsymbol{x} \to h^{\boldsymbol{x}}$ can be shown to be an enriched functor as follows. An enriched functor $f = h^{\boldsymbol{x}} : \mathcal{L} \to \hat{\mathcal{L}}$ satisfies the property that $\mathcal{L}(\boldsymbol{x},\boldsymbol{y}) \leq \hat{\mathcal{L}}(f(\boldsymbol{x}),f(\boldsymbol{y}))$. Here, we define $\hat{\mathcal{L}}$ to be the category which contains all the enriched functors. To show this, we need to first define the function $\hat{\mathcal{L}}(f(\boldsymbol{x}),f(\boldsymbol{y}))$.

$$\hat{\mathcal{L}}(f(\boldsymbol{x}),f(\boldsymbol{y})) = \begin{cases} 1, & \text{if } \boldsymbol{x} \leq \boldsymbol{y}, \\ \dfrac{f(\boldsymbol{x})}{f(\boldsymbol{y})}, & \text{if } \boldsymbol{x} > \boldsymbol{y}. \end{cases} \tag{A.2}$$

For $\boldsymbol{x} \leq \boldsymbol{y}$, $\mathcal{L}(\boldsymbol{x},\boldsymbol{y}) \leq 1 = \hat{\mathcal{L}}(f(\boldsymbol{x}),f(\boldsymbol{y}))$. In the case, when $\boldsymbol{x} > \boldsymbol{y}$, $\mathcal{L}(\boldsymbol{x},\boldsymbol{y}) = 0 \leq \frac{f(x)}{f(y)}$, with $f(\boldsymbol{x}) \leq f(\boldsymbol{y})$. Hence, $f = h^{\boldsymbol{x}}$ is an enriched functor. Hence, all the representable copresheaves $h^{\boldsymbol{x}}, \forall \boldsymbol{x} \in \mathcal{L}$ within the context c forms a [0,1] enriched category which we define as $\hat{\mathcal{L}}$ (representing all possible worlds that follows from $\mathcal{L}$). We denote the codomain of $h^{\boldsymbol{x}}$ as $\mathcal{H}^{\boldsymbol{x}}$. Further, for completion, we can define the hom object between two representable copresheaves as follows, $\hat{\mathcal{L}}(h^{\boldsymbol{x}}, h^{\boldsymbol{y}}) = \sum_{\boldsymbol{d} \in \mathcal{H}^{\boldsymbol{y}}} h^{\boldsymbol{x}}(\boldsymbol{d})$. For any three objects $\boldsymbol{x}, \boldsymbol{y}, \boldsymbol{z}$, it satisfies $\hat{\mathcal{L}}(h^{\boldsymbol{x}}, h^{\boldsymbol{z}}) = \sum_{\boldsymbol{d}_1 \in \mathcal{H}^{\boldsymbol{x}}} \sum_{\boldsymbol{d}_2 \in \mathcal{H}^{\boldsymbol{z}}} \pi(\boldsymbol{d}_1 \mid \boldsymbol{x}) \pi(\boldsymbol{d}_2 \mid \boldsymbol{y}) = \hat{\mathcal{L}}(h^{\boldsymbol{x}}, h^{\boldsymbol{y}}) \hat{\mathcal{L}}(h^{\boldsymbol{y}}, h^{\boldsymbol{z}})$. Hence, associative property of enriched categories is satisfied. $\hat{\mathcal{L}}(h^{\boldsymbol{x}}, h^{\boldsymbol{x}}) = 1$ satisfies the identity morphism. Thus, $\hat{\mathcal{L}}$ is a $[0, 1]$-enriched category of representable copresheaves of the objects in $\mathcal{L}$.

Foundations of Semantic Communication Networks, First Edition.
Edited by Walid Saad, Christina Chaccour, Christo Kurisummoottil Thomas, and Merouane Debbah.

A.2 Proof of Theorem 11.1

We consider that the cost of transmission is negligible, $c(\boldsymbol{u}) = 0$ and assume that the normalization constraints are satisfied with equality, hence $\boldsymbol{\alpha}_s = 0$. By using the inequality, $\sum_x p(x) \log q(x) \leq \sum_x p(x) \log p(x)$ (non-negativity of KL divergence), we can write (where p, q are π_s, π_l, respectively) $F_s(\pi_l, \pi_s) \geq \mathbb{E}\left[S_s(\boldsymbol{z}; \boldsymbol{u} \mid \pi_s, c)\right] - \lambda E\left[\log \pi(\boldsymbol{z} \mid \boldsymbol{u})\right]$. Substituting for $\pi(\boldsymbol{z} \mid \boldsymbol{u})$ using Bayes rule, we get

$$\begin{aligned}
&\mathbb{E}\left[S_s(\boldsymbol{z}; \boldsymbol{u} \mid \pi_s, c)\right] + \lambda_s \mathbb{E}\left[V(\pi_s, \pi_l)\right] \\
&\geq \mathbb{E}\left[S_s(\boldsymbol{z}; \boldsymbol{u} \mid \pi_s, c)\right] - \lambda_s \mathbb{E}\left[\log \frac{\pi(\boldsymbol{u} \mid \boldsymbol{z})\pi(\boldsymbol{z} \mid c)\pi(c)}{\pi(\boldsymbol{u})}\right] \\
&= (1 - \lambda_s)\mathbb{E}\left[S_s(\boldsymbol{z}; \boldsymbol{u} \mid \pi_s, c)\right] + \lambda_s H_s(\boldsymbol{z}, c),
\end{aligned} \tag{A.3}$$

where $H_s(\boldsymbol{z})$ is entropy of $\boldsymbol{z}$. For $\lambda_s \in [0, 1]$, the minimum is attained at $\mathbb{E}\left[S_s(\boldsymbol{z}; \boldsymbol{u} | \pi_s, c)\right] = 0$ and hence, $\mathbb{E}\left[S_s(\boldsymbol{z}; \boldsymbol{u} | \pi_s, c)\right] + \lambda_s \mathbb{E}[V(\pi_s, \pi_l)] \geq \lambda_s H_s(\boldsymbol{z}, c)$. It can be shown that the zero for $\mathbb{E}\left[S_s(\boldsymbol{z}; \boldsymbol{u} | \pi_s, c)\right]$ can be achieved when $\pi_s(\boldsymbol{u} \mid \boldsymbol{z}) = 1/K$, $\pi_l(\boldsymbol{z} \mid \boldsymbol{u}) = \pi(\boldsymbol{z})$, and $H_s(\boldsymbol{u}) = \log K$, where $\mathcal{U} = K$. This implies that the minimum can be achieved with a noninformative speaker when the constraint on the sender utility is weak. When $\lambda_s > 1$, the $F_s(\pi_l, \pi_s)$ is minimized when $\mathbb{E}\left[S_s(\boldsymbol{z}; \boldsymbol{u} \mid \pi_s, c)\right]$ is maximized. $\mathbb{E}\left[S_s(\boldsymbol{z}; \boldsymbol{u} \mid \pi_s, c)\right]$ is maximum when there is a bijection that preserves the semantics, that is $\psi : \mathcal{W} \to \mathcal{U}$ such that $\pi_s(\boldsymbol{u} \mid \boldsymbol{z}, c) = \delta_{\boldsymbol{u}, \psi(s)}$, $\pi_l(\boldsymbol{z} \mid \boldsymbol{u}, c) = \delta_{\boldsymbol{u}, \psi(s)}$. In this case, the listener can map accurately the received signal (assuming error-free communication) to the speaker intended semantics and thus extracting the maximum semantic information.

In the partial pooling case, where $|\mathcal{W}| > |\mathcal{U}|$, we first look at how the listener side extracts the semantic state. For simplicity of analysis, we assume that communication is error-free, which implies $\hat{\boldsymbol{u}} = \boldsymbol{u}$. We can write $F_l(\pi_l, \pi_s) \leq \mathbb{E}_{\boldsymbol{u}_t}\left[S_l(\boldsymbol{z}; \boldsymbol{u} \mid \pi_s, \pi_l)\right]$, from which it is clear that the maximum is attained when there is a bijective mapping between $\boldsymbol{z}$ and $\boldsymbol{u}$, that contradicts our assumption for partial pooling case. Hence, the expression

$$\pi_l^* = \arg\max_{\pi_l} S_l(\hat{\boldsymbol{z}}; \boldsymbol{u} \mid \boldsymbol{z}, \pi_l), \tag{A.4}$$

is maximized only when the overlap between the copresheaves $h^{\boldsymbol{z}}$ and $h^{\hat{\boldsymbol{z}}}$ are maximized. Further, defining a semantic distance metric as $\|\boldsymbol{z} - \hat{\boldsymbol{z}}\|_S = \sum_{Y \in \mathcal{H}^{\boldsymbol{z}} - (\mathcal{H}^{\boldsymbol{z}} \cap \mathcal{H}^{\hat{\boldsymbol{z}}})} S_s(\boldsymbol{Y}; \boldsymbol{z})$. Intuitively, it can be interpreted as a weighted ($S_s(\boldsymbol{Y}; \boldsymbol{z})$ are the weights) Hamming distance (HD) between the copresheaves represented by $\mathcal{H}^{\boldsymbol{z}}$ and $\mathcal{H}^{\hat{\boldsymbol{z}}}$. Hence, at equilibrium, the listener chooses the expression $\hat{\boldsymbol{z}}$ that is semantically closer to $\boldsymbol{z}$ or, in other words, according to the optimization (equivalent to (A.4)), $\hat{\boldsymbol{z}}^* = \arg\min_{\hat{\boldsymbol{z}}} \|\boldsymbol{z} - \hat{\boldsymbol{z}}\|_S$. Given that the listener chooses HD-based decoding, how should the speaker select the encoded signal to maximize the received semantic information (or minimize the HD)? The optimal encoded signal can be understood using a similar analysis as in Jäger et al. [2011]. The speaker can partition $\mathcal{W}$ into $|\mathcal{U}|$ of them, with each denoted $\mathcal{W}_k$ and every state in $\mathcal{W}_k$ mapped to transmit signal $\boldsymbol{u}_k$. Let the listener's decoded state be $\hat{\boldsymbol{z}}_k$ corresponding to $\boldsymbol{u}_k$. The speaker wants the decoded expression to be as close to the transmitted one as possible. Hence, the speaker will transmit $\boldsymbol{u}_k$ whenever the decoded state $\hat{\boldsymbol{z}}_k$ is semantically closest to all the states in $\mathcal{W}_k$. Such a partition of the $\mathcal{W}$ is called a *Voronoi tessellation* of $\mathcal{W}$.

A.3 Proof of Proposition 11.5

Given a listener policy $\pi_{l,t}$, we optimize the speaker policy by maximizing the Lagrangian.

$$\begin{aligned}
\pi^*_{s,t} &= \arg\max_{\pi_{s,t}} F_s(\pi_{s,t}, \pi_{l,t}), \text{ where,} \\
F_s(\pi_{s,t}, \pi_{l,t}) &= -\mathbb{E}_{\boldsymbol{u}_t}\left[S_s(\boldsymbol{z}_t; \boldsymbol{u}_t \mid \pi_{s,t}, c, [\boldsymbol{z}_{t-1}])\right] \\
&\quad - \lambda_s \mathbb{E}[V(\pi_{s,t}, \pi_{l,t})] + \sum_{\boldsymbol{z}_t} \alpha_{\boldsymbol{z}_t}\left[\sum_{\boldsymbol{u}_t} \pi(\boldsymbol{u}_t \mid \boldsymbol{z}_t)\right].
\end{aligned} \tag{A.5}$$

It can be validated that $F_s(\pi_{s,t}, \lambda_s)$ is concave in $\pi_{s,t}$ (follows from showing that term of the form $x \log \frac{x}{a}$ in $F_s(\pi_{s,t}, \lambda_s)$ is convex), hence the global maximum is obtained at the point where derivative of $F_s(\pi_{s,t}, \pi_{l,t})$ is zero. λ_s represents the Lagrange multiplier for the inequality constraint associated with V. $\alpha_{\boldsymbol{z}_t}$ represents the Lagrange multiplier for the normalization of the distribution $\pi(\boldsymbol{u}_t \mid \boldsymbol{z}_t)$. $\boldsymbol{\alpha}_s$ represents the vector of all $\alpha_{\boldsymbol{z}_t}$. Further, taking derivative w.r.t $\pi_{s,t}$, we obtain

$$\begin{aligned}
&-\pi(\boldsymbol{z}_t)S_s(\boldsymbol{z}_t)\left(\log \pi_{s,t} - \log \pi(\boldsymbol{u}_t) + 1\right) - \lambda_s\left(c(\boldsymbol{u}_t) - \log \pi_{l,t}\right) + \alpha_{\boldsymbol{z}_t} = 0, \implies \\
&\pi_{s,t} = \pi(\boldsymbol{u}_t) \\
&\exp\left(\frac{-1}{S_s(\boldsymbol{z}_t \mid [\boldsymbol{z}_{t-1}])\pi(\boldsymbol{z}_t \mid c)}(\lambda_s(c(\boldsymbol{u}_t) - \log \pi_{l,t}) - \alpha_{\boldsymbol{z}_t}) - 1\right).
\end{aligned} \tag{A.6}$$

$$\begin{aligned}
\pi^*_{l,t} &= \arg\max_{\pi_{l,t}} F_l(\pi_{l,t}, \pi_{s,t}), \\
F_l(\pi^*_{l,t}, \pi_{s,t}) &= \underbrace{\sum_{\boldsymbol{u}_t} \pi\left(\boldsymbol{u}_t \mid \boldsymbol{z}_t\right)\left[\sum_{\hat{\boldsymbol{z}}_t} \pi(\hat{\boldsymbol{z}}_t \mid \boldsymbol{u}_t) \log \frac{\pi(\hat{\boldsymbol{z}}_t \mid \boldsymbol{u}_t)}{\pi(\hat{\boldsymbol{z}}_t)} Z_{\hat{\boldsymbol{z}}_t \boldsymbol{z}_t}\right] S_s(\boldsymbol{z}_t \mid c, [\boldsymbol{z}_{t-1}])}_{S_l(\hat{\boldsymbol{z}}_t; \boldsymbol{u}_t \mid \pi_{s,t}, \pi_{l,t}, [\hat{\boldsymbol{z}}_{t-1}])} \\
&\quad - \sum_{\boldsymbol{u}_t} \alpha_{\boldsymbol{u}_t} \sum_{\boldsymbol{z}_t} \pi_{l,t}.
\end{aligned} \tag{A.7}$$

$$\pi^*_{l,t} = \arg\max_{\pi_{l,t}} \hat{F}_l(\pi_{l,t}, \pi_{s,t}),$$
$$\hat{F}_l(\pi_{l,t}, \pi_{s,t}) = \sum_{\boldsymbol{u}_t} \hat{S}_l\left(\hat{\boldsymbol{z}}_t; \boldsymbol{u}_t \mid \pi_{s,t}, \pi_{l,t}, [\hat{\boldsymbol{z}}_{t-1}]\right) - \sum_{\boldsymbol{u}_t} \alpha_{\boldsymbol{u}_t} \sum_{\boldsymbol{z}_t} \pi(\hat{\boldsymbol{z}}_t \mid \boldsymbol{u}_t). \tag{A.8}$$

Similarly, for a fixed $\pi_{s,t}$, we can obtain $\pi_{l,t}$ by maximizing the Lagrangian (A.7). Equation (A.7) is nonconcave since it is a summation of a convex and a concave function. Since a linear function is simultaneously convex and concave, consider the first order Taylor series expansion of $\pi_{l,t} \log \pi_{l,t}$, using which we can rewrite $S_l(\hat{\boldsymbol{z}}_t; \boldsymbol{u}_t \mid \pi_{s,t}, \pi_{l,t}, [\hat{\boldsymbol{z}}_{t-1}])$ as

$$\hat{S}_l\left(\hat{\boldsymbol{z}}_t; \boldsymbol{u}_t \mid \pi_{s,t}, \pi_{l,t}, [\hat{\boldsymbol{z}}_{t-1}]\right)$$
$$= \pi\left(\boldsymbol{u}_t \mid \boldsymbol{z}_t\right) S_s(\boldsymbol{z}_t \mid c, [\boldsymbol{z}_{t-1}]) \sum_{\hat{\boldsymbol{z}}_t, \hat{c}} \pi_{l,t}\left[1 + \log \pi_{l,t-1} - \log \pi(\hat{\boldsymbol{z}}_t)\right] Z_{\hat{\boldsymbol{z}}_t \boldsymbol{z}_t} + \text{const.}$$

Using the resulting approximate function obtained above which can be shown as a lower bound to the original $S_l(\hat{\boldsymbol{z}}_t; \boldsymbol{u}_t \mid \pi_{s,t}, \pi_{l,t}, [\hat{\boldsymbol{z}}_{t-1}])$, we can solve the approximate Lagrangian function (A.8). We can now derive (A.9) w.r.t $\pi_{l,t}$

$$\pi\left(\boldsymbol{u}_t \mid \boldsymbol{z}_t\right) S_s(\boldsymbol{z}_t \mid c, [\boldsymbol{z}_{t-1}]) Z_{\hat{\boldsymbol{z}}_t \boldsymbol{z}_t}$$
$$\times \left(\left(1 + \log \pi_{l,t-1}\right) - \log \pi(\hat{\boldsymbol{z}}_t) - \frac{\pi_{l,t}}{\pi(\hat{\boldsymbol{z}}_t)} \pi(\boldsymbol{u}_t)\right) = \alpha_{\boldsymbol{u}_t} \implies \tag{A.9}$$
$$\pi_{l,t} \propto \frac{1}{\pi(\boldsymbol{u}_t)} \exp\left(\frac{-\alpha_{\boldsymbol{u}_t}}{\pi\left(\boldsymbol{u}_t \mid \boldsymbol{z}_t\right) Z_{\hat{\boldsymbol{z}}_t \boldsymbol{z}_t} S_s(\boldsymbol{z}_t \mid c, [\boldsymbol{z}_{t-1}])} + \log \pi_{l,t-1}\right).$$

To arrive at $\pi_{l,t}$ above, we used the approximation $\pi(\hat{\boldsymbol{z}}_t) \approx \pi_{l,t}\pi(\boldsymbol{u}_t)$, since the probabilities are expected to be higher near the peak of $\pi_{l,t}$ and zero for all other $\pi(\hat{\boldsymbol{z}}_t \mid \boldsymbol{u}'_t)$, with $\boldsymbol{u}'_t \neq \boldsymbol{u}_t$.

A.4 Proof of Theorem 11.3

In our envisioned ESC model, we have the following set of inequality that mimics the data processing inequality in information theory [Cover and Thomas, 1991], $S_s(\boldsymbol{\Phi}; \boldsymbol{z}) \geq S_s(\boldsymbol{\Phi}; \boldsymbol{u}) \geq S_s(\boldsymbol{\Phi}; \hat{\boldsymbol{z}})$. Here $\boldsymbol{\Phi}$ (logical formulas) represents the semantic category space represented by $\boldsymbol{z}$. This follows from the Markov Chain $\boldsymbol{\Phi} \to \boldsymbol{z} \to \boldsymbol{u} \to \hat{\boldsymbol{u}} \to \hat{\boldsymbol{z}} \to \hat{\boldsymbol{\Phi}}$. In a classical system, the state description represented as $\boldsymbol{z}_c$ does not consider the hidden semantics and may contain irrelevant information. In such a case, where the semantic category space is excluded, the corresponding Markov Chain would be $\boldsymbol{z}_c \to \boldsymbol{u} \to \hat{\boldsymbol{u}} \to \hat{\boldsymbol{z}}_c$. Moreover, $H(\boldsymbol{z}_c \mid \hat{\boldsymbol{z}}_c) \geq H(\boldsymbol{z}_c \mid \boldsymbol{u}) \geq H(\boldsymbol{z} \mid \boldsymbol{u})$, assuming the same representation space for ESC and classical systems. Hence, we have the following Fano's inequality [Cover and Thomas, 1991], for a classical system, $P_e = p(\hat{\boldsymbol{z}}_c \neq \boldsymbol{z}_c) \geq \frac{H(\hat{\boldsymbol{z}}_c \mid \boldsymbol{z}_c) - 1}{\log |\mathcal{W}|}$. Next, we look at how the semantic probability error measure differs from the classical error quantity above. We define $\boldsymbol{e}$ as the random variable that captures the error in semantics represented by $\boldsymbol{z}$ and $\hat{\boldsymbol{z}}$. $\boldsymbol{e}$ denotes the set $\mathcal{H}^{\boldsymbol{z}} - (\mathcal{H}^{\boldsymbol{z}} \cap \mathcal{H}^{\hat{\boldsymbol{z}}})$, that represents the semantic space that cannot be represented by $\hat{\boldsymbol{z}}$. Moreover, consider $H(\boldsymbol{e}, \boldsymbol{\Phi} \mid \hat{\boldsymbol{z}}) = H(\boldsymbol{\Phi} \mid \hat{\boldsymbol{z}}) + H(\boldsymbol{e} \mid \boldsymbol{\Phi}, \hat{\boldsymbol{z}}) = H(\boldsymbol{\Phi} \mid \hat{\boldsymbol{z}}) = H(\boldsymbol{z} \mid \hat{\boldsymbol{z}})$. Further, we proceed following similar derivation as for Fano's inequality. Rewriting $H(\boldsymbol{e}, \boldsymbol{\Phi} \mid \hat{\boldsymbol{z}}) = H(\boldsymbol{e} \mid \hat{\boldsymbol{z}}) + H(\boldsymbol{\Phi} \mid \boldsymbol{e}, \hat{\boldsymbol{z}})$. With probability $p(\mathcal{H}^{\boldsymbol{z}} = \mathcal{H}^{\hat{\boldsymbol{z}}})$, we have $H(\boldsymbol{\Phi} \mid \boldsymbol{e}, \hat{\boldsymbol{z}}) = 0$. With probability $S_e = 1 - p(\mathcal{H}^{\boldsymbol{z}} = \mathcal{H}^{\hat{\boldsymbol{z}}})$, we have $H(\boldsymbol{\Phi} \mid \boldsymbol{e}, \hat{\boldsymbol{z}}) = H(\boldsymbol{z} \mid \boldsymbol{e}, \hat{\boldsymbol{z}}) \leq H(\boldsymbol{z}) \leq \log |\Omega|$. Further, we are able to conclude as

$$\begin{aligned} &H(\boldsymbol{z} \mid \hat{\boldsymbol{z}}) \leq H(\boldsymbol{e} \mid \hat{\boldsymbol{z}}) + S_e \log |\Omega| \\ &\implies S_e \geq \frac{H(\boldsymbol{z} \mid \hat{\boldsymbol{z}}) - H(\boldsymbol{e} \mid \hat{\boldsymbol{z}})}{\log |\mathcal{W}|}, \end{aligned} \tag{A.10}$$

where $H(\boldsymbol{e}) \geq 1$, with equal to happens only when the intersection of the semantic space represented by $\boldsymbol{z}$ and $\hat{\boldsymbol{z}}$ is a null set. Moreover, $H(\boldsymbol{z}_c \mid \hat{\boldsymbol{z}}_c) \geq H(\boldsymbol{z} \mid \hat{\boldsymbol{z}})$. Combining these two, we can conclude that even with some bit errors in the physical transmission, the semantics represented by $\hat{\boldsymbol{z}}$ can be similar (with its semantic space having significant intersection to $\mathcal{H}^{\boldsymbol{z}}$).

A.5 Proof of Theorem 11.2

Assume that the length of each codeword $\boldsymbol{u}_i \in \mathcal{U}$ is l_i. Also, consider that the listener is able to disambiguate the ambiguous signaling strategy from the sender by inferring the context. Then the average semantic representation length for the vocabulary $\mathcal{U}$ can be lower bounded as (using the Kraft's inequality [Cover and Thomas, 1991]) $\sum_{\boldsymbol{u}_i \in \mathcal{U}} \pi(\boldsymbol{u}_i) l_i \geq \sum_{c \in C} \pi(c) H(\boldsymbol{z}_i \mid c)$. Moreover, we can upper bound the codeword length for the semantic representation as follows.

$$\sum_{\boldsymbol{u}_i \in \mathcal{U}} \pi(\boldsymbol{u}_i) l_i \leq -\sum_{c \in C} \pi(c) \sum_{\boldsymbol{z}_i} \pi(\boldsymbol{z}_i \mid c) \lceil \log \pi(\boldsymbol{z}_i \mid c) \rceil. \tag{A.11}$$

For a causal reasoning system, where the contextual information is not taken into account $\sum_{\boldsymbol{u}_i \in \mathcal{U}} \pi(\boldsymbol{u}_i) l_i \geq H(\boldsymbol{z}_i)$, and the upperbound is $\sum_{\boldsymbol{u}_i \in \mathcal{U}} \pi(\boldsymbol{u}_i) l_i \leq \max_{c \in C} \sum_{\boldsymbol{z}_i} \pi(\boldsymbol{z}_i \mid c) \lceil \log \pi(\boldsymbol{z}_i \mid c) \rceil$. Further, we consider a classical system with $H_c(\boldsymbol{z}_i)$ representing the entropy. Classical systems are designed without a causal reasoning component and directly encodes all the entities. The bounds in this case are

$$\sum_{\boldsymbol{u}_i \in \mathcal{U}} \pi(\boldsymbol{u}_i) l_i \geq H_c(\boldsymbol{z}_i), \text{ where, } H_c(\boldsymbol{z}_i) \overset{(a)}{>} H(\boldsymbol{z}_i), \tag{A.12}$$

with, $H_c(\boldsymbol{z}_i)$ representing the entropy for classical wireless system. Statement (a) holds because the ESC system utilizes a uniform encoding for all states that convey equivalent meanings from the listener's standpoint (correspond to logical conclusions). In contrast, conventional wireless systems employ data statistics of $\mathbf{z}_i^0$ for encoding and do not account for the listener's perspective. This leads to a larger state space $\mathcal{W}$ for conventional systems as opposed to ESC, which employs a reduced state space. Given that the conditional entropy obeys the relation $H(\mathbf{z}_i \mid c) \leq H(\mathbf{z}_i) < H_c(\mathbf{z}_i^0)$, the lower bound for semantic representation gets reduced further when we take into account the communication context, compared to a classical communication system. Further, we look at the upper bound which is obtained as

$$\sum_{\mathbf{u}_i \in \mathcal{U}} \pi(\mathbf{u}_i) l_i \leq \max_{c \in C} \sum_{\mathbf{z}_i} \pi(\mathbf{z}_i \mid c) \left[\sum_{\mathbf{s}_i \in \mathbf{z}_i} \lceil \log \pi(\mathbf{s}_i \mid c) \rceil \right]. \tag{A.13}$$

The upper bound (A.13) follows from the reasoning that the classical system should be designed considering the worst-case scenario such that the listener can decode properly with an arbitrarily low probability of error. This results in a higher upper bound for classical system compared to ESC.

Bibliography

T. M. Cover and J. A. Thomas. *Elements of Information Theory*. Wiley, 1991.

G. Jäger, L. P. Metzger, and F. Riedel. Voronoi languages: Equilibria in cheap-talk games with high-dimensional types and few signals. *Games and Economic Behavior*, 73(2):517–537, Nov. 2011.

Index

Foundations of Semantic Communication Networks, First Edition.
Edited by Walid Saad, Christina Chaccour, Christo Kurisummoottil Thomas, and Merouane Debbah.

d

e

Printed and bound by CPI Group (UK) Ltd, Croydon, CR0 4YY
19/09/2026
14965091-0002